MyWorkBook

Beverly Fusfield

Beginning Algebra
Eleventh Edition

Margaret L. Lial
American River College

John Hornsby
University of New Orleans

Terry McGinnis

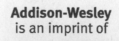
Addison-Wesley
is an imprint of

PEARSON

Copyright © 2012 Pearson Education, Inc.
Publishing as Pearson Addison-Wesley, 75 Arlington Street, Boston, MA 02116.

ISBN-13: 978-0-321-70251-7
ISBN-10: 0-321-70251-4

5 6 7 EBM 14 13 12

Addison-Wesley
is an imprint of

www.pearsonhighered.com

CONTENTS

CHAPTER 1   THE REAL NUMBER SYSTEM..1

CHAPTER 2   LINEAR EQUATIONS AND INEQUALITIES IN ONE
             VARIABLE...51

CHAPTER 3   LINEAR EQUATIONS AND INEQUALITIES IN
             TWO VARIABLES; FUNCTIONS..115

CHAPTER 4   SYSTEMS OF LINEAR EQUATIONS AND INEQUALITIES .. 163

CHAPTER 5   EXPONENTS AND POLYNOMIALS 205

CHAPTER 6   FACTORING AND APPLICATIONS...247

CHAPTER 7   RATIONAL EXPRESSIONS AND APPLICATIONS.................289

CHAPTER 8   ROOTS AND RADICALS 349

CHAPTER 9   QUADRATIC EQUATIONS 393

ANSWERS 427

Chapter 1 THE REAL NUMBER SYSTEM

1.1   Fractions

Learning Objectives
Learning Objectives
1   Learn the definition of *factor*.
2   Write fractions in lowest terms.
3   Multiply and divide fractions.
4   Add and subtract fractions.
5   Solve applied problems that involve fractions.
6   Interpret data in a circle graph.

Key Terms
Use the vocabulary terms listed below to complete each statement in exercises 1–21.

natural (counting) numbers	**whole numbers**	**fraction**
numerator	**denominator**	**proper fraction**
improper fraction	**mixed number**	**factor**
product	**prime**	**composite**
prime factors	**basic principle of fractions**	
lowest terms	**greatest common factor**	
reciprocals	**quotient**	**sum**
least common denominator (LCD)		**difference**

1.   A fraction in which the numerator is less than the denominator is called a(n) _____ .

2.   A(n) _____ of a given number is any number that divides evenly (without remainder) into the given number.

3.   The set of _____ consists of the numbers used for counting.

4.   Given several denominators, the smallest number that is divisible by all the denominators is called the _____ .

5.   The _____ of a list of integers is the largest common factor of those integers.

6.   A fraction in which the numerator is greater than the denominator is called a(n) _____ .

7.   The set of _____ is {0, 1, 2, 3, …}.

8.   The _____ states that, if the numerator and denominator of a fraction are multiplied or divided by the same nonzero number, the value of the fraction is not changed.

9.   Pairs of numbers whose product is 1 are called _____ of each other.

10. The answer to an addition problem is called the _____.

11. A(n) _____ includes a whole number and a fraction written together and is understood to be the sum of the whole number and the fraction.

12. In the fraction $\frac{2}{9}$, the 2 is the _____.

13. A(n) _____ number has at least one factor other than itself and 1.

14. The _____ of 66 are 2, 3, and 11.

15. The answer to a subtraction problem is called the _____.

16. A fraction is in _____ when there are no common factors in the numerator and denominator.

17. The answer to a division problem is called the _____.

18. The answer to a multiplication problem is called the _____.

19. A natural number (except 1) is _____ if it has only 1 and itself as factors.

20. One way to write a division problem is as a _____.

21. The _____ of a fraction shows the number of equal parts in a whole.

Name: Date:

Instructor: Section:

Guided Examples

Review this example for Objective 1:

1. Write each number as a product of prime factors.

   a. 21

   Write 21 as the product of the prime factors 3 and 7, or as $3 \cdot 7$.

   b. 104

   We show a factor tree on the right. The prime factors are circled.

   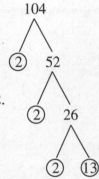

   Divide by the least prime factor of 104, 2.
   $104 = 2 \cdot 52$.

   Now divide 52 by 2 to find two factors of 52.
   $52 = 2 \cdot 26$.

   Now factor 26 as
   $26 = 2 \cdot 13$.
   All factors are prime.

Now Try:

1. Write each number as a product of prime factors.

   a. 35   _____

   b. 120   _____

Review this example for Objective 2:

2. Write each fraction in lowest terms.

   a. $\dfrac{10}{25}$

   First write the numerator and the denominator as the product of prime factors. Then divide the numerator and the denominator by the **greatest common factor**, which is the product of all factors common to both.

   $$\frac{10}{25} = \frac{2 \cdot 5}{5 \cdot 5} = \frac{2 \cdot 1}{5 \cdot 1} = \frac{2}{5}$$

   b. $\dfrac{28}{168} = \dfrac{2 \cdot 2 \cdot 7}{2 \cdot 2 \cdot 2 \cdot 3 \cdot 7} = \dfrac{1 \cdot 1 \cdot 1}{2 \cdot 1 \cdot 1 \cdot 3 \cdot 1} = \dfrac{1}{6}$

Now Try:

2. Write each fraction in lowest terms.

   a. $\dfrac{30}{36}$ _____

   b. $\dfrac{48}{150}$ _____

Review this example for Objective 3:

3. Find the product or quotient and write it in lowest terms.

a. $\dfrac{10}{75} \cdot \dfrac{5}{4} = \dfrac{10 \cdot 5}{75 \cdot 4}$ Multiply numerators. Multipy denominators.

$= \dfrac{2 \cdot 5 \cdot 5}{3 \cdot 5 \cdot 5 \cdot 2 \cdot 2}$ Factor the numerator and the denominator.

$= \dfrac{1}{3 \cdot 2}$ Divide numerator and denominator by $2 \cdot 5 \cdot 5$, or 50.

$= \dfrac{1}{6}$ Lowest terms

b. $2\dfrac{5}{8} \div \dfrac{3}{4} = \dfrac{21}{8} \div \dfrac{3}{4}$ Write the mixed number as an improper fraction.

$= \dfrac{21}{8} \cdot \dfrac{4}{3}$ Multiply by the reciprocal of the second fraction.

$= \dfrac{21 \cdot 4}{8 \cdot 3}$ Multiply numerators. Multipy denominators.

$= \dfrac{3 \cdot 7 \cdot 2 \cdot 2}{2 \cdot 2 \cdot 2 \cdot 3}$ Factor the numerator and the denominator.

$= \dfrac{7}{2}$ Lowest terms

Now Try:

3. Find the product or quotient and write it in lowest terms.

a. $\dfrac{15}{36} \cdot \dfrac{9}{25}$ _____

b. $\dfrac{5}{4} \div 2\dfrac{7}{24}$ _____

Review this example for Objective 4:

4. Find the sum or difference and write it in lowest terms.

   a. $\dfrac{5}{12} + \dfrac{3}{8}$

   First find the least common denominator by factoring both denominators
   $12 = 2 \cdot 2 \cdot 3$ and $8 = 2 \cdot 2 \cdot 2$
   Since 2 and 3 are the factors and 2 is a factor of 8 three times, the LCD is
   $2 \cdot 2 \cdot 2 \cdot 3 = 24$.
   Write each fraction with 24 as the denominator.

   $\dfrac{5}{12} = \dfrac{5 \cdot 2}{12 \cdot 2} = \dfrac{10}{24}$ and $\dfrac{3}{8} = \dfrac{3 \cdot 3}{8 \cdot 3} = \dfrac{9}{24}$

   Add the two equivalent fractions.

   $\dfrac{10}{24} + \dfrac{9}{24} = \dfrac{19}{24}$

   b. $4\dfrac{5}{12} - 1\dfrac{11}{16} = \dfrac{53}{12} - \dfrac{27}{16}$ Write each mixed number as an improper fraction.

   $= \dfrac{53 \cdot 4}{48} - \dfrac{27 \cdot 3}{48}$ Find a common denominator.

   $= \dfrac{212}{48} - \dfrac{81}{48}$

   $= \dfrac{131}{48}$ or $2\dfrac{35}{48}$ Subtract. Write as a mixed number.

Now Try:

4. Find the sum or difference and write it in lowest terms.

   a. $\dfrac{5}{12} + \dfrac{5}{18}$ _____

   b. $2\dfrac{5}{8} - 1\dfrac{15}{32}$ _____

Review this example for Objective 5:

5. *Solve*: A triangle has sides of length $\dfrac{1}{2}$ foot, $1\dfrac{1}{4}$ feet, and $1\dfrac{1}{8}$ feet. What is the distance around this triangle?

   Solution:
   This is an addition problem.

   $\dfrac{1}{2} + 1\dfrac{1}{4} + 1\dfrac{1}{8} = \dfrac{4}{8} + 1\dfrac{2}{8} + 1\dfrac{1}{8} = 2\dfrac{7}{8}$

   The distance around the triangle is $2\dfrac{7}{8}$ feet.

Now Try:

5. *Solve*: Saul sold $\dfrac{3}{5}$ bushel of potatoes, $\dfrac{2}{5}$ bushel of apples, $\dfrac{3}{4}$ bushel of pears, $\dfrac{1}{4}$ bushel of peppers, and $1\dfrac{1}{4}$ bushels of tomatoes. How many bushels of fruits and vegetables did he sell?

   _____

Name: Date:
Instructor: Section:

Review this example for Objective 6:

6. Use the circle graph to answer the question.

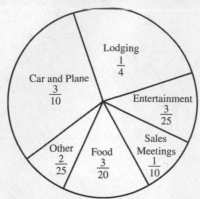

The circle graph shows the expenses involved in keeping a sales force on the road. Each expense item is expressed as a fraction of the total sales force cost of $950,000. How much was spent on car and plane expenses?

Solution:

We must find $\frac{3}{10}$ of $950,000.

$$\frac{3}{10} \cdot \frac{950,000}{1} = \frac{2,850,000}{10} = 285,000$$

$285,000 was spent on car and plane expenses.

Now Try:

6. Use the circle graph to answer the question.

The circle graph shows the expenses involved in keeping a sales force on the road. Each expense item is expressed as a fraction of the total sales force cost of $750,000. How much was spent on food expenses?

_____

Objective 1 Learn the definition of *factor*.

For extra help, see Example 1 on page 3 of your text, the Section Lecture video for Section 1.1, and Exercise Solution Clip 13.

Identify each numbers as prime, composite, *or* neither. *If the number is composite, write the number as the product of prime factors.*

1. 48

2. 127

1. _____

2. _____

Objective 2 Write fractions in lowest terms.
For extra help, see Example 2 on page 4 of your text, the Section Lecture video for Section 1.1, and Exercise Solution Clip 29.

Write the fraction in lowest terms.

3. $\dfrac{42}{15}$

3. _____

4. $\dfrac{144}{324}$

4. _____

Objective 3 Multiply and divide fractions.
For extra help, see Examples 3 and 4 on pages 4–6 of your text, the Section Lecture video for Section 1.1, and Exercise Solution Clips 43 and 53.

Find the product or quotient and write it in lowest terms.

5. $4\dfrac{3}{8} \cdot 5\dfrac{3}{7}$

5. _____

6. $\dfrac{12}{13} \div 6$

6. _____

7. $9\dfrac{5}{8} \div 3\dfrac{1}{2}$

7. _____

Objective 4 Add and subtract fractions.
For extra help, see Examples 5–7 on pages 6–8 of your text, the Section Lecture video for Section 1.1, and Exercise Solution Clips 71, 73, and 83.

Find the sum or difference and write it in lowest terms.

8. $6\dfrac{3}{5} + 5\dfrac{1}{2}$

8. _____

9. $\dfrac{7}{15} - \dfrac{3}{10}$

9. _____

10. $11\frac{1}{6} - 2\frac{2}{3}$ 10. _____

Objective 5 Solve applied problems that involve fractions.
For extra help, see Example 8 on page 9 of your text, the Section Lecture video for Section 1.1, and Exercise Solution Clip 99.

Solve the problem.

11. Pete has $12\frac{2}{3}$ cords of firewood for sale. If he sells the   11. _____

firewood in face cord lots (a face cord equals $\frac{1}{3}$ of a

cord), how many face cords does he have for sale?

12. A cake recipe calls for $1\frac{2}{3}$ cups of sugar. A caterer has   12. _____

20 cups of sugar on hand. How many cakes can she
make?

13. Marissa bought 8 yards of fabric. She used $1\frac{3}{4}$ yards for   13. _____

a blouse and $3\frac{5}{8}$ yards for a skirt. How many yards
were left?

Name: Date:

Instructor: Section:

Objective 6 Interpret data in a circle graph.

For extra help, see Example 9 on pages 9–10 of your text and the Section Lecture video for Section 1.1.

Use the circle graph to answer the questions.

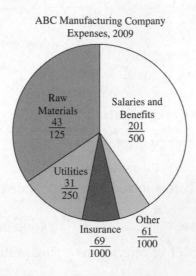

ABC Manufacturing Company
Expenses, 2009

Raw Materials $\frac{43}{125}$

Salaries and Benefits $\frac{201}{500}$

Utilities $\frac{31}{250}$

Insurance $\frac{69}{1000}$

Other $\frac{61}{1000}$

14. What fractional part of the total expenses came from raw materials?

14. _____

15. If the expenses were $2.7 billion:
    (a) How much more was spent on salaries and benefits than on raw materials?
    (b) How much was spent on areas other than insurance?
    (c) How much more was spent on insurance than on "other" expenses?

15. _____

Chapter 1 THE REAL NUMBER SYSTEM

1.2   Exponents, Order of Operations, and Inequality

Learning Objectives
1   Use exponents.
2   Use the rules for order of operations.
3   Use more than one grouping symbol.
4   Know the meanings of $\neq$, $<$, $>$, $\leq$, and $\geq$.
5   Translate word statements to symbols.
6   Write statements that change the direction of inequality symbols.

Key Terms
Use the vocabulary terms listed below to complete each statement in exercises 1–6.

exponent (power)	**base**	**exponential expression**
grouping symbols	**order of operations**	**inequality**

1.   A number or letter (variable) written with an exponent is a(n) _____.

2.   The _____ is the number that is a repeated factor when written with an exponent.

3.   The _____ is used to evaluate expressions containing more than one operation.

4.   A(n) _____ is a number that indicates how many times a factor is repeated.

5.   A(n) _____ is a statement that two expressions are not equal.

6.   _____ are parentheses ( ), brackets [ ], or fraction bars.

Name: _____   Date: _____

Instructor: _____   Section: _____

Guided Examples

Review this example for Objective 1:

1. Find the value of each exponential expression.

   a. 3^4

     Read 3^4 as "3 to the fourth power."

     $3^4 = \underline{3 \cdot 3 \cdot 3 \cdot 3} = 81$

          3 is used as a factor four times.

   b. $\left(\dfrac{2}{5}\right)^3 = \dfrac{2}{5} \cdot \dfrac{2}{5} \cdot \dfrac{2}{5} = \dfrac{8}{125}$

Now Try:

1. Find the value of each exponential expression.

   a. 4^3 _____

   b. $\left(\dfrac{4}{3}\right)^5$ _____

Review this example for Objective 2:

2. Find the value of each expression.

   a. $7 + 2 \cdot 4$

     Multiply first, then add.

     $7 + 2 \cdot 4 = 7 + 8$

            $= 15$

   b. $8(6-3) - 7 \cdot 3 = 8(3) - 7 \cdot 3$

                  $= 24 - 21$

                  $= 3$

   c. $6^2 \div 3^2 - 4 \cdot 3 + 2 \cdot 5 = 36 \div 9 - 4 \cdot 3 + 2 \cdot 5$

                        $= 4 - 4 \cdot 3 + 2 \cdot 5$

                        $= 4 - 12 + 10$

                        $= 2$

Now Try:

2. Find the value of each expression.

   a. $18 - 15 \div 3$ _____

   b. $7(5+3) - 4 \cdot 6$ _____

   c. $7 \cdot 3 - 4^2 \div 2^3 + 6 \cdot 5$ _____

Name:
Instructor:

Date:
Section:

Review this example for Objective 3:
3. Simplify each expression.

a. $8\left[14+3^2(9-4)\right]$
   $= 8\left[14+9(9-4)\right]$
   Apply exponent.
   $= 8\left[14+9(5)\right]$
   Subtract inside parentheses.
   $= 8\left[14+45\right]$
   Multiply.
   $= 8\left[59\right]$
   Add inside brackets.
   $= 472$
   Multiply

b. $\dfrac{6\cdot 10+9\cdot 2}{3(4-2)} = \dfrac{60+18}{3(2)}$
   Simplify numerator and denominator separately.
   $= \dfrac{78}{6}$
   $= 13$

Now Try:
3. Simplify each expression.

a. $3^3\left[(6+5)-2^2\right]$ _____

b. $\dfrac{4\cdot 10-2\cdot 2}{2^2(9-6)}$ _____

Review this example for Objective 4:
4. Determine whether each statement is *true* or *false*.

a. $7 \neq 21 \div 3$
   This statement is *false* because $7 = 21 \div 3$.

b. $3\cdot 6 > 12-9$
   This statement is *true* because $18 > 3$.

c. $25 \leq 30$
   This statement is *true* because $25 < 30$.

d. $\dfrac{5}{8} < \dfrac{4}{9}$
   First convert the fractions to fractions with common denominators. Then compare the fractions.
   $\dfrac{5}{8} = \dfrac{45}{72}; \ \dfrac{4}{9} = \dfrac{32}{72}; \ \dfrac{45}{72} < \dfrac{32}{72}$
   This statement is false.

Now Try:
4. Determine whether each statement is *true* or *false*.

a. $24 \neq 8\cdot 3$ _____

b. $18 \div 6 < 4+5$ _____

c. $65 \geq 9\cdot 7$ _____

d. $\dfrac{13}{4} \geq \dfrac{88}{24}$ _____

Name: _____   Date: _____
Instructor: _____   Section: _____

Review this example for Objective 5:	Now Try:
5. Write each word statement in symbols.	5. Write each word statement in symbols.
a. Eight is not equal to thirteen minus four. $8 \neq 13 - 4$.	**a.** Six is less than or equal to six. _____
b. Twenty-one is greater than fifteen. $21 > 15$.	**b.** Seven is greater than the quotient of fifteen and five. _____

Review this example for Objective 6:	Now Try:
6. Write the statement as another true statement with the inequality symbol reversed. $12 \leq 21$ **Solution:** $21 \geq 21$	6. Write the statement as another true statement with the inequality symbol reversed. $17 > 5$ _____

Objective 1 Use exponents.

For extra help, see Example 1 on page 15 of your text, the Section Lecture video for Section 1.2, and Exercise Solution Clip 9.

Find the value of each exponential expression.

1. 3^4

1. _____

2. $\left(\dfrac{2}{3}\right)^4$

2. _____

3. $(0.4)^2$

3. _____

Objective 2 Use the rules for order of operations.

For extra help, see Example 2 on page 16 of your text, the Section Lecture video for Section 1.2, and Exercise Solution Clips 29 and 51.

Find the value of each expression.

4. $3 \cdot 15 + 10^2$

4. _____

5. $20 \div 5 - 3 \cdot 1$ 5. _____

6. $\left(\dfrac{5}{6}\right)\left(\dfrac{3}{2}\right) - \left(\dfrac{1}{3}\right)^2$ 6. _____

Objective 3 Use more than one grouping symbol.
For extra help, see Example 3 on page 17 of your text, the Section Lecture video for Section 1.2, and Exercise Solution Clips 29 and 51.

Find the value of each expression.

7. $4\left[3 + 2(9 - 2)\right]$ 7. _____

8. $19 + 3\left[8(5 - 2) + 6\right]$ 8. _____

9. $3^3\left[(6 + 5) - 2^2\right]$ 9. _____

Objective 4 Know the meanings of $\neq$, $<$, $>$, $\leq$, and $\geq$.
For extra help, see Example 4 on page 18 of your text, the Section Lecture video for Section 1.2, and Exercise Solution Clip 57.

Tell whether each statement is true *or* false.

10. $3 \cdot 4 \div 2^2 \neq 3$ 10. _____

11. $\dfrac{5 + 4 \cdot 5}{14 - 2 \cdot 3} \geq 2$ 11. _____

Objective 5 Translate word statements to symbols.
For extra help, see Example 5 on page 18 of your text, the Section Lecture video for Section 1.2, and Exercise Solution Clip 87.

Write each word statement in symbols.

12. Seven equals twelve minus five.   12. _____

13. The sum of nine and thirteen is greater than twenty-one.   13. _____

Objective 6 Write statements that change the direction of inequality symbols.
For extra help, see Example 6 on page 19 of your text and the Section Lecture video for Section 1.2.

Write each statement with the inequality symbol reversed.

14. $\dfrac{2}{3} < \dfrac{3}{4}$ 14. _____

15. $0.921 \le 0.922$ 15. _____

Chapter 1 THE REAL NUMBER SYSTEM

1.3   Variables, Expressions, and Equations

Learning Objectives
1   Evaluate algebraic expressions, given values for the variables.
2   Translate word phrases to algebraic expressions.
3   Identify solutions of equations.
4   Identify solutions of equations from a set of numbers.
5   Distinguish between *expressions* and *equations*.

Key Terms
Use the vocabulary terms listed below to complete each statement in exercises 1–6.

variable	algebraic expression	equation
solution	set	elements

1. A(n) _____ is a symbol, usually a letter, used to represent an unknown number.

2. A(n) _____ is a statement that two algebraic expressions are equal.

3. A(n) _____ is a collection of objects.

4. A(n) _____ of an equation is any replacement for the variable that makes the equation true.

5. _____ are the objects that belong to a set.

6. A(n) _____ is a sequence of numbers, variables, operation symbols, and/or grouping symbols (such as parentheses) formed according to the rules of algebra.

Guided Examples

Review these examples for Objective 1:
1. Find the value of each algebraic expression for $m = 4$.

   a. $7m$
       $7m = 7 \cdot m = 7 \cdot 4 = 28$

   b. $2m^3$
       $2m^3 = 2 \cdot 4^3 = 2 \cdot 64 = 128$

Now Try:
1. Find the value of each algebraic expression for $n = 3$.

   a. $6n$ _____

   b. $4n^4$ _____

2. Find the value of each algebraic expression for $x = 2$ and $y = 4$.

   a. $2x + 3y$

$$2x + 3y = 2(2) + 3(4) \quad x = 2,\ y = 4$$
$$= 4 + 12 \qquad \text{Multiply.}$$
$$= 16 \qquad \text{Add.}$$

   b. $\dfrac{2x + 3y}{3x - y + 2}$

$$\frac{2x + 3y}{3x - y + 2} = \frac{2(2) + 3(4)}{3(2) - 4 + 2} \quad x = 2,\ y = 4$$
$$= \frac{4 + 12}{6 - 4 + 2} \qquad \text{Multiply.}$$
$$= \frac{16}{4} = 4 \qquad \begin{array}{l}\text{Simplify numerator}\\\text{and denominator,}\\\text{then divide.}\end{array}$$

   c. $3x^4 - 2y^2$

$$3x^4 - 2y^2 = 3 \cdot 2^4 - 2 \cdot 4^2 \quad x = 2,\ y = 4$$
$$= 3 \cdot 16 - 2 \cdot 16 \qquad \text{Apply the exponents.}$$
$$= 48 - 32 \qquad \text{Multiply.}$$
$$= 16 \qquad \text{Subtract.}$$

2. Find the value of each algebraic expression for $x = 4$ and $y = 2$.

   a. $3x + 2y$ _____

   b. $\dfrac{3x - 2y}{2x + 3y + 2}$ _____

   c. $\dfrac{1}{2}x^4 + \dfrac{1}{4}y^2$ _____

Review this example for Objective 2:

3. Write each word phrase as an algebraic expression, using x as the variable.

   a. The sum of four and a number.
     $x + 4$, or $4 + x$

   b. The quotient of a number and 8.
     $\dfrac{x}{8}$, or $x \div 8$

   c. The difference between twice a number and 7
     $2x - 7$

Now Try:

3. Write each word phrase as an algebraic expression, using x as the variable.

   a. The sum of eight and a number _____

   b. The quotient of 8 and a number _____

   c. Ten times a number, added to 21 _____

Review this example for Objective 3:

4. Decide whether the given number is a solution of the equation.

$2x + 4 = 16;\ 6$

$2 \cdot 6 + 4 \overset{?}{=} 16 \quad x = 6$

$12 + 4 \overset{?}{=} 16 \quad$ Multiply.

$16 = 16 \quad$ True. The right side of the equation equals the left side.

Now Try:

4. Decide whether the given number is a solution of the equation.

$8b + 6 = 14;\ 2$ _____

Review this example for Objective 4:

5. Write the word statement as an equation. Then find all solutions of the equation from the set $\{1, 3, 5, 7, 9\}$

The sum of twice a number and four is ten.

Start with $5x$, and   The word *is*
then add 4 to it.   translates as = .   ten
$\downarrow$ $\downarrow$ $\downarrow$
$2x + 4$ $=$ 10

Substitute each of the given numbers. The solution is 3 since $2 \cdot 3 + 4 = 10$ is true.

Now Try:

5. Write the word statement as an equation. Then find all solutions of the equation from the set $\{1, 3, 5, 7, 9\}$

The quotient of thirty-five and a number is seven.

_____

Review this example for Objective 5:

6. Decide whether the following is an *equation* or an *expression*.

$5x^2 + 3$

There is no equals symbol, so this is an expression.

Now Try:

6. Decide whether the following is an *equation* or an *expression*.

$9x + 2y = 2$

_____

Objective 1 Evaluate algebraic expressions, given values for the variables.

For extra help, see Examples 1 and 2 on page 23 of your text and the Section Lecture video for Section 1.3.

Find the value of each expression if $x = 2$ and $y = 4$.

1. $8x^2 - 6x$ 1. _____

2. $\dfrac{x^2 + y}{x + 1}$ 2. _____

3. $\dfrac{3y^2 + 2x^2}{5x + y^2}$

3. _____

Objective 2 Translate word phrases to algebraic expressions.

For extra help, see Example 3 on page 23 of your text, the Section Lecture video for Section 1.3, and Exercise Solution Clip 39.

*Change the word phrases to algebraic expressions. Use **x** as the variable.*

4. The product of four less than a number and two

4. _____

5. The sum of a number and 4 is divided by twice the number

5. _____

6. Half a number is subtracted from two-thirds of the number

6. _____

Objective 3 Identify solutions of equations.

For extra help, see Example 4 on page 24 of your text, the Section Lecture video for Section 1.3, and Exercise Solution Clip 57.

Determine whether the given number is a solution of the equation.

7. $5 + 3x^2 = 19; \ 2$

7. _____

8. $\dfrac{x^2 - 7}{x} = 6; \ 2$

8. _____

9. $x^2 + 2x + 1 = 9; \ 2$

9. _____

Objective 4 Identify solutions of equations from a set of numbers.
For extra help, see Example 5 on page 25 of your text, the Section Lecture video for
Section 1.3, and Exercise Solution Clip 67.

Change the word statement to an equation. Use ***x*** *as the variable. Then find the solution
of the equation from the set* {0, 2, 4, 6, 8, 10}.

10. The sum of three times a number and five is 23.   **10.** _____

11. Three times a number is equal to two more than twice   **11.** _____
the number.

12. 10 divided by a number is three more than the number.   **12.** _____

Objective 5 Distinguish between *expressions* and *equations*.
For extra help, see Example 6 on page 25 of your text, the Section Lecture video for
Section 1.3, and Exercise Solution Clip 75.

Decide whether each of the following is an equation or an expression.

13. $3x + 2y$ **13.** _____

14. $y^2 - 4y - 3$ **14.** _____

15. $\dfrac{x+3}{15} = 2x$ **15.** _____

Chapter 1 THE REAL NUMBER SYSTEM

1.4   Real Numbers and the Number Line

Learning Objectives

1   Classify numbers and graph them on number lines.
2   Tell which of two real numbers is less than the other.
3   Find the additive inverse of a real number.
4   Find the absolute value of a real number.
5   Interpret the meanings of real numbers from a table of data.

Key Terms

Use the vocabulary terms listed below to complete each statement in exercises 1–11.

number line	**integers**	**signed numbers**
rational numbers	**set-builder notation**	**graph**
coordinate	**irrational number**	**real numbers**
additive inverses	**absolute value**	

1.   Each number on a number line is called the _____ of the point that it labels.

2.   The _____ of a number is the distance between 0 and the number on a number line.

3.   A(n) _____ is a line with a scale that is used to show how numbers relate to each other.

4.   Two numbers are called _____ if their sum is equal to zero.

5.   _____ is used to describe a set of numbers without actually having to list all the elements.

6.   Rational and irrational numbers together form the set of _____.

7.   _____ are numbers that can be written with a positive or negative sign.

8. $\{\ldots, -3, -2, -1, 0, 1, 2, 3, \ldots\}$ is the set of _____.

9.   _____ can be written as the quotient of two integers, with denominator not 0.

10. The point on a number line that corresponds to a number is its _____.

11. An _____ cannot be written as the quotient of two integers but can be represented by a point on the number line.

Guided Examples

Review these examples for Objective 1:

1. Use an integer to express the number in boldface italics in the following statement.

The record low temperature in the United States was *80* degrees below zero at Prospect Creek Camp in Alaska on January. 23, 1971. (*Source*: USAToday.com)

Answer: -80

Now Try:

1. Use an integer to express the number in boldface italics in the following statement.

Between 2000 and 2009, the population of Arizona increased by *1,465,146*. _____ (*Source*: factfinder.census.gov)

2. List the numbers in the following set that belong to each set of numbers.

$$\left\{-8, -\tfrac{2}{3}, 0, \sqrt{2}, 3.5, \pi\right\}$$

a. Whole numbers

b. Integers

c. Rational numbers

d. Irrational numbers

Answers:
  a. $\{0\}$
  b. $\{-8, 0\}$
  c. $\left\{-8, -\tfrac{2}{3}, 0, 3.5\right\}$
  d. $\left\{\sqrt{2}, \pi\right\}$

2. List the numbers in the following set that belong to each set of numbers.

$$\left\{-9, -\tfrac{4}{3}, 0, 0.\overline{3}, \sqrt{3}, \pi\right\}$$

a. Natural numbers _____

b. Integers _____

c. Irrational numbers _____

d. Real numbers _____

Review this example for Objective 2:

3. Determine whether the statement is *true* or *false*.

$-5 > -1$

Answer: False

Now Try:

3. Determine whether the statement is *true* or *false*.

$\tfrac{2}{3} \le -\tfrac{1}{2}$ _____

Name: Date:
Instructor: Section:

Review this example for Objective 3:	Now Try:
4. Find the additive inverse of −1.5.	**4.** Find the additive inverse of 2.35.
Answer: 1.5	_____

Review this example for Objective 4:	Now Try:
5. Simplify by finding the absolute value.	**5.** Simplify by finding the absolute value.
a. $\|7\|$	**a.** $\|-2\|$ _____
b. $\|-7\|$	**b.** $-\|-2\|$ _____
c. $-\|-7\|$	**c.** $\|2\|$ _____
Answers:	
a. 7	
b. 7	
c. −7	

Review this example for Objective 5:

6. The table below shows the changes in population for five cities.

City	1980–1990	1990–2000	2000–2009
New York	250,925	685,714	383,603
Los Angeles	518,548	209,422	137,048
Chicago	−221,346	112,290	−44,748
Houston	35,415	323,078	304,295
Philadelphia	−102,633	−68,027	29,747

Source: factfinder.census.gov

Which city had the greatest change in population? During which years?

Answer: New York from 1990–2000.

Now Try:

6. Use the table shown at the left to determine which city has the least change in population and in which years. _____

Objective 1 Classify numbers and graph them on number lines.
For extra help, see Examples 1 and 2 on pages 29–31 of your text, the Section Lecture video for Section 1.4, and Exercise Solution Clip 27.

List all the sets among the following to which the number belongs: natural numbers, whole numbers, integers, rational numbers, irrational numbers, real numbers.

1. −2.6

1. _____

2. $\sqrt{7}$

2. _____

Graph each group of rational numbers on a number line.

3. $\frac{1}{2}, 0, -3, -\frac{5}{2}$

3.

Objective 2 Tell which of two real numbers is less than the other.
For extra help, see Example 3 on page 31 of your text, the Section Lecture video for Section 1.4, and Exercise Solution Clip 65.

Select the smaller number in each pair.

4. $-0.802, -0.820$

4. _____

*Decide whether each statement is **true** or **false**.*

5. $-76 < 45$

5. _____

6. $-5 > -5$

6. _____

Objective 3 Find the additive inverse of a real number.
For extra help, see the Section Lecture video for Section 1.4.

Find the additive inverse of the number.

7. -15

7. _____

8. $\frac{5}{8}$

8. _____

9. $-2\frac{5}{8}$

9. _____

Objective 4 Find the absolute value of a real number.

For extra help, see Example 4 on page 33 of your text, the Section Lecture video for Section 1.4, and Exercise Solution Clip 43.

Simplify by removing absolute value symbols.

10. $-|95|$

10. _____

11. $-|-25|$

11. _____

12 $|-7.52|$

12. _____

Objective 5 Interpret the meanings of real numbers from a table of data.

For extra help, see Example 5 on page 33 of your text and the Section Lecture video for Section 1.4.

Use the table below of counties in the state of Indiana to answer the questions.

County Name	July 1, 2008 Estimated Population	July 1, 2007 Estimated Population	Numeric Population Change 2007–2008
Allen	348,791	350,523	−1732
Benton	8769	8774	−5
Delaware	114,685	115,167	−482
Elkhart	199,137	197,639	1498
Henry	47,162	46,991	171
Lake	493,800	491,238	2562
Monroe	128,992	127,625	1367
Union	7157	7177	−20

Source: U.S. Census

13. Which county experienced the greatest population growth?

13. _____

14. Which county experienced the greatest decline in population?

14. _____

15. By how many more people did the population of Elkhart County change compared with Henry County?

15. _____

Chapter 1 THE REAL NUMBER SYSTEM

1.5   Adding and Subtracting Real Numbers

Learning Objectives
1   Add two numbers with the same sign.
2   Add two numbers with different signs.
3   Use the definition of subtraction.
4   Use the rules for order of operations with real numbers.
5   Translate words and phrases involving addition and subtraction.
6   Use signed numbers to interpret data.

Key Terms
Use the vocabulary terms listed below to complete each statement in exercises 1–2.

   minuend   **subtrahend**

1.   In the subtraction 256 – 187, 256 is called the _____.

2.   In the subtraction $a - b$, b is called the _____.

Guided Examples

Review these examples for Objective 1:
1. Use a number line to find each sum.

    a. $2 + 4$
    Start at 0 and draw an arrow 2 units to the *right*. From the right end of that arrow, draw another arrow 4 units to the right. The number below the end of this second arrow is 6, so $2 + 4 = 6$.

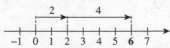

    b. $-1 + (-3)$
    Start at 0 and draw an arrow 1 unit to the *left*. From the left end of that arrow, draw another arrow 3 units to the left. The number below the end of this second arrow is -4, so $-1 + (-3) = -4$.

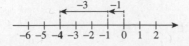

Now Try:
1. Use a number line to find each sum.

    a. $1 + 5$ _____

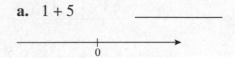

    b. $(-2) + (-5)$ _____

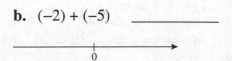

2. Find the sum.

$-12 + (-8)$

To add two numbers with the same sign, add the absolute values of the numbers. The sum has the same sign as the numbers being added.

$-12 + (-8) = -20$

2. Find the sum.

$(-7) + (-16)$ _____

Review these examples for Objective 2:

3. Use a number line to find the sum.

$4 + (-9)$

Start at 0 and draw an arrow 4 units to the *right*. From the right end of that arrow, draw another arrow 9 units to the *left*. The number below the end of this second arrow is −5, so $4 + (-9) = -5$.

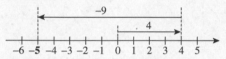

Now Try:

3. Use a number line to find the sum.

$2 + (-5)$ _____

---

4. Find the sum.

$-14 + 6$

Find the absolute value of each number, then find the difference between these absolute values.

$|-14| = 14; \ |6| = 6$

$14 - 6 = 8$

The sum will be negative since $|-14| > |6|$.

$-14 + 6 = -8$

4. Find the sum.

$-16 + 7$ _____

---

5. Check each answer by adding mentally.

a. $-\dfrac{4}{7} + \dfrac{3}{5} = \dfrac{1}{35}$

Find a common denominator.

$-\dfrac{4}{7} + \dfrac{3}{5} = -\dfrac{20}{35} + \dfrac{21}{35} = \dfrac{1}{35}$

The answer is correct.

5. Check each answer by adding mentally.

a. $-\dfrac{2}{9} + \dfrac{2}{5} = \dfrac{8}{45}$ _____

b. $1.6 + (-3.2) = -1.6$
Find the absolute value of each number, then find the difference between these absolute values. $|1.6| = 1.6$; $|-3.2| = 3.2$
$3.2 - 1.6 = 1.6$
The sum will be negative since $|-3.2| > |1.6|$.
The answer is correct.

b. $2.4 + (-5.2) = -2.8$

_____

Review this example for Objective 3:

6. Subtract.

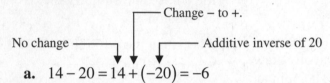

a. $14 - 20 = 14 + (-20) = -6$

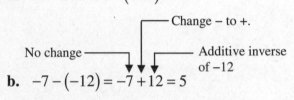

b. $-7 - (-12) = -7 + 12 = 5$

c. $\dfrac{3}{8} - \left(-\dfrac{1}{2}\right) = \dfrac{3}{8} + \dfrac{1}{2} = \dfrac{3}{8} + \dfrac{4}{8} = \dfrac{7}{8}$

Now Try:

6. Subtract.

a. $22 - (-24)$ _____

b. $-7.2 - 8.9$ _____

c. $\dfrac{1}{10} - \dfrac{1}{2}$ _____

Review this example for Objective 4:

7. Perform each indicated operation.

a. $3 - \left[-4 + (11 - 19)\right]$
$= 3 - \left[-4 + (-8)\right]$

      Start within the innermost parenthesis.

$= 3 - \left[-12\right]$ Add.
$= 3 + 12$ Definition of subtraction.
$= 15$ Add.

b. $-2 + 4\left|(-12 + 10) - (-6 + 2)\right|$

      Start within the innermost parenthesis.

$= -2 + 4\left|-2 - (-4)\right|$ Add.
$= -2 + 4\left|-2 + 4\right|$ Definition of subtraction.
$= -2 + 4\left|2\right|$ Add.
$= -2 + 4 \cdot 2$ Evaluate absolute value.
$= -2 + 8$ Multiply.
$= 6$ Add.

Now Try:

7. Perform each indicated operation.

a. $-9 + \left[5 + (21 - 30)\right]$

_____

b. $\left|-6 + 9\right| - 2\left|-4 + (11 - 19)\right|$

_____

Name: Date:

Instructor: Section:

Review these examples for Objective 5:

8. Write a numerical expression for the phrase and simplify the expression.

The sum of –14 and –29, increased by 27

Answer: [–14 + (–29)] + 27

9. Write a numerical expression for the phrase and simplify the expression.

The sum of –4 and 12, decreased by 9

Answer: (–4 + 12) – 9

10. A scientist runs an experiment at –43.3°C. He then lowers the temperature by 7.9°C. What is the new temperature for the experiment?

–43.3 – 7.9 = –51.2

The new temperature for the experiment is –51.2° C.

Review this example for Objective 6:

11. The bar graph below gives the Consumer Price Index (CPI) for footwear between 2002 and 2007. Use a signed number to represent the change in the CPI from 2005 to 2007.

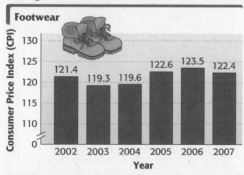

Source: U.S. Bureau of Labor Statistics.

Start with the index number for 2007. Subtract the index number for 2005 from it.

122.4 – 122.6 = 122.4 + (–122.6) = –0.2

The CPI decreased by 0.2 from 2005 to 2007.

Now Try:

8. Write a numerical expression for the phrase and simplify the expression.

–10 added to the sum of 20 and –4

_____

9. Write a numerical expression for the phrase and simplify the expression.

2 less than the difference between 10 and –4

_____

10. The highest point in a country has an elevation of 1408 meters. The lowest point is 396 meters below sea level. Using zero as sea level, find the difference between the two elevations.

_____

Now Try:

11. Refer to the bar graph at the left. Use a signed number to represent the change in the CPI from 2003 to 2006.

_____

    29

Objective 1 Add two numbers with the same sign.
For extra help, see Examples 1 and 2 on pages 37–38 of your text, the Section Lecture video for Section 1.5, and Exercise Solution Clip 11.

Find the sum.

1. $8 + 7$ 1. _____

2. $-4 + (-6)$ 2. _____

3. $-7 + (-11)$ 3. _____

Objective 2 Add two numbers with different signs.
For extra help, see Examples 3–5 on pages 38–39 of your text, the Section Lecture video for Section 1.5, and Exercise Solution Clip 31.

Find the sum.

4. $9 + (-16)$ 4. _____

5. $3\frac{5}{8} + \left(-2\frac{1}{4}\right)$ 5. _____

6. $14.1 + (-14.1)$ 6. _____

Objective 3 Use the definition of subtraction.
For extra help, see Example 6 on page 40 of your text, the Section Lecture video for Section 1.5, and Exercise Solution Clip 47.

Find each difference.

7. $13 - (-9)$ 7. _____

8. $4.5 - (-2.8)$ 8. _____

9. $3\frac{3}{4} - \left(-2\frac{1}{8}\right)$

9. _____

Objective 4 Use the rules for order of operations with real numbers.
For extra help, see Example 7 on pages 40–41 of your text, the Section Lecture video for Section 1.5, and Exercise Solution Clip 69.

Find the sum.

10. $\left[(-7)+14\right]+\left[(-16)+3\right]$

10. _____

11. $-7.6+\left[5.2+(-11.4)\right]$

11. _____

12. $-\frac{4}{5}+\left[\frac{1}{4}+\left(-\frac{2}{3}\right)\right]$

12. _____

Objective 5 Translate words and phrases involving addition and subtraction.
For extra help, see Examples 8–10 on pages 41–43 of your text, the Section Lecture video for Section 1.5, and Exercise Solution Clips 95 and 99.

Write a numerical expression for the phrase, and then simplify the expression.

13. 6 more than -2, increased by 8

13. _____

14. -6 subtracted from the sum of 2 and -3

14. _____

Objective 6 Use signed numbers to interpret data.
For extra help, see Example 11 on page 43 of your text, the Section Lecture video for Section 1.5, and Exercise Solution Clip 111.

Solve the problem by writing a sum or difference of real numbers and adding or subtracting. No variables are needed.

15. A mountain climber starts to climb at an altitude of 4325 feet. He climbs so that he gains 208 feet in altitude. Then he finds that, because of an obstruction, he must descend 25 feet. Then he climbs 58 feet up. What is his final altitude?

15. _____

Chapter 1 THE REAL NUMBER SYSTEM

1.6 Multiplying and Dividing Real Numbers

Learning Objectives
1   Find the product of a positive number and a negative number.
2   Find the product of two negative numbers.
3   Identify factors of integers.
4   Use the reciprocal of a number to apply the definition of division.
5   Use the rules for order of operations when multiplying and dividing signed numbers.
6   Evaluate expressions involving variables.
7   Translate words and phrases involving multiplication and division.
8   Translate simple sentences into equations.

Key Terms

Use the vocabulary terms listed below to complete each statement in exercises 1–3.

   Multiplicative inverse (reciprocal)   **dividend**   **divisor**

1. In the division $\frac{27}{4}$, 27 is the _____ .

2. In the division $a \div b$, b is the _____ .

3. The _____ of a nonzero real number a is $\frac{1}{a}$.

Guided Examples

Review this example for Objective 1:
1. Find each product.

   a. $(-80)(4)$
      $(-80)(4) = -(80 \cdot 4) = -320$

   b. $\left(\frac{1}{5}\right)\left(-\frac{2}{3}\right)$

     $\left(\frac{1}{5}\right)\left(-\frac{2}{3}\right) = -\left(\frac{1}{5} \cdot \frac{2}{3}\right) = -\frac{2}{15}$

Now Try:
1. Find each product.

   a. $(-3.2)(4.1)$ _____

   b. $\left(-\frac{3}{8}\right)\left(\frac{14}{9}\right)$ _____

Review this example for Objective 2:
  2. Find the product.

  $-12(-8)$

  The product of two negative numbers is positive, so $-12(-8) = 12(8) = 96$.

Now Try:
  2. Find the product.

  $(-7)(-16)$ _____

Review these examples for Objective 4:

3. Find each quotient, using the definition of division.

a. $\dfrac{48}{-8}$

$\dfrac{48}{-8} = 48\left(-\dfrac{1}{8}\right) = -6 \qquad \dfrac{x}{y} = x \cdot \dfrac{1}{y}$

b. $-\dfrac{1}{5} \div \left(-\dfrac{2}{3}\right)$

$-\dfrac{1}{5} \div \left(-\dfrac{2}{3}\right) = -\dfrac{1}{5} \cdot \left(-\dfrac{3}{2}\right) = \dfrac{3}{10}$

4. Find each quotient.

a. $\dfrac{10}{0}$

Division by zero is not allowed, so the quotient is undefined.

b. $5.5 \div (-2.2)$

$5.5 \div (-2.2) = \dfrac{5.5}{-2.2} = -\dfrac{5}{2}$, or -2.5

Review this example for Objective 5:

5. Perform each indicated operation.

a. $-4\big[(-2)(7) - 2\big]$

$= -4\big[-14 - 2\big]$ Multiply inside brackets

$= -4\big[-16\big]$ Subtract.

$= 64$ Multiply.

b. $\dfrac{-7(2) - (-3)}{5 + (-3)}$

$= \dfrac{-14 + 3}{2}$ Simplify the numerator and denominator separately.

$= \dfrac{-11}{2} = -\dfrac{11}{2}$

Now Try:

3. Find each quotient, using the definition of division.

a. $\dfrac{-120}{-20}$ _____

b. $-\dfrac{3}{16} \div \dfrac{9}{8}$ _____

4. Find each quotient, using the definition of division.

a. $\dfrac{0}{15}$ _____

b. $-\dfrac{5}{8} \div \left(-\dfrac{25}{16}\right)$ _____

Now Try:

5. Perform each indicated operation.

a. $-7\big[-4 - (-2)(-3)\big]$

_____

b. $\dfrac{9(-4) - (-5 - 1)}{-6 - 4(-2)}$

_____

Review this example for Objective 6:

6. Evaluate $-x^2 + 2a^2 - 3y$ if $x = -3$, $y = 2$, and $a = 4$.

   $-x^2 + 2a^2 - 3y$

   $= -(-3)^2 + 2(4)^2 - 3(2)$ Substitute the given values for the variables.

   $= -9 + 2 \cdot 16 - 3(2)$ Apply the exponents.

   $= -9 + 32 - 6$ Multiply.

   $= 17$ Add, then subtract.

Now Try:

6. Evaluate $\dfrac{4a - 3x}{(y-5)^2}$ if $x = -3$, $y = 2$, and $a = 4$.

   _____

Review these examples for Objective 7:

7. Write a numerical expression for each phrase, and simplify the expression.

   a. 70% of the sum of 20 and –4

   $0.7[20 + (-4)]$ simplifies to 11.2

   b. –34 subtracted from two-thirds of the sum of 16 and –10

   $\frac{2}{3}[16 + (-10)] - (-34)$ simplifies to 38.

Now Try:

7. Write a numerical expression for each phrase, and simplify the expression.

   a. The product of 10 and –2, subtracted from –2

   _____

   b. 30% of the difference between 50 and –10, subtracted from 85

   _____

---

8. Write a numerical expression for the phrase and simplify the expression.

   The sum of –12 and the quotient of 49 and –7

   $-12 + \dfrac{49}{-7}$ simplifies to –19.

8. Write a numerical expression for the phrase and simplify the expression.

   The product of 40 and –3, divided by the difference between 5 and –10

   _____

Review this example for Objective 8:

9. Write each sentence as an equation, using x as the variable. Then find the solution from the list of integers between -12 and 12, inclusive.

a. The quotient of a number and -4 is 2.

$$\frac{x}{-4} = 2$$

Since $\frac{-8}{-4} = 2$, the solution of the equation is -8.

b. Two-thirds of a number is -6.

$$\frac{2}{3}x = -6$$

Since $\frac{2}{3}(-9) = -6$, the solution of the equation is -9.

Now Try:

9. Write each sentence as an equation, using x as the variable. Then find the solution from the list of integers between -12 and 12, inclusive.

a. The product of -2 and a number is 12.

_____

b. The sum of 5 and a number is -3.

_____

Objective 1 Find the product of a positive number and a negative number.
For extra help, see Example 1 on page 49 of your text and Section Lecture video for Section 1.6.

Find the product.

1. $7(-4)$

1. _____

2. $7(-2.5)$

2. _____

Objective 2 Find the product of two negative numbers.
For extra help, see Example 2 on page 50 of your text and Section Lecture video for Section 1.6.

Find the product.

3. $\left(-\frac{2}{7}\right)\left(-\frac{14}{5}\right)$

3. _____

Objective 3 Identify factors of integers.

For extra help, see the Section Lecture video for Section 1.6, and Exercise Solution Clip 31.

Find all the integer factors of the given number.

4. 40   **4.** _____

5. 36   **5.** _____

Objective 4 Use the reciprocal of a number to apply the definition of division.

For extra help, see Examples 3 and 4 on pages 51–52 of your text, the Section Lecture video for Section 1.6, and Exercise Solution Clips 35 and 41.

Use the definition of division to find each quotient.

6. $\dfrac{-120}{-20}$ **6.** _____

7. $-\dfrac{3}{16} \div \dfrac{9}{8}$ **7.** _____

Objective 5 Use the rules for order of operations when multiplying and dividing signed numbers.

For extra help, see Example 5 on page 53 of your text, the Section Lecture video for Section 1.6, and Exercise Solution Clip 55.

Perform the indicated operations.

8. $(-4)(9) - (-5)(4)$ **8.** _____

Simplify the numerators and denominators separately. Then find the quotient.

9. $\dfrac{-4\left[8-(-3+7)\right]}{-6\left[3-(-2)\right]-3(-3)}$ **9.** _____

Objective 6 Evaluate expressions involving variables.
For extra help, see Example 6 on page 54 of your text, the Section Lecture video for Section 1.6, and Exercise Solution Clip 83.

Evaluate the following expressions if $x = 3,\ y = 2,\ and\ a = 4$.

10. $(-2y + 4a)(3x + y)$ 10. _____

11. $\dfrac{2x^2 - 3y}{4a}$ 11. _____

Objective 7 Translate words and phrases involving multiplication and division.
For extra help, see Examples 7 and 8 on pages 54–55 of your text, the Section Lecture video for Section 1.6, and Exercise Solution Clips 95 and 101.

Write a numerical expression for each phrase and simplify.

12. The product of 7 and –2, added to 4   12. _____

13. The product of –4 and 7, divided by the sum of –3 and 14   13. _____

Objective 8 Translate simple sentences into equations.
For extra help, see Example 9 on pages 55–56 of your text, the Section Lecture video for Section 1.6, and Exercise Solution Clip 113.

Write each sentence as an equation, using x as the variable. Then find the solution from the set of integers between −12 and 12, inclusive.

14. The quotient of a number and –2 is –4.   14. _____

15. When 8 is added to a number, the result is 6.   15. _____

Chapter 1 THE REAL NUMBER SYSTEM

1.7   Properties of Real Numbers

Learning Objectives
1   Use the commutative properties.
2   Use the associative properties.
3   Use the identity properties.
4   Use the inverse properties.
5   Use the distributive property.

Key Terms
Use the vocabulary terms listed below to complete each statement in exercises 1–7.

commutative property   associative property   identity property

identity element for addition (additive identity)

identity element for multiplication (multiplicative identity)

inverse property   distributive property

1.   For any real numbers a, b, and c, the _____ states that
      $a(b+c) = ab + ac$ and $(b+c)a = ba + ca$.

2.   The _____ states that the sum of 0 and any number equals the
      number, and the product of 1 and any number equals the number.

3.   The _____ states that the way in which numbers being added (or
      multiplied) are grouped does not change the sum (or product).

4.   Since multiplying a number by 1 does not change the number, 1 is called the
      _____.

5.   The _____ states that a number added to its opposite is 0 and a
      number multiplied by its reciprocal is 1.

6.   The _____ states that the order of the numbers in an addition
      problem (or multiplication problem) can be changed without changing the sum (or
      product).

7.   Since adding 0 to a number does not change the number, 0 is called the
      _____.

Guided Examples

Review this example for Objective 1:

1. Use a commutative property to complete each statement.

    a. $y + 4 = \underline{\hspace{1cm}} + y$

        $y + 4 = \underline{4} + y$ Notice that the "order" changed.

    b. $5(2) = \underline{\hspace{1cm}}(5)$

        $5(2) = \underline{2}(5)$

Now Try:

1. Use a commutative property to complete each statement.

    a. $3 + (-4) = -4 + \underline{\hspace{1cm}}$

        _____

    b. $-4(p + 9) = \underline{\hspace{1cm}}(-4)$

        _____

Review these examples for Objective 2:

2. Use an associative property to complete each statement.

    a. $4r + (3s + 14t) = \underline{\hspace{1cm}} + 14t$

        $4r + (3s + 14t) = \underline{(4r + 3s)} + 14t$

                The "order" is the same. The "grouping" changed.

    b. $(4 \cdot 5)(-7) = \underline{\hspace{1cm}}\left[5(-7)\right]$

        $(4 \cdot 5)(-7) = \underline{4} \cdot \left[5(-7)\right]$

3. Is $(2 + 4) \cdot 6 = 6 \cdot (2 + 4)$ an example of the associative property or the commutative property?

    Answer: commutative

4. Find each sum or product.

    a. $8 + (-17) + 5 + 12 + 7$

        $= (8 + 12) + (-17 + 7) + 5$

        $= 20 + (-10) + 5$

        $= 15$

    b. $5(-19)(2) = (5 \cdot 2)(-19)$

               $= 10(-19)$

               $= -190$

Now Try:

2. Use an associative property to complete each statement.

    a. $\left[-4 + (-2)\right] + y$

        $= \underline{\hspace{1cm}} + (-2 + y)$

        _____

    b. $(2m)(-7) = (2)\underline{\hspace{1cm}}$

        _____

3. Is $(2 + 4) + 6 = (4 + 2) + 6$ an example of the associative property or the commutative property?

     _____

4. Find each sum or product.

    a. $23 + 86 + 25 + (-3) + (-16)$

        _____

    b. $\dfrac{1}{4}(13)(-8)$

        _____

Name: _____ Date: _____

Instructor: _____ Section: _____

Review these examples for Objective 3:

5. Use an identity property to complete each statement.

a. $\dfrac{3}{4} +$ _____ $= \dfrac{3}{4}$

$\dfrac{3}{4} + \underline{0} = \dfrac{3}{4}$

b. $-7 \cdot$ _____ $= -7$

$-7 \cdot \underline{1} = -7$

6. Simplify.

a. $\dfrac{16}{72}$

$\dfrac{16}{72} = \dfrac{2 \cdot 8}{9 \cdot 8}$ Factor.

$= \dfrac{2}{9} \cdot \dfrac{8}{8}$ Write as a product.

$= \dfrac{2}{9} \cdot 1$ Divide.

$= \dfrac{2}{9}$ Identity property

b. $\dfrac{4}{5} - \dfrac{3}{25}$

$\dfrac{4}{5} - \dfrac{3}{25} = \dfrac{4}{5} \cdot 1 - \dfrac{3}{25}$ Identity property

$= \dfrac{4}{5} \cdot \dfrac{5}{5} - \dfrac{3}{25}$ Use $1 = \dfrac{5}{5}$ to get a common denominator.

$= \dfrac{20}{25} - \dfrac{3}{25}$ Multiply.

$= \dfrac{17}{25}$ Subtract.

Now Try:

5. Use an identity property to complete each statement.

a. _____ $\cdot \dfrac{2}{3} = \dfrac{2}{3}$ _____

b. $-2 +$ _____ $= -2$ _____

6. Simplify.

a. $\dfrac{36}{84}$ _____

b. $\dfrac{5}{6} + \dfrac{2}{9}$ _____

Review these examples for Objective 4:

7. Use an inverse property to complete each statement.

   a. $\dfrac{3}{4} + \underline{\hspace{1cm}} = 0$

   $\dfrac{3}{4} + \left(-\dfrac{3}{4}\right) = 0$

   b. $-7 \cdot \underline{\hspace{1cm}} = 1$

   $-7 \cdot \left(-\dfrac{1}{7}\right) = 1$

Now Try:

7. Use an inverse property to complete each statement.

   a. $\underline{\hspace{1cm}} \cdot \dfrac{2}{3} = 1$ $\underline{\hspace{2cm}}$

   b. $-2 + \underline{\hspace{1cm}} = 0$ $\underline{\hspace{1.5cm}}$

8. Simplify. $4x + 7 - 4x$.

   $\begin{aligned}
   4x + 7 &- 4x \\
   &= (4x + 7) - 4x && \text{Order of operations} \\
   &= (7 + 4x) - 4x && \text{Commutative property} \\
   &= 7 + (4x - 4x) && \text{Associative property} \\
   &= 7 + \left[4x + (-4x)\right] && \text{Definition of subtraction} \\
   &= 7 + 0 && \text{Inverse property} \\
   &= 7 && \text{Identity property}
   \end{aligned}$

8. Simplify.

   $-\dfrac{3}{5}x - 12 + \dfrac{3}{5}x$ $\underline{\hspace{2cm}}$

Review these examples for Objective 5:

9. Use the distributive property to rewrite each property.

   a. $2(7y - 3z)$

   $2(7y - 3z) = 2(7y) + 2(-3z)$

   $ \text{Distributive property}$

   $= 14y - 6z \quad \text{Multiply.}$

   b. $n(2a - 4b + 6c)$

   $n(2a - 4b + 6c)$

   $= n(2a) + n(-4b) + n(6c)$

   $ \text{Distributive property}$

   $= 2an - 4bn + 6cn$

   $ \text{Commutative property}$

Now Try:

9. Use the distributive property to rewrite each property.

   a. $-4(2x + 9y)$ $\underline{\hspace{2cm}}$

   b. $4(12x - 9y - 11z)$

   $\underline{\hspace{3cm}}$

c. $6y + 7y$

$6y + 7y = y(6 + 7)$ Distributive property in reverse

$= y(13)$ Add.

$= 13y$ Commutative property

10. Write each expression without parentheses.

a. $-(-10 + n)$

$-(-10 + n)$

$= -1 \cdot (-10 + n)$ $-a = -1 \cdot a$

$= -1 \cdot (-10) + (-1) \cdot n$ Distributive property

$= 10 - n$ Multiply.

b. $-(-2a - 3b + c)$

$-(-2a - 3b + c)$

$= -1 \cdot (-2a - 3b + c)$

$= -1(-2a) - 1(-3b) - 1(c)$

$= 2a + 3b - c$

c. $3a - 3b$ _____

10. Write each expression without parentheses.

a. $-(-6x + 2)$ _____

b. $-(5x - 3y - z)$

_____

Objective 1 Use the commutative properties.

For extra help, see Example 1 on page 61 of your text and the Section Lecture video for Section 1.7.

Complete each statement. Use a commutative property.

1. $(ab)(2) = (2)$_____

2. $10\left(\frac{1}{4} \cdot 2\right) = $_____$(10)$

3. $2 + [10 + (-9)] = $_____$+ 2$

1. _____

2. _____

3. _____

Objective 2 Use the associative properties.

For extra help, see Examples 2–4 on pages 61–62 of your text, the Section Lecture video for Section 1.7, and Exercise Solutions Clip 27 and 45.

Complete each statement. Use an associative property.

4. $x(9y) = $_____$(y)$

4. _____

5. $(-r)\left[(-p)(-q)\right] = $ _____ $(-q)$

5. _____

6. $\left[x+(-4)\right]+3y = x+$ _____

6. _____

Objective 3 Use the identity properties.

For extra help, see Examples 5 and 6 on pages 62–63 of your text, the Section Lecture video for Section 1.7, and Exercise Solution Clip 39.

Simplify.

7. $-7+0 =$

7. _____

8. $1(-4) =$

8. _____

9. $7(1) =$

9. _____

Objective 4 Use the inverse properties.

For extra help, see Examples 7 and 8 on pages 63–64 of your text, the Section Lecture video for Section 1.7, and Exercise Solution Clip 55.

Complete the statements so that they are examples of either an identity property or an inverse property. Identify which property is used.

10. $-4+$ _____ $= 0$

10. _____

11. $\frac{2}{7} \cdot$ _____ $= 1$

11. _____

12. $-\frac{3}{5} \cdot$ _____ $= 1$

12. _____

Objective 5 Use the distributive property.

For extra help, see Examples 9 and 10 on pages 65–66 of your text, the Section Lecture video for Section 1.7, and Exercise Solution Clips 67 and 89.

Use the distributive property to rewrite each expression. Simplify the result if necessary.

13. $10r-4r$

13. _____

14. $-2(5y-9z)$

14. _____

15. $-14x+(-14y)$

15. _____

Chapter 1 THE REAL NUMBER SYSTEM

1.8   Simplifying Expressions

Learning Objectives
1   Simplify expressions.
2   Identify terms and numerical coefficients.
3   Identify like terms.
4   Combine like terms.
5   Simplify expressions from word phrases.

Key Terms

Use the vocabulary terms listed below to complete each statement in exercises 1–5.

    term   **numerical coefficient**   **like terms**

    unlike terms   **combining like terms**

1.   The numerical factor in a term is its _____.

2.   _____ is a method of adding or subtracting like terms by using the properties of real numbers.

3.   A(n) _____ is a number, a variable, or the product or quotient of a number and one or more variables raised to powers.

4.   Terms with exactly the same variables raised to exactly the same powers are called _____.

5.   _____ are terms that do not have the same variable or terms with the same variables but whose variables are not raised to the same powers.

Guided Examples

Review this example for Objective 1:
1. Simplify each expression.

    a. $3(2d - 8y)$

        $3(2d - 8y) = 3(2d) + 3(-8y)$

                 Distributive property

        $= (3 \cdot 2)d + \left[3 \cdot (-8)\right]y$

                 Associative property

        $= 6d - 24y$ Multiply.

Now Try:
1. Simplify each expression.

    a. $-9(-5x + 8y)$

        _____

b. $-2(a-12)-4$

$-2(a-12)-4=-2a-2(-12)-4$

Distributive property

$=-2a+24-4$ Multiply.

$=-2a+20$ Add.

b. $-2(-5x+2)+7$

_____

Review these examples for Objective 4:

2. Combine like terms in each expression.

   a. $3x+x-7x$

     $3x+x-7x=(3+1-7)x$

          $=-3x$

   b. $12-4x-2-7x$

     $12-4x-2-7x=(12-2)+(-4x-7x)$

          $=10-11x$

   c. $0.8y^2-0.2xy-0.3xy+0.9y^2$

     $0.8y^2-0.2xy-0.3xy+0.9y^2$

       $=0.8y^2+0.9y^2+(-0.2xy-0.3xy)$

       $=1.7y^2-0.5xy$

3. Simplify each expression.

   a. $2(3x+5)-8x$

     $2(3x+5)-8x$

      $=2(3x)+2(5)-8x$ Distributive property

      $=6x+10-8x$ Multiply.

      $=-2x+10$ Combine like terms.

   b. $7r+6-(2r+4)$

     $7r+6-(2r+4)$

      $=7r+6-1(2r+4)$ $-a=-1\cdot a$

      $=7r+6-1(2r)-1(4)$ Distributive property

      $=7r+6-2r-4$ Multiply.

      $=5r+2$ Combine like terms.

Now Try:

2. Combine like terms in each expression.

   a. $5a-a-4a$ _____

   b. $16+7a+9-12a$

     _____

   c. $12y-7y^2+4y-3y^2$

     _____

3. Simplify each expression.

   a. $2.5(3y+1)-4.5y$

     _____

   b. $2(4x-1)-(5x+2)$

     _____

Review this example for Objective 5:

4. Translate the phrase into a mathematical expression and simplify.

   Twelve times the difference between 4 and twice a number, subtracted from 10

   $10 - 12(4 - 2x)$ simplifies to $24x - 38$

Now Try:

4. Translate the phrase into a mathematical expression and simplify.

   The difference between five times a number and 3, added to four times the sum of the number and 2

   _____

Objective 1 Simplify expressions.
For extra help, see Example 1 on pages 69–70 of your text, the Section Lecture video for Section 1.8, and Exercise Solution Clips 7, 41, and 61.

Simplify each expression.

1. $14 + 3y - 8$ 1. _____

2. $11 - (d - 2) + (-6)$ 2. _____

3. $4(-6p - 2) + 2 - 4$ 3. _____

Objective 2 Identify terms and numerical coefficients.
For extra help, see the Section Lecture video for Section 1.8 and Exercise Solution Clip 13.

Give the numerical coefficient of each term.

4. $4x$ 4. _____

5. $0.3a^2b$ 5. _____

6. $-\frac{5}{9}v^6w^4$ 6. _____

Name: _____   Date: _____

Instructor: _____   Section: _____

Objective 3 Identify like terms.

For extra help, see the Section Lecture video for Section 1.8 and Exercise Solution Clip 27.

*Identify each group of terms as **like** or **unlike**.*

7. $-7q^2, 2q^2$ **7.** _____

8. $4x, -10x^2, -9x^2$ **8.** _____

9. $2, -4, 16$ **9.** _____

Objective 4 Combine like terms.

For extra help, see Examples 2 and 3 on pages 71–72 of your text, the Section Lecture video for Section 1.8, and Exercise Solution Clips 41, and 61.

Simplify each expression by combining like terms.

10. $4a^2 - 4a^3 - 2a^2 + 7a^3$ **10.** _____

11. $\frac{7}{10}r + \frac{3}{10}s - \frac{2}{5}r - \frac{4}{5}s$ **11.** _____

Use the distributive property and combine like terms to simplify the following expressions.

12. $-6(a+2) + 4(2a-1)$ **12.** _____

Objective 5 Simplify expressions from word phrases.

For extra help, see Example 4 on page 72 of your text, the Section Lecture video for Section 1.8, and Exercise Solution Clip 81.

*Write each phrase as a mathematical expression and simplify by combining like terms. Use **x** as the variable.*

13. The sum of seven times a number and 2, subtracted   **13.** _____
from three times the number

    49

14. The sum of ten times a number and 7, subtracted from the difference between 2 and nine times the number

14. _____

15. Four times the difference between twice a number and –10, subtracted from three times the sum of –7 and five times the number

15. _____

Chapter 2 LINEAR EQUATIONS AND INEQUALITIES IN ONE VARIABLE

2.1 The Addition Property of Equality

Learning Objectives
1   Identify linear equations.
2   Use the addition property of equality.
3   Simplify and then use the addition property of equality.

Key Terms

Use the vocabulary terms listed below to complete each statement in exercises 1–5.

    equation   **linear equation in one variable**   **solution**

    solution set   **equivalent equations**

1. A(n) _____ can be written in the form $Ax + B = C$, where A, B, and C are real numbers and $A \neq 0$.

2. The _____ is the set of all solutions to a particular equation.

3. _____ are equations that have the same solution set.

4. A _____ of an equation is a number that makes the equation true when it replaces the variable.

5. A statement that asserts that two algebraic expressions are equal is a(n) _____.

Guided Examples

Review these examples for Objective 2:
1. Solve $x - 4 = 16$.

$$x - 4 = 16$$
$$x - 4 + 4 = 16 + 4 \quad \text{Add 4 to each side.}$$
$$x = 20 \qquad \text{Combine like terms.}$$

The solution set is $\{20\}$.

2. Solve $x - 3.8 = -7.2$.

$$x - 3.8 = -7.2$$
$$x - 3.8 + 3.8 = -7.2 + 3.8 \quad \text{Add 4 to each side.}$$
$$x = -3.4 \qquad \text{Combine like terms.}$$

The solution set is $\{-3.4\}$.

Now Try:
1. Solve $x - 9 = -12$.

_____

2. Solve $x - 9.1 = 8.4$.

_____

3. Solve $-9 = x + 13$.

$$-9 = x + 13$$
$$-9 - 13 = x + 13 - 13 \quad \text{Subtract 13 from each side.}$$
$$-22 = x \quad\quad\quad\quad\quad \text{Combine terms.}$$

The solution set is $\{-22\}$.

3. Solve $-15 = x + 8$.

_____

4. Solve $\dfrac{2}{3}t - 5 = \dfrac{5}{3}t$.

$$\frac{2}{3}t - 5 = \frac{5}{3}t$$

$$\frac{2}{3}t - 5 - \frac{2}{3}t = \frac{5}{3}t - \frac{2}{3}t \quad \text{Subtract } \frac{2}{3}t \text{ from each side.}$$

$$-5 = 1t \quad\quad \begin{array}{l} \frac{2}{3}t - \frac{2}{3}t = 0; \\ \frac{5}{3}t - \frac{2}{3}t = \frac{3}{3}t = 1 \end{array}$$

$$-5 = t \quad\quad \text{Multiplicative identity}$$

The solution set is $\{-5\}$.

4. Solve $\dfrac{9}{8}p - \dfrac{1}{2} = \dfrac{1}{8}p$.

_____

5. Solve $4x + 2 = 5x + 8$.

$$4x + 2 = 5x + 8$$

$$4x + 2 - 4x = 5x + 8 - 4x \quad \begin{array}{l}\text{Subtract } 4x \text{ from} \\ \text{each side.}\end{array}$$

$$2 = x + 8 \quad\quad\quad \text{Combine like terms.}$$

$$2 - 8 = x + 8 - 8 \quad\quad \begin{array}{l}\text{Subtract 8 from} \\ \text{each side.}\end{array}$$

$$-6 = x \quad\quad\quad\quad \text{Combine like terms.}$$

The solution set is $\{-6\}$.

5. Solve $-5x - 3 = -4x + 7$.

_____

Review these examples for Objective 3:

6. Solve $6x - 3x + 2 = 11x - 2 - 7x$.

$$6x - 3x + 2 = 11x - 2 - 7x$$

$$3x + 2 = 4x - 2 \quad\quad\quad \text{Combine like terms.}$$

$$3x + 2 - 3x = 4x - 2 - 3x \quad \begin{array}{l}\text{Subtract } 3x \text{ from} \\ \text{each side.}\end{array}$$

$$2 = x - 2 \quad\quad\quad\quad \text{Combine like terms.}$$

$$2 + 2 = x - 2 + 2 \quad\quad\quad \text{Add 2 to each side.}$$

$$4 = x \quad\quad\quad\quad\quad\quad \text{Combine like terms.}$$

The solution set is $\{4\}$.

Now Try:

6. Solve

$$10x + 4x - 11x - 3 = -4x - 1 + 8x$$

_____

7. Solve $3(t+3)-(2t+7)=9$.

$$3(t+3)-(2t+7)=9$$

$$3(t+3)-1(2t+7)=9$$

$$3t+3(3)-1(2t)-1(7)=9$$

$$3t+9-2t-7=9$$

$$t+2=9$$

$$t+2-2=9-2$$

$$t=7$$

$-(2t+7)$
$\quad = -1(2t+7)$

Distributive property

Multiply.

Combine like terms.

Subtract 2 from each side.

Combine like terms.

The solution set is $\{7\}$.

7. Solve
$$2(4t+6)-3(2t-3)=17+t$$

_____

Objective 1 Identify linear equations.

For extra help, see page 86 of your text and the Section Lecture video for Section 2.1.

Tell whether each of the following is a linear equation.

1. $9x+2=0$

2. $3x^2+4x+3=0$

3. $7x^2=10$

4. $\dfrac{5}{x}-\dfrac{3}{2}=0$

5. $4x-2=12x+9$

1. _____

2. _____

3. _____

4. _____

5. _____

Objective 2 Use the addition property of equality.

For extra help, see Examples 1–5 on pages 87–89 of your text, the Section Lecture video for Section 2.1, and Exercise Solution Clips 7, 19, 25, and 39.

Solve each equation by using the addition property of equality. Check each solution.

6. $6y=7y-1$

6. _____

7. $p - \dfrac{2}{3} = \dfrac{5}{6}$ 7. _____

8. $y + 4\dfrac{1}{2} = 3\dfrac{3}{4}$ 8. _____

9. $5.7x + 12.8 = 4.7x$ 9. _____

10. $9.5y - 2.4 = 10.5y$ 10. _____

Objective 3 Simplify, and then use the addition property of equality.

For extra help, see Examples 6 and 7 on pages 89–90 of your text, the Section Lecture video for Section 2.1, and Exercise Solution Clips 55 and 63.

Solve each equation. First simplify each side of the equation as much as possible. Check each solution.

11. $6x + 3x - 7x + 4 = 10$ 11. _____

12. $-4(5g - 7) + 3(8g - 3) = 15 - 4 + 3g$ 12. _____

13. $4(3a - 2) - 7(2 + a) = 4(a - 5)$ 13. _____

14. $\dfrac{5}{12} + \dfrac{7}{6}s - \dfrac{1}{6} = \dfrac{5}{6}s + \dfrac{1}{4} - \dfrac{2}{3}s$

14. _____

15. $3.6p + 4.8 + 4.0p = 8.6p - 3.1 + 0.7$

15. _____

Chapter 2 LINEAR EQUATIONS AND INEQUALITIES IN ONE VARIABLE

2.2 The Multiplication Property of Equality

Learning Objectives
1 Use the multiplication property of equality.
2 Simplify, and then use the multiplication property of equality.

Key Terms

Use the vocabulary terms listed below to complete each statement in exercises 1–3.

multiplication property of equality   **reciprocal**   **coefficient**

1. The _____ states that the same nonzero number can be multiplied by (or divided into) both sides of an equation to obtain an equivalent equation.

2. A _____ is the numerical factor of a term.

3. To eliminate a fractional coefficient, it is usually easier to multiply by the _____ of the fraction.

Guided Examples

Review these examples for Objective 1:
1. Solve $6x = 24$.

$$6x = 24$$
$$\frac{6x}{6} = \frac{24}{6} \quad \text{Divide each side by 6.}$$
$$x = 4 \qquad \frac{6x}{6} = \frac{6}{6}x = 1x = x$$

The solution set is $\{4\}$.

2. Solve $18x = -75$.

$$18x = -75$$
$$\frac{18x}{18} = \frac{-75}{18} \quad \text{Divide each side by 18.}$$
$$x = -\frac{25}{6} \quad \text{Write in lowest terms.}$$

The solution set is $\left\{-\frac{25}{6}\right\}$.

Now Try:
1. Solve $12x = 84$.

_____

2. Solve $16x = -42$.

_____

3. Solve $-7.5x = 61.5$

$-7.5x = 61.5$

$\dfrac{-7.5x}{-7.5} = \dfrac{61.5}{7.5}$ Divide each side by -7.5.

$x = 8.2$ Divide.

The solution set is $\{-8.2\}$.

4. Solve $\dfrac{b}{5} = -4$.

$\dfrac{b}{5} = -4$

$\dfrac{1}{5}b = -4$ $\dfrac{b}{5} = \dfrac{1}{5}b$

$5 \cdot \dfrac{1}{5}b = -4 \cdot 5$ Multiply each side by 5, the reciprocal of $\dfrac{1}{5}$.

$\phantom{5 \cdot \dfrac{1}{5}}b = -20$ Multiplicative inverse property; Multiplicative identity property

The solution set is $\{-20\}$.

5. Solve $-\dfrac{3}{4}r = -27$.

$-\dfrac{3}{4}r = -27$

$\left(-\dfrac{4}{3}\right) \cdot \left(-\dfrac{3}{4}\right)r = -27 \cdot \left(-\dfrac{4}{3}\right)$

       Multiply each side by $-\dfrac{4}{3}$,

       the reciprocal of $-\dfrac{3}{4}$.

$1 \cdot r = -27 \cdot \left(-\dfrac{4}{3}\right)$

       Multiplicative inverse property

$r = 36$

       Multiplicative identity property; Multiply fractions.

The solution set is $\{36\}$.

3. Solve $-1.5x = 8.1$

_____

4. Solve $\dfrac{b}{-2} = 21$

_____

5. Solve $\dfrac{3}{7}p = -6$

_____

6. Solve $-x = -25$.

$$-x = -25$$
$$-1x = -25 \quad -x = -1x$$
$$-1(-1x) = -1(-25)$$

Multiply each side by -1.

$$1 \cdot x = 25 \quad \text{Multiply.}$$
$$x = 25$$

The solution set is $\{25\}$.

6. Solve $-p = 16$

_____

Review this example for Objective 2:

7. Solve $7y - 2y = 45$.

$$7y - 2y = 45$$
$$5y = 45 \quad \text{Combine like terms.}$$
$$\frac{5y}{5} = \frac{45}{5} \quad \text{Divide each side by 5.}$$
$$y = 9 \quad \text{Multiplicative identity property; divide.}$$

The solution set is $\{9\}$.

Now Try:

7. Solve $7a - 10a = -24$.

_____

Objective 1 Use the multiplication property of equality.

For extra help, see Examples 1–6 on pages 93–95 of your text, the Section Lecture video for Section 2.2, and Exercise Solution Clips 29, 35, 41, 45, and 49.

Solve each equation and check your solution.

1. $-3w = 42$

1. _____

2. $-16a = -48$

2. _____

3. $-h = \frac{7}{4}$

3. _____

4. $-\frac{7}{2}t = -4$

4. _____

5. $\frac{6}{7}y = \frac{2}{3}$

5. _____

6. $2.1a = 9.03$ **6.** _____

7. $-5.9y = -21.24$ **7.** _____

Objective 2 Simplify, and then use the multiplication property of equality.
For extra help, see Example 7 on page 95 of your text, the Section Lecture video for
Section 2.2, and Exercise Solution Clip 57.

Solve each equation and check your solution.

8. $4r + 3r = 63$ **8.** _____

9. $8f + 4f - 3f = 72$ **9.** _____

10. $-3b - 4b = 8$ **10.** _____

11. $9p - 10p = -18$ **11.** _____

12. $3w - 7w = 20$ **12.** _____

13. $7q - 10q = -24$ **13.** _____

14. $4x - 8x + 2x = 16$ **14.** _____

15. $-11h - 6h + 14h = -21$ **15.** _____

Chapter 2 LINEAR EQUATIONS AND INEQUALITIES IN ONE VARIABLE

2.3 More on Solving Linear Equations

Learning Objectives
1   Learn and use the four steps for solving a linear equation.
2   Solve equations with fractions or decimals as coefficients.
3   Solve equations with no solution or infinitely many solutions.
4   Write expressions for two related unknown quantities.

Key Terms

Use the vocabulary terms listed below to complete each statement in exercises 1–4.

conditional equation   identity   contradiction   empty set

1.   A(n) _____ is true for some replacements of the variable and false for others.

2.   A(n) _____ is an equation that is never true. It has no solution.

3.   The _____, denoted by { } or Ø, is the set containing no elements.

4.   A(n) _____ is an equation that is true for all replacements of the variable. It has an infinite number of solutions.

Guided Examples

Review these examples for Objective 1:
1.   Solve $6 - 3r = -12$

   Step 1 **Simplify each side separately.** There are no parentheses, fractions, or decimals in this equation, so this step is not necessary.

   Step 2 **Isolate the variable term on one side.**
   $$6 - 3r = -12$$
   $6 - 3r - 6 = -12 - 6$ Subtract 6 from each side.
   $$-3r = -18$$ Combine like terms.

   Step 3 **Isolate the variable.**
   $$\frac{-3r}{-3} = \frac{-18}{-3}$$ Divide each term by -3.
   $$r = 6$$

Now Try:
1.   Solve $-8s + 14 = 70$.

   _____

Step 4 **Check.** Check by substituting 6 for r in the original equation.

$$6 - 3r = -12 \quad \text{Original equation}$$

$$6 - 3(6) \overset{?}{=} -12 \quad \text{Let } r = 6.$$

$$6 - 18 \overset{?}{=} -12 \quad \text{Multiply.}$$

$$-12 = -12 \checkmark \quad \text{True}$$

The solution set is $\{6\}$.

2. Solve $7j + 1 = 10j - 29$.

Step 1 **Simplify each side separately.** There are no parentheses, fractions, or decimals in this equation, so this step is not necessary.

Step 2 **Isolate the variable term on one side.**

$$7j + 1 = 10j - 29$$

$$7j + 1 - 7j = 10j - 29 - 7j \quad \text{Subtract } 7j \text{ from each side.}$$

$$1 = 3j - 29 \quad \text{Combine like terms.}$$

$$1 + 29 = 3j - 29 + 29 \quad \text{Add 29 to each side.}$$

$$30 = 3j \quad \text{Combine like terms.}$$

Step 3 **Isolate the variable.**

$$\frac{30}{3} = \frac{3j}{3} \quad \text{Divide each side by 3.}$$

$$10 = j \quad \text{Divide.}$$

Step 4 **Check.** Check by substituting 10 for j in the original equation.

$$7j + 1 = 10j - 29 \quad \text{Original equation}$$

$$7(10) + 1 \overset{?}{=} 10(10) - 29 \quad \text{Let } j = 10.$$

$$70 + 1 \overset{?}{=} 100 - 29 \quad \text{Multiply.}$$

$$71 = 71 \checkmark \quad \text{True}$$

The solution set is $\{10\}$.

2. Solve $-5r + 6 = 20 - 3r$.

_____

3. Solve $-3a + 4(a-4) = -2a - 4$.

Step 1 Clear parentheses using the distributive property.

$$-3a + 4(a-4) = -2a - 4$$
$$-3a + 4a + 4(-4) = -2a - 4 \quad \text{Distributive property}$$
$$-3a + 4a - 16 = -2a - 4 \quad \text{Multiply.}$$
$$a - 16 = -2a - 4 \quad \text{Combine like terms.}$$

Step 2

$$a - 16 + 2a = -2a - 4 + 2a$$

Add $2a$ to each side.

$$3a - 16 = -4 \quad \text{Combine like terms.}$$
$$3a - 16 + 16 = -4 + 16 \quad \text{Add 16 to each side.}$$
$$3a = 12 \quad \text{Combine like terms.}$$

Step 3

$$\frac{3a}{3} = \frac{12}{3} \quad \begin{array}{l}\text{Divide each side}\\ \text{by 3.}\end{array}$$
$$a = 4 \quad \text{Divide.}$$

Step 4 **Check.**

$$-3a + 4(a-4) = -2a - 4 \quad \text{Original equation}$$
$$-3(4) + 4(4-4) \overset{?}{=} -2(4) - 4 \quad \text{Let } a = 4.$$
$$-12 + 4(0) \overset{?}{=} -8 - 4$$
$$-12 = -12 \checkmark \quad \text{True}$$

The solution set is $\{4\}$.

4. Solve $3t - (2t + 7) = 8t + 2$.

Step 1
$$3t - (2t + 7) = 8t + 2$$
$$3t - 1(2t + 7) = 8t + 2 \quad \begin{array}{l}\text{Multiplicative identity}\\ \text{property}\end{array}$$
$$3t - 2t - 7 = 8t + 2 \quad \text{Distributive property}$$
$$t - 7 = 8t + 2 \quad \text{Combine like terms.}$$

Step 2
$$t - 7 - t = 8t + 2 - t$$

Subtract t from each side.

$$-7 = 7t + 2 \quad \text{Combine like terms.}$$
$$-7 - 2 = 7t + 2 - 2$$

Subtract 2 from each side.

$$-9 = 7t \quad \text{Combine like terms.}$$

3. Solve $-w + 3(w-7) = -4w + 9$.

_____

4. Solve $7x - (5x - 3) = 5 - 6x$.

_____

Step 3

$$\frac{-9}{7} = \frac{7t}{7} \qquad \text{Divide each side by 7.}$$

$$-\frac{9}{7} = t$$

Step 4 Check that $-\frac{9}{7}$ is the solution.

The solution set is $\left\{ -\frac{9}{7} \right\}$.

5. Solve $4(3a - 2) - 12 = 5(2a - 4) + 6a$.

   Step 1

   $$4(3a - 2) - 12 = 5(2a - 4) + 6a$$
   $$12a - 8 - 12 = 10a - 20 + 6a$$
   $$\qquad\qquad\qquad\qquad \text{Distributive property}$$
   $$12a - 20 = 16a - 20 \quad \text{Combine like terms.}$$

   Step 2
   $$12a - 20 - 12a = 16a - 20 - 12a$$
   $$\qquad\qquad\qquad\qquad \text{Subtract } 12a.$$
   $$-20 = 4a - 20 \quad \text{Combine like terms.}$$
   $$-20 + 20 = 4a - 20 + 20$$
   $$\qquad\qquad\qquad\qquad \text{Add 20.}$$
   $$0 = 4a \qquad \text{Combine like terms.}$$

   Step 3

   $$\frac{0}{4} = \frac{4a}{4} \qquad \text{Divide by 4.}$$
   $$0 = a$$

   Step 4 **Check**.

   $$4(3a - 2) - 12 = 5(2a - 4) + 6a$$
   $$4[3(0) - 2] - 12 = 5[2(0) - 4] + 6(0)$$
   $$\qquad\qquad\qquad\qquad \text{Let } a = 0.$$

   $$4(0 - 2) - 12 \overset{?}{=} 5(0 - 4) + 0 \quad \text{Multiply.}$$

   $$4(-2) - 12 \overset{?}{=} 5(-4) + 0 \qquad \text{Add.}$$
   $$-20 = -20 \checkmark \qquad \text{True}$$

The solution set is {0}.

5. Solve
   $$8t - 3(2t - 7) = -3(3t - 4) + 4$$

   _____

Review these examples for Objective 2:

6. Solve $\dfrac{2}{3}y - \dfrac{1}{4}y = -\dfrac{5}{12}y + \dfrac{5}{2}$.

Step 1 The LCD of all the fractions in the equation is 12.

$$\frac{2}{3}y - \frac{1}{4}y = -\frac{5}{12}y + \frac{5}{2}$$

$$12\left(\frac{2}{3}y - \frac{1}{4}y\right) = 12\left(-\frac{5}{12}y + \frac{5}{2}\right)$$

Multiply each side by the LCD, 12.

$$12\left(\frac{2}{3}y\right) + 12\left(-\frac{1}{4}y\right) = 12\left(-\frac{5}{12}y\right) + 12\left(\frac{5}{2}\right)$$

Distributive property

$$8y - 3y = -5y + 30$$

Multiply.

$$5y = -5y + 30$$

Combine like terms.

Step 2

$$5y + 5y = -5y + 30 + 5y$$

Add $5y$.

$$10y = 30 \quad \text{Combine like terms.}$$

Step 3

$$\frac{10y}{10} = \frac{30}{10} \quad \text{Divide by 10.}$$

$$y = 3$$

Step 4 **Check.**

$$\frac{2}{3}y - \frac{1}{4}y = -\frac{5}{12}y + \frac{5}{2}$$

$$\frac{2}{3}(3) - \frac{1}{4}(3) = -\frac{5}{12}(3) + \frac{5}{2} \quad \text{Let } y = 3.$$

$$2 - \frac{3}{4} \overset{?}{=} -\frac{5}{4} + \frac{5}{2} \quad \text{Multiply.}$$

$$\frac{5}{4} = \frac{5}{4} \ \checkmark \quad \text{True}$$

The solution set is {3}.

Now Try:

6. Solve $\dfrac{1}{2}r + \dfrac{5}{14}r = r - \dfrac{4}{7}$.

_____

Name: _____   Date: _____

Instructor: _____   Section: _____

7. Solve $\dfrac{5}{6}(r-2)-\dfrac{2}{9}(r+4)=\dfrac{1}{2}$.

Step 1

$$\dfrac{5}{6}(r-2)-\dfrac{2}{9}(r+4)=\dfrac{1}{2}$$

$$18\left[\dfrac{5}{6}(r-2)-\dfrac{2}{9}(r+4)\right]=18\left(\dfrac{1}{2}\right)$$

Multiply each side by
the LCD, 18.

$$18\left(\dfrac{5}{6}(r-2)\right)+18\left(-\dfrac{2}{9}(r+4)\right)=18\left(\dfrac{1}{2}\right)$$

Distributive property

$$15(r-2)-4(r+4)=9$$

Multiply.

$$15r-30-4r-16=9$$

Distributive property

$$11r-46=9$$

Combine like terms.

Step 2

$$11r-46+46=9+46 \quad \text{Add 46.}$$

$$11r=55$$

Combine like terms.

Step 3

$$\dfrac{11r}{11}=\dfrac{55}{11}$$

Divide by 11.

$$r=5$$

Step 4 Check to confirm that {5} is the solution set.

8. Solve

$$0.03(x+20)-0.09(-x+10)=(0.01)(210).$$

Step 1 Multiply by the least power of 10 to eliminate the decimals. Here, we use $10^2 = 100$.

$$0.03(x+20)-0.09(-x+10)$$
$$=(0.01)(210)$$

$$100\left[0.03(x+20)-0.09(-x+10)\right]$$
$$=100\left[(0.01)(210)\right]$$

Multiply each side
by 100.

7. Solve $\dfrac{4}{3}(2-4b)-\dfrac{3}{2}(5-b)=\dfrac{17}{6}$.

_____

8. Solve

$$0.24x-0.38(x+2)=-0.34(x+4).$$

_____

$$100\left[0.03(x+20)\right]+100\left[-0.09(-x+10)\right]$$
$$=100\left[(0.01)(210)\right]$$

Distributive property

$$3(x+20)-9(-x+10)=1(210)$$

Multiply.

$$3x+3(20)-9(-x)-9(10)=210$$

Distributive property

$$3x+60+9x-90=210$$

Multiply.

$$12x-30=210$$

Combine like terms.

Step 2

$$12x-30+30=210+30$$

Add 30.

$$12x=240$$

Combine like terms.

Step 3

$$\frac{12x}{12}=\frac{240}{12}$$

Divide by 12.

$$x=20$$

Step 4 Check to confirm that $\{20\}$ is the solution set.

Review these examples for Objective 3:

9. Solve $22-10h=12+5(2-2h)$.

$$22-10h=12+5(2-2h)$$
$$22-10h=12+10-10h$$

Distributive property

$$22-10h=22-10h \quad \text{Combine like terms.}$$
$$22-10h+10h=22-10h+10h$$

Add 10h.

$$22=22 \quad \text{Combine like terms.}$$

Since the last statement ($22=22$) is true, *any* real number is a solution. The solution set is $\{$all real numbers$\}$.

Now Try:

9. Solve $3(r-2)-r+12=2r+6$

_____

10. Solve $16d - 3(7d + 12) = -5d - 20$.

$$16d - 3(7d + 12) = -5d - 20$$
$$16d - 21d - 36 = -5d - 20$$

Distributive property

$$-5d - 36 = -5d - 20$$

Combine like terms.

$$-5d - 36 + 5d = -5d - 20 + 5d$$

Add $5d$.

$$-36 = -20 \text{ False}$$

Since the last statement $(-36 = -22)$ is false, the equation has no solution. The solution set is $\varnothing$.

10. Solve $-1 - (2 + y) = -(-4 + y)$

_____

Review this example for Objective 4:

11. Shirley is x years old. Her mother is 28 years older. Write an expression for her mother's age.

Answer: $x + 28$

Now Try:

11. The length of a rectangle is x inches. The sum of the length and width is 12 inches. Write an expression for the width of the rectangle.

_____

Objective 1 Learn and use the four steps for solving a linear equation.

For extra help, see Examples 1–5 on pages 98–100 of your text, the Section Lecture video for Section 2.3, and Exercise Solution Clips 21, 25, 27, 29, 35, and 37.

Solve each equation and check your solution.

1. $7x + 11 = 9x + 25$

1. _____

2. $4 + x = -(x + 6)$

2. _____

3. $4(z - 2) - (3z - 1) = 2z - 6$

3. _____

4. $3(x+4) = 6 - 2(x-8)$ 4. _____

Objective 2 Solve equations with fractions or decimals as coefficients.
For extra help, see Examples 6–8 on pages 101–102 of your text, the Section Lecture video for Section 2.3, and Exercise Solution Clip 39.

Solve each equation.

5. $\dfrac{3}{8}x - \dfrac{1}{3}x = \dfrac{1}{12}$ 5. _____

6. $\dfrac{1}{5}(z-5) = \dfrac{1}{3}(z+2)$ 6. _____

7. $0.12x + 0.24(x-5) = 0.56x$ 7. _____

8. $0.45a - 0.35(20-a) = 0.02(50)$ 8. _____

Objective 3 Solve equations with no solution or infinitely many solutions.
For extra help, see Examples 9 and 10 on page 103 of your text and the Section Lecture video for Section 2.3.

Solve each equation.

9. $3(6x-7)=2(9x-6)$ 9. _____

10. $6y-3(y+2)=3(y-2)$ 10. _____

11. $4(2p-3)-3(3p+1)=-18-p+3$ 11. _____

12. $7y-11=6(2y+3)-5y$ 12. _____

Objective 4 Write expressions for two related unknown quantities.
For extra help, see Example 11 on page 104 of your text, the Section Lecture video for Section 2.3, and Exercise Solution Clip 65.

Write an expression for the two related unknown quantities.

13. Two numbers have a sum of 36. One is *m*. Find the   13. _____
     other number.

14. The product of two numbers is 17. One number is p. What is the other number?

14. _____

15. A cashier has q dimes. Find the value of the dimes in cents.

15. _____

Chapter 2 LINEAR EQUATIONS AND INEQUALITIES IN ONE VARIABLE

2.4 An Introduction to Applications of Linear Equations

Learning Objectives
1 Learn the six steps for solving applied problems.
2 Solve problems involving unknown numbers.
3 Solve problems involving sums of quantities.
4 Solve problems involving consecutive integers.
5 Solve problems involving supplementary and complementary angles.

Key Terms

Use the vocabulary terms listed below to complete each statement in exercises 1–6.

consecutive integers	degree	complementary
right angle	**supplementary**	**straight angle**

1. _____ angles are angles whose measures have a sum of 90°.

2. One _____ is a basic unit of measure for angles equal to $\frac{1}{360}$ of a complete revolution.

3. _____ angles are angles whose measures have a sum of 180°.

4. A _____ measures 90°.

5. A _____ measures 180°.

6. Two integers that differ by 1 are called _____.

Guided Examples

Review this example for Objectives 1 and 2:
1. If 2 is subtracted from four times a number, the result is 4 more than six times the number. What is the number?

Step 1 **Read** the problem carefully. We are asked to find a number.

Step 2 **Assign a variable** to represent the unknown quantity.
Let x = the number.

Now Try:
1. If –2 is multiplied by the difference between 4 and a number, the result is 24. Find the number.

_____

Step 3 **Write an equation.**

$4x - 2 = 4 + 6x$

Step 4 **Solve** the equation.

$$4x - 2 = 4 + 6x$$

$4x - 2 - 4x = 4 + 6x - 4x$	Subtract $4x$.
$-2 = 4 + 2x$	Combine like terms.
$-2 - 4 = 4 + 2x - 4$	Subtract 4.
$-6 = 2x$	Combine like terms.
$\dfrac{-6}{2} = \dfrac{2x}{2}$	Divide by 2.
$-3 = x$	

Step 5 **State the answer.** The number is -3.

Step 6 **Check.** When 2 is subtracted from 4 times -3, we get -14. Four more than 6 times -3 is also -14, so the answer is correct.

Review these examples for Objective 3:

2. George and Al were opposing candidates in the school board election. George received 21 more votes than Al, with 439 votes cast. How many votes did Al receive?

Step 1 **Read** the problem carefully. We are given information about the total number of votes and are asked to find the number of votes each received.

Step 2 **Assign a variable.**

Let x = the number of votes Al received.
Then $x + 21$ = the number of votes George received.

Step 3 **Write an equation.**

$439 = x + (x + 21)$

Step 4 **Solve** the equation.

$439 = x + (x + 21)$	
$439 = 2x + 21$	Combine like terms.
$439 - 21 = 2x + 21 - 21$	Subtract 21.
$418 = 2x$	Combine like terms.
$\dfrac{418}{2} = \dfrac{2x}{2}$	Divide by 2.
$209 = x$	

Step 5 **State the answer.** The variable x represents the number of votes that Al received, so Al received 209 votes. George received $209 + 21 = 230$ votes.

Now Try:

2. On a psychology test, the highest grade was 38 points more than the lowest grade. The sum of the two grades was 142. Find the lowest grade.

Step 6 **Check.**
Since Al received 209 votes and George received 230 votes, the total number of votes was 209 + 230 = 439. Because 230 − 21 = 209, George received 21 more votes than Al. This information agrees with what is given in the problem, so the answer checks.

3. Jose had a Big Burger and a can of cola for lunch. The number of calories in the can of cola was $\frac{3}{8}$ the number of calories in the Big Burger. If Jose's lunch had 550 calories in all, how many calories were in the can of cola?

   Step 1 **Read the problem carefully.** It asks for the number of calories in the Big Burger.

   Step 2 **Assign a variable.** Because of the way the problem is stated, let the variable represent the number of calories in the Big Burger. Let x = the number of calories in the Big Burger.
   Then $\frac{3}{8}x$ = the number of calories in the can of cola.

   Step 3 **Write an equation.** Use the fact that the total number of calories was 550.

   $$x + \frac{3}{8}x = 550$$

   Step 4 **Solve.**

   $$x + \frac{3}{8}x = 550$$
   $$\frac{11}{8}x = 550 \qquad \text{Combine like terms.}$$
   $$\frac{8}{11} \cdot \frac{11}{8}x = 550 \cdot \frac{8}{11} \qquad \text{Multiply by } \frac{8}{11}.$$
   $$x = 400$$

   Step 5 **State the answer.** In this problem x does not represent the quantity that we are asked to find. The number of calories in the can of cola was $\frac{3}{8}x = \frac{3}{8}(400) = 150$.

3. Charlene had a Cobb salad and a bottle of raspberry iced tea for lunch. The number of calories in the bottle of iced tea was $\frac{4}{7}$ the number of calories in the Cobb salad. If Charlene's lunch had 440 calories in all, how many calories were in the bottle of iced tea?

   _____

Step 6 **Check.**
The number of calories in the can of cola, 150,

is $\frac{3}{8}$ the number of calories in the Big Burger,

400, and 400 + 150 = 550. This agrees with the
information given in the problem, so the
answer is correct.

4. U.S. quarters are made from a combination of
nickel and copper. For every 1 pound of nickel
used, 12 pounds of copper are used. How many
pounds of each are needed to make 338 pounds
of quarters? *Source*: The United States Mint.

Step 1 **Read** the problem carefully. We must
find how many pounds of nickel and copper
are needed.

Step 2 **Assign a variable.**
Let x = the number of pounds of nickel.
Then $12x$ = the number of pounds of copper.

Step 3 **Write an equation.**
$x + 12x = 338$

Step 4 **Solve.**

$x + 12x = 338$

$13x = 338$ Combine like terms.

$\dfrac{13x}{13} = \dfrac{338}{13}$ Divide by 13.

$x = 26$

Step 5 **State the answer.** 26 pounds of nickel
are needed and $12 \cdot 26 = 312$ pounds of copper
are needed.

Step 6 **Check.**
Since 26 + 312 = 338 and 312 is 12 times 26,
the answer checks.

5. A piece of fabric is 180 in. long. It is cut into
three pieces. The longest piece is 25 in. more
than the middle-sized piece, and the shortest
piece measures half the middle-sized piece.
Find the lengths of the three pieces.

Step 1 **Read** the problem carefully. There will
be three answers.

4. For every adult ticket sold to a
magic show, three children's
tickets were sold. If a total of 96
tickets were sold, how many adult
tickets and how many children's
tickets were sold?

_____

5. One winter, the heating bill for
the Jones family was $360 more
than that of the Sanchez family.
The heating bill for the Kwan
family was $150 less than twice
that of the Sanchez family.
Altogether, the three families paid
$3910 for heating. What was the
heating bill for each family?

_____

Step 2 **Assign a variable.** Since the middle-sized piece appears in both pairs of comparisons, let x represent the length, in inches of the middle-sized piece.

Let $x =$ the length of the middle-sized piece.

Then $x + 25 =$ the length of the longest piece

and $\frac{1}{2}x =$ the length of the shortest piece.

Step 3 **Write an equation.**

$$x + x + 25 + \frac{1}{2}x = 180$$

Step 4 **Solve.**

$$x + x + 25 + \frac{1}{2}x = 180$$

$$25 + \frac{5}{2}x = 180 \qquad \text{Combine like terms.}$$

$$25 + \frac{5}{2}x - 25 = 180 - 25 \quad \text{Subtract 25.}$$

$$\frac{5}{2}x = 155 \qquad \text{Combine like terms.}$$

$$\frac{2}{5} \cdot \frac{5}{2}x = 155 \cdot \frac{2}{5} \quad \text{Multiply by } \frac{2}{5}.$$

$$x = 62$$

Step 5 **State the answer.** The middle-sized piece is 62 in., the longest piece is $62 + 25 = 87$ in., and the shortest piece is $\frac{1}{2} \cdot 62 = 31$ in.

Step 6 **Check.**
Since $62 + 87 + 31 = 180$, the answer checks.

Review these examples for Objective 4:

6. The sum of the page numbers of two back-to-back pages in a book is 135. What are the page numbers?

Step 1 **Read** the problem carefully. Because the pages are back-to-back, they must have page numbers that are consecutive integers.

Step 2 **Assign a variable.**
Let $x =$ the lesser page number
Then $x + 1 =$ the greater page number.

Step 3 **Write an equation.**
$x + x + 1 = 135$

Now Try:

6. The sum of the page numbers of two back-to-back pages in a book is 151. What are the page numbers?

_____

Step 4 **Solve.**

$$x + x + 1 = 135$$
$$2x + 1 = 135 \qquad \text{Combine like terms.}$$
$$2x + 1 - 1 = 135 - 1 \qquad \text{Subtract 1.}$$
$$2x = 134 \qquad \text{Combine like terms.}$$
$$\frac{2x}{2} = \frac{134}{2} \qquad \text{Divide by 2.}$$
$$x = 67$$

Step 5 **State the answer.** The lesser page number is 67 and the greater page number is $67 + 1 = 68$.

Step 6 **Check.**
Since $67 + 68 = 135$, the answer checks.

7. Find two consecutive even integers such that the smaller, added to twice the larger, is 292.

Step 1 **Read** the problem carefully. There will be two answers.

Step 2 **Assign a variable.**
Let x = the lesser integer.
Then $x + 2$ = the greater integer.

Step 3 **Write an equation.**

$$x + 2(x + 2) = 292$$

Step 4 **Solve.**

$$x + 2(x + 2) = 292$$
$$x + 2x + 4 = 292 \qquad \text{Distributive property}$$
$$3x + 4 = 292 \qquad \text{Combine like terms.}$$
$$3x + 4 - 4 = 292 - 4 \qquad \text{Subtract 4.}$$
$$3x = 288 \qquad \text{Combine like terms.}$$
$$\frac{3x}{3} = \frac{288}{3} \qquad \text{Divide by 3.}$$
$$x = 96$$

Step 5 **State the answer.** The two integers are 96 and 98.

Step 6 **Check.**
Since $96 + 2(98) = 292$, the answer checks.

7. Find two consecutive odd integers such that the larger, added to eight times the smaller, equals 119.

_____

Name: _____   Date: _____

Instructor: _____   Section: _____

Review these examples for Objective 5:

8. Find the measure of an angle whose complement is 9° more than twice its measure.

Step 1 **Read** the problem carefully. We must find the measure of an angle, given information about the measure of its complement.

Step 2 **Assign a variable.**
Let x = the degree measure of the angle.
Then $90 - x$ = the degree measure of its complement.

Step 3 **Write an equation.**
$90 - x = 9 + 2x$

Step 4 **Solve.**

$$90 - x = 9 + 2x$$
$$90 - x + x = 9 + 2x + x \quad \text{Add } x.$$
$$90 = 9 + 3x \quad \text{Combine like terms.}$$
$$90 - 9 = 9 + 3x - 9 \quad \text{Subtract 9.}$$
$$81 = 3x \quad \text{Combine like terms.}$$
$$\frac{81}{3} = \frac{3x}{3} \quad \text{Divide by 3.}$$
$$27 = x$$

Step 5 **State the answer.** The measure of the angle is 27°.

Step 6 **Check.**
If the angle measures 27°, then its complement measures $90° - 27° = 63°$.
Since $63° = 9° + 2(27°)$, the answer checks.

9. Find the measure of an angle whose supplement measures 6° more than 7 times its complement.

Step 1 **Read** the problem carefully. We must find the measure of an angle, given information about the measure of its complement and its supplement.

Step 2 **Assign a variable.**
Let x = the degree measure of the angle.
Then $90 - x$ = the degree measure of its complement and $180 - x$ = the degree measure of its supplement.

Step 3 **Write an equation.**
$180 - x = 6 + 7(90 - x)$

Now Try:

8. Find the measure of an angle such that the difference between the measures of an angle and its complement is 20°.

_____

9. Find the measure of an angle whose supplement measures 20° more than twice its complement.

_____

Step 4 **Solve.**

$$180 - x = 6 + 7(90 - x)$$
$$180 - x = 6 + 7(90) + 7(-x)$$

Distributive property

$$180 - x = 636 - 7x \qquad \text{Combine like terms.}$$
$$180 - x + 7x = 636 - 7x + 7x \qquad \text{Add } 7x.$$
$$180 + 6x = 636 \qquad \text{Combine like terms.}$$
$$180 + 6x - 180 = 636 - 180 \qquad \text{Subtract 180.}$$
$$6x = 456$$
$$\frac{6x}{6} = \frac{456}{6} \qquad \text{Divide by 6.}$$
$$x = 76$$

Step 5 **State the answer.** The measure of the angle is 76°.

Step 6 **Check.**
If the angle measures 76°, then its complement measures $90° - 76° = 14°$ and its supplement measures $180° - 76° = 104°$.
Since $104° = 6° + 7(14°)$, the answer checks.

Objective 1 Learn the six steps for solving applied problems.
Objective 2 Solve problems involving unknown numbers.
For extra help, see Example 1 on page 108 of your text, the Section Lecture video for Section 2.4, and Exercise Solution Clip 7.

Write an equation for each of the following and then solve the problem. Use x as the variable.

1. Six times the difference between a number and 4 equals the product of the number and –2. Find the number.

1. _____

2. If four times a number is added to 7, the result is five less than six times the number. Find the number.

2. _____

3. When the difference between a number and 4 is multiplied by –3, the result is two more than –5 times the number. Find the number.

3. _____

4. If seven times a number is added to 3, the result is two less than eight times the number. Find the number.

4. _____

Objective 3 Solve problems involving sums of quantities.

For extra help, see Example 2–5 on pages 109–112 of your text, the Section Lecture video for Section 2.4, and Exercise Solution Clips 23, 29 and 33.

Solve each problem.

5. A rope 116 inches long is cut into three pieces. The middle-sized piece is 10 inches shorter than twice the shortest piece. The longest piece is $\frac{5}{3}$ as long as the shortest piece. What is the length of the shortest piece?

5. _____

6. Mount McKinley in Alaska is 5910 feet higher than Mount Rainier in Washington. Together, their heights total 34,730 feet. How high is each mountain?

6. _____

7. Penny is making punch for a party. The recipe requires twice as much orange juice as cranberry juice and 8 times as much ginger ale as cranberry juice. If she plans to make 176 ounces of punch, how much of each ingredient should she use?

7. _____

8. Pablo, Joaquin, and Mark swim at a public pool each day for exercise. One day Pablo swam five more than three times as many laps as Mark, and Joaquin swam four times as many laps as Mark. If the men swam 29 laps altogether, how many laps did each one swim?

8. _____

Objective 4 Solve problems involving consecutive integers.
For extra help, see Examples 6 and 7 on pages 112–113 of your text and the Section Lecture video for Section 2.4.

Solve each problem.

9. Find two consecutive even integers whose sum is 154.   9. _____

10. Find two consecutive integers such that the larger,   10. _____
    added to three times the smaller, is 109.

11. Find two consecutive odd integers such that if three   11. _____
    times the smaller is added to twice the larger, the sum is
    69.

Objective 5 Solve problems involving supplementary and complementary angles.
For extra help, see Example 8 and 9 on pages 114–115 of your text, the Section Lecture video for Section 2.4, and Exercise Solution Clip 55.

Solve each problem.

12. Find the measure of an angle if the measure of the angle   12. _____
    is 8° less than three times the measure of its supplement.

13. Find the measure of an angle such that the difference   13. _____
    between the measure of its supplement and twice the
    measure of its complement is 49°.

Name: Date:

Instructor: Section:

14. Find the measure of an angle if its supplement measures   **14.**_____
4° less than three times its complement.

15. Find the measure of an angle such that the difference   **15.**_____
between the measure of the angle and the measure of its
complement is 28°.

81

Chapter 2 LINEAR EQUATIONS AND INEQUALITIES IN ONE VARIABLE

2.5 Formulas and Applications from Geometry

Learning Objectives
1   Solve a formula for one variable, given values of the other variables.
2   Use a formula to solve an applied problem.
3   Solve problems involving vertical angles and straight angles.
4   Solve a formula for a specified variable.

Key Terms

Use the vocabulary terms listed below to complete each statement in exercises 1–5.

formula   **area**   **perimeter**

vertical angles   **volume**

1.   A(n) _____ is an equation in which letters are used to describe relationships.

2.   _____ is a measure of the surface covered by a two-dimensional (flat) figure.

3.   When two intersecting lines are drawn, the angles that lie opposite each other have the same measure and are called _____.

4.   The _____ of a two-dimensional figure is the measure of the distance around the outside edges of the figure – that is, the sum of the lengths of its sides.

5.   The _____ of a three-dimensional figure is a measure of the space occupied by the figure.

Name: Date:

Instructor: Section:

Guided Examples

Review this example for Objective 1:

1. Find the value of the remaining variable.
   $I = prt$; $I = 288$, $r = 0.04$, $t = 3$

   $I = prt$
   $288 = p(0.04)(3)$ Let $r = 0.04$ and $t = 3$.
   $288 = 0.12p$ Multiply.
   $\dfrac{288}{0.12} = \dfrac{0.12p}{0.12}$ Divide by 0.12.
   $2400 = p$

Now Try:

1. Find the value of the remaining variable.

   $V = \frac{1}{3}Bh$; $B = 27$, $V = 63$

   _____

Review these examples for Objective 2:

2. Find the dimensions of a rectangular garden if its perimeter is 96 feet and its length is three times its width.

   Step 1 **Read** the problem carefully. We must find the dimensions of the garden.

   Step 2 **Assign a variable.**
   Let w = the width of the garden.
   Then $3w$ = the length of the garden.

   Step 3 **Write an equation.**
   $P = 2L + 2W$
   $96 = 2(3w) + 2w$ Substitute $3w$ for L and w for W.

   Step 4 **Solve** the equation.
   $96 = 2(3w) + 2w$
   $96 = 6w + 2w$ Multiply.
   $96 = 8w$ Combine like terms.
   $\dfrac{96}{8} = \dfrac{8w}{8}$ Divide by 8.
   $12 = w$

   Step 5 **State the answer.** The width is 12 feet and the length is 3(12) = 36 feet.

   Step 6 **Check.**
   If the width is 12 ft and the length is 36 ft, then the perimeter is 2(12) + 2(36) = 96 ft as required.

Now Try:

2. Find the dimensions of a rectangular room if its perimeter is 74 feet and its length is 8 feet less than twice its width.

   _____

3. The longest side of a triangle is 2 cm less than 3 times the shortest side. The medium side is 2 cm more than twice the shortest side. If the perimeter of the triangle is 30 cm, what are the lengths of the three sides?

Step 1 **Read** the problem carefully. We must find the lengths of the sides of a triangle..

Step 2 **Assign a variable.**
Let s = the length of the shortest side.
Then $2s + 2$ = the length of the medium side and $3s - 2$ = the length of the longest side.

Step 3 **Write an equation.**

$P = a + b + c$ Perimeter of a triangle.
$30 = s + (2s + 2) + (3s - 2)$ Substitute.

Step 4 **Solve.**

$30 = s + (2s + 2) + (3s - 2)$
$30 = 6s$ Combine like terms.
$\dfrac{30}{6} = \dfrac{6s}{6}$ Divide by 6.
$5 = s$

Step 5 **State the answer.** The shortest side has length 5 cm. The medium side has length $2(5) + 2 = 12$ cm, and the longest side has length $3(5) - 2 = 13$ cm.

Step 6 **Check.**
The perimeter is $5 + 12 + 13 = 30$ cm, as required.

4. The area of a triangular banner is 48 in^2. The base is 12 in. Find the height of the triangle.

Step 1 **Read** the problem carefully. We must find the height of the triangle.

Step 2 **Assign a variable.**
Let h = the height of the triangle.

Step 3 **Write an equation.**

$A = \dfrac{1}{2}bh$

$48 = \dfrac{1}{2}(12)h$ Substitute $b = 12$.

3. The medium side of a triangle is 10 cm more than twice the shortest side. The longest side of the triangle is 4 cm more than three times the shortest side. If the perimeter of the triangle is 560 cm, what are the lengths of the three sides?

_____

4. The area of a triangle is 108 in^2. The height is 12 in. Find the base of the triangle.

_____

Step 4 **Solve.**

$$48 = \frac{1}{2}(12)h$$

$48 = 6h$ Combine like terms.

$$\frac{48}{6} = \frac{6h}{6}$$ Divide by 6.

$8 = h$

Step 5 **State the answer.** The height of the triangle is 8 in.

Step 6 **Check.**

$$48 \overset{?}{=} \frac{1}{2} \cdot 12 \cdot 8$$

$48 = 48$ ✓

Review this example for Objective 3:	**Now Try:**

5. Find the measure of each marked angle in the figure.

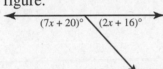

5. Find the measure of each marked angle in the figure.

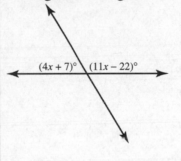

Since the marked angles form a straight angle, they are supplements of each other, and the sum of their measures in $180°$.

$$(7x + 20) + (2x + 16) = 180$$

$9x + 36 = 180$ Combine like terms.

$9x + 36 - 36 = 180 - 36$ Subtract 36.

$9x = 144$ Combine like terms.

$$\frac{9x}{9} = \frac{144}{9}$$ Divide by 9.

$x = 16$

Replace *x* with 16 in the expression for the measure of each angle.

$7(16) + 20 = 132°$; $2(16) + 16 = 48°$.

The two angle measures are $132°$ and $48°$.

Review these examples for Objective 4:	**Now Try:**

6. Solve $V = LWH$ for H.

$$V = LWH$$

$$\frac{V}{LW} = H$$ Isolate H by dividing each side by LW.

6. Solve $i = prt$ for p.

7. Solve $P = 2b + a + c$ for b.

$$P = 2b + a + c$$

$$p - a - c = 2b + a + c - a - c \qquad \text{Subtract } a; \text{ subtract } c.$$

$$p - a - c = 2b \qquad \text{Combine like terms.}$$

$$\frac{p - a - c}{2} = \frac{2b}{2} \qquad \text{Divide by 2.}$$

$$\frac{p - a - c}{2} = b$$

8. Solve $m = \dfrac{y_2 - y_1}{x_2 - x_1}$ for y_1.

$$m = \frac{y_2 - y_1}{x_2 - x_1}$$

$$m(x_2 - x_1) = y_2 - y_1 \qquad \text{Multiply by } (x_2 - x_1).$$

$$m(x_2 - x_1) - y_2 = y_2 - y_1 - y_2 \qquad \text{Subtract } y_2.$$

$$m(x_2 - x_1) = -y_1 \qquad \text{Combine like terms.}$$

$$(-1)m(x_2 - x_1) = (-1)(-y_1) \qquad \text{Multiply by } -1.$$

$$-m(x_2 - x_1) = y_1$$

9. $a_n = a_1 + (n-1)d$ for n

$$a_n = a_1 + (n-1)d$$

$$a_n - a_1 = (n-1)d \qquad \text{Subtract } a_1.$$

$$\frac{a_n - a_1}{d} = n - 1 \qquad \text{Divide by } d.$$

$$\frac{a_n - a_1}{d} + 1 = n - 1 + 1 \qquad \text{Add 1.}$$

$$\frac{a_n - a_1}{d} + 1 = n \qquad \text{Combine like terms.}$$

7. Solve $P = 3s + 2a$ for s.

_____

8. Solve $V = h^2 + \dfrac{1}{3}bh$ for b.

_____

9. Solve $A = R \cdot \dfrac{x+1}{w}$ for x.

_____

Name: _____   Date: _____
Instructor: _____   Section: _____

Objective 1 Solve a formula for one variable, given values of the other variables.
For extra help, see Example 1 on page 120 of your text, the Section Lecture video for
Section 2.5, and Exercise Solution Clip 19.

In the following exercises, a formula is given, along with the values of all but one of the
variables in the formula. Find the value of the variable that is not given.

1. $A = \pi r^2$; $r = 3$, $\pi = 3.14$ 1. _____

2. $C = \frac{5}{9}(F - 32)$; $F = 104$ 2. _____

3. $A = \frac{1}{2}(b + B)h$; $b = 6$, $B = 16$, $A = 132$ 3. _____

Objective 2 Use a formula to solve an applied problem.
For extra help, see Examples 2–4 on pages 121–122 of your text, the Section Lecture video
for Section 2.5, and Exercise Solution Clips 41, 43, and 51.

Use a formula to write an equation for each of the following applications; then solve
the application. Formulas are found on the inside covers of your text.

4.   A water tank is a right circular cylinder. The tank has a   4. _____
     radius of 6 meters and a volume of 1356.48 cubic
     meters. Find the height of the tank. (Use 3.14 as an
     approximation for π.)

5.   A spherical balloon has a radius of 9 centimeters. Find   5. _____
     the amount of air required to fill the balloon. (Use 3.14
     as an approximation for π.)

6. Linda invests $5000 at 6% simple interest and earns   **6.** _____
$450. How long did Linda invest her money?

7. Find the height of an ice cream cone if the diameter is 6   **7.** _____
centimeters and the volume is 37.68 cubic centimeters.
(Use 3.14 as an approximation for π .)

Objective 3 Solve problems involving vertical angles and straight angles.
For extra help, see Example 5 on page 123 of your text, the Section Lecture video for
Section 2.5, and Exercise Solution Clip 59.

Find the measure of each marked angle.

8.   **8.** _____

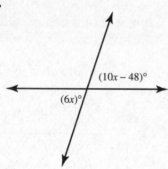

$(10x - 48)°$

$(6x)°$

9.   **9.** _____

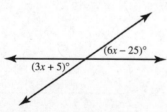

$(6x - 25)°$

$(3x + 5)°$

10.   **10.** _____

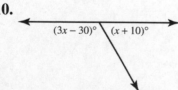

$(3x - 30)°$ $(x + 10)°$

11.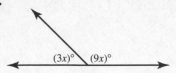

11. _____

Objective 4 Solve a formula for a specified variable.
For extra help, see Examples 6–9 on pages 124–125 of your text, the Section Lecture video for Section 2.5, and Exercise Solution Clips 67, 83, 85, and 89.

Solve each formula for the specified variable.

12. $A = p(1 + rt)$ for p

12. _____

13. $S = \dfrac{a}{1-r}$ for r

13. _____

14. $S = 2\pi rh + 2\pi r^2$ for h

14. _____

15. $C = \frac{5}{9}(F - 32)$ for F

15. _____

Chapter 2 LINEAR EQUATIONS AND INEQUALITIES IN ONE VARIABLE

2.6 Ratios, Proportion, and Percent

Learning Objectives
1   Write ratios.
2   Solve proportions.
3   Solve applied problems by using proportions.
4   Find percents and percentages.

Key Terms

Use the vocabulary terms listed below to complete each statement in exercises 1–6.

ratio	proportion	terms
extremes	means	cross products

1. The _____ of the proportion $\frac{a}{b} = \frac{c}{d}$ are the a and d terms.

2. The _____ of the proportion $\frac{a}{b} = \frac{c}{d}$ are the b and c terms.

3. A _____ is a comparison of two quantities with the same units.

4. A _____ is a statement that two ratios are equal.

5. The _____ in the proportion $\frac{a}{b} = \frac{c}{d}$ are ad and bc.

6. The _____ of the proportion $\frac{a}{b} = \frac{c}{d}$ are $a, b, c,$ and d.

Guided Examples

Review these examples for Objective 1:
1. Write a ratio for each word phrase.

    a.   7 cm to 12 cm

$$\frac{7 \text{ cm}}{12 \text{ cm}} = \frac{7}{12}$$

Now Try:
1. Write a ratio for each word phrase.

    a.   5 months to 8 months

    _____

b. 10 days to 2 weeks

We must convert 2 weeks to days.

$$\frac{10 \text{ days}}{2 \text{ weeks}} = \frac{10 \text{ days}}{14 \text{ days}} = \frac{10}{14} = \frac{5}{7}$$

b. 23 yards to 10 feet

_____

2. A supermarket charges the following prices for a certain brand of trash bags.

Box size	Price
10-count	$1.25
15-count	$1.60
20-count	$1.99
25-count	$2.49

Which size is the best buy? What is the unit cost for that size?

To find the best buy, write ratios comparing the price for each box size to the number of units (bags) per box. Then divide to obtain the price per unit (bag).

Box size	Unit Cost (dollars per bag)
10-count	$\frac{\$1.25}{10} = \$0.125$
15-count	$\frac{\$1.60}{15} = \$0.107$
20-count	$\frac{\$1.99}{20} = \$0.0995$
25-count	$\frac{\$2.49}{25} = \$0.0996$

Because the 20-count box produces the lowest unit cost, it is the best buy. The unit cost is $0.0995 per bag.

2. A supermarket charges the following prices for a certain brand of tomato catsup.

Size	Price
14 oz.	$0.93
32 oz.	$1.92
44 oz.	$2.59
64 oz.	$3.45

Which size is the best buy? What is the unit cost for that size?

_____

Review these examples for Objective 2:

3. Decide whether each proportion is *true* or *false*.

   a. $\dfrac{4}{7} = \dfrac{16}{28}$ b. $\dfrac{2}{9} = \dfrac{10}{36}$

   Check to see whether the cross products are equal.

   a. $\dfrac{4}{7} \diagdown\!\!\!\!\diagup \dfrac{16}{28}$ $4 \cdot 28 = 112$ $7 \cdot 16 = 112$

   The cross products are equal, so the proportion is true.

   b. $\dfrac{2}{9} \diagdown\!\!\!\!\diagup \dfrac{10}{36}$ $2 \cdot 36 = 72$ $9 \cdot 10 = 90$

   The cross products are not equal, so the proportion is false.

4. Solve the proportion. $\dfrac{z}{20} = \dfrac{25}{125}$

   $\dfrac{z}{20} = \dfrac{25}{125}$

   $125z = 20 \cdot 25$ Cross products are equal.

   $125z = 500$ Multiply.

   $\dfrac{125z}{125} = \dfrac{500}{125}$ Divide by 125.

   $z = 4$

   Check by substituting 4 for z in the proportion. The solutions set is $\{4\}$.

5. Solve the equation. $\dfrac{w+4}{6} = \dfrac{w+10}{8}$

   $\dfrac{w+4}{6} = \dfrac{w+10}{8}$

   $8(w+4) = 6(w+10)$ Cross products

   $8w + 32 = 6w + 60$ Distributive property

   $2w + 32 = 60$ Subtract $6w$.

   $2w = 28$ Subtract 32.

   $w = 14$ Divide by 2.

   The solution set is $\{14\}$.

Now Try:

3. Decide whether each proportion is *true* or *false*.

   a. $\dfrac{5}{16} = \dfrac{16}{56}$ b. $\dfrac{48}{7} = \dfrac{240}{35}$

   a. _____

   b. _____

4. Solve the proportion. $\dfrac{25}{3} = \dfrac{125}{x}$

   _____

5. Solve the equation. $\dfrac{4}{z+1} = \dfrac{2}{z+7}$

   _____

Review this example for Objective 3:

6. Ginny can type 8 pages of her term paper in 30 minutes. How long will it take her to type the paper if it has 20 pages?

Set up a proportion with number of minutes in the numerator and number of pages in the denominator. Let x = the number of pages needed to type 20 pages.

$$\frac{30 \text{ min.}}{8 \text{ pages}} = \frac{x \text{ min.}}{20 \text{ pages}}$$

$\quad 20 \cdot 30 = 8x \qquad$ Cross products

$\qquad 600 = 8x \qquad$ Multiply.

$\qquad\;\; 75 = x \qquad$ Divide by 8.

It will take Ginny 75 minutes to type 20 pages.

Now Try:

6. A certain lawn mower uses 5 tanks of gas to cut 18 acres of lawn. How many acres could be cut using 12 tanks of gas?

_____

Review these examples for Objective 4:

7.a. Write 6% as a decimal.

$\quad 6\% = 6 \cdot 1\% = 6 \cdot 0.01 = 0.06$

b. Write 1.26 as a percent.

$\quad 1.26 = 126 \cdot 0.01 = 126 \cdot 1\% = 126\%$

8. Solve each problem.

a. What is 12% of 85?

$\quad$ What   is   12%   of   85?

$\qquad \downarrow \quad \downarrow \quad \downarrow \quad \downarrow \quad \downarrow$

$\quad n \;\; = 0.12 \;\; \cdot \;\; 85 \quad$ Write the percent equation.

$\quad n \;\; = 10.2 \qquad\qquad$ Multiply.

Thus, 12% of 85 is 10.2.

b. 165% of what number is 363?

$\quad$ 165%   of   what number   is   363?

$\qquad \downarrow \qquad \downarrow \qquad \downarrow \qquad\quad \downarrow \quad \downarrow$

$\quad 1.65 \quad \cdot \qquad n \qquad\; = 363 \quad$ Write the percent equation.

$$n = \frac{363}{1.65} = 220 \quad \text{Divide by 1.65; simplify.}$$

Thus, 165% of 220 is 363.

Now Try:

7.a. Write 32% as a decimal.

_____

b. Write 0.893 as a percent.

_____

8. Solve each problem.

a. What is 2.5% of 3500?

_____

b. 2.75% of what number is 20.625?

_____

c. 108.8 is what percent of 128?

108.8   is what percent   of   128?

$\downarrow$ $\downarrow$ $\downarrow$ $\downarrow$ $\downarrow$

$108.8 =$ p $\cdot$ 128 Write the percent equation.

$\dfrac{108.8}{128} = p$ Divide by 128.

$0.85 = p$, or $85\% = p$ Simplify. Write 0.85 as a percent.

Thus, 108.8 is 85% of 128.

c. What percent of 5200 is 104?

_____

9. An advertisement for a Blu-Ray disc player gives a sale price of $125. The regular price is $180. Find the percent of the regular price was the savings?

The savings amounted to $180 − $125 = $55. Now restate the problem: What percent of $180 is $55? (Round your answer to the nearest whole percent.)

What percent   of   180   is $55?

$\downarrow$ $\downarrow$ $\downarrow$ $\downarrow$ $\downarrow$

p $\cdot$ $180 =$ 55 Write the percent equation.

$p = \dfrac{55}{180}$ Divide by 180.

$p \approx 0.31$, or $p \approx 31\%$ Simplify. Write 0.31 as a percent.

The sale price represents a 31% savings.

9. Jasmine solved 40 problems correctly on a test, giving her a score of 62.5%. How many problems were on the test?

_____

Objective 1 Write ratios.
For extra help, see Examples 1 and 2 on pages 130–131 of your text, the Section Lecture video for Section 2.6, and Exercise Solution Clips 3, 7, and 15.

Write a ratio for each word phrase. Write fractions in lowest terms.

1. 9 dollars to 48 quarters

1. _____

A supermarket was surveyed and the following prices were charged for items in various sizes. Find the best buy (based on price per unit) for each of the following items.

2. Corn oil
    18-ounce bottle: $1.27
    32-ounce bottle: $2.40
    48-ounce bottle: $3.19
    64-ounce bottle: $4.43

2. _____

3. Applesauce
    8-ounce jar: $ .59
    16-ounce jar: $ .96
    24-ounce jar: $1.31
    48-ounce jar: $1.99

3. _____

Objective 2 Solve proportions.
For extra help, see Examples 3–5 on pages 132–133 of your text, the Section Lecture video for Section 2.6, and Exercise Solution Clips 23, 29, and 37.

Solve each equation.

4. $\dfrac{5}{m} = \dfrac{12}{5}$

4. _____

5. $\dfrac{m}{5} = \dfrac{m-2}{2}$

5. _____

6. $\dfrac{6y-4}{y} = \dfrac{11}{5}$

6. _____

7. $\dfrac{3x+4}{x-2} = \dfrac{1}{3}$

7. _____

Objective 3 Solve applied problems by using proportions.
For extra help, see Example 6 on page 133 of your text and the Section Lecture video for Section 2.6.

Solve the following problems involving proportions.

8. If 50 DVD-Rs cost $26.98, how much will 15 DVD-Rs cost? (Round your answer to the nearest cent.)

8. _____

9. On a road map, 6 inches represents 50 miles. How many inches would represent 125 miles?

9. _____

10. If 3 ounces of medicine must be mixed with 10 ounces of water, how many ounces of medicine must be mixed with 15 ounces of water?

10. _____

11. A certain lawn mower uses 7 tanks of gas to cut 15 acres of lawn. How many tanks of gas are needed to cut 30 acres of lawn?

11. _____

Objective 4 Find percents and percentages.
For extra help, see Examples 7–9 on pages 134–135 of your text and the Section Lecture video for Section 2.6.

Solve the following problems involving proportions.

12. 35% of the students at Mountainview Community College are married. How many of the 3260 students enrolled this year are married?

12. _____

13. 782 members on an organization are between 25 and 45 years of age. If this group makes up 92% of the total membership, find the total membership.

13. _____

Name: Date:

Instructor: Section:

14. Vera's Antique Shoppe says that of its 5100 items in stock, 4233 are just plain junk, while the rest are antiques. What percent of the items in stock are antiques?

14. _____

15. A pair of shoes with a regular price of $57 is on sale at 25% off. Find the amount of the discount and the sale price of the shoes.

15. discount _____

sale price _____

Chapter 2 LINEAR EQUATIONS AND INEQUALITIES IN ONE VARIABLE

2.7 Further Applications of Linear Equations

Learning Objectives
1 Use percent in solving problems involving rates.
2 Solve problems involving mixtures.
3 Solve problems involving simple interest.
4 Solve problems involving denominations of money.
5 Solve problems involving distance, rate, and time.

Key Terms

Use the vocabulary terms listed below to complete each statement in exercises 1–4.

mixture problem   **simple interest**   **denomination**   **distance problems**

1. The formula $I = prt$ is used to calculate _____, where p is the principal, r is the rate, and t is time (usually in years).

2. A(n) _____ typically involves mixing different concentrations of a substance.

3. Most _____ are solved using the formula $d = rt$.

4. The _____ of a coin or bill gives the monetary value of the coin or bill.

Guided Examples

Review this example for Objective 1:
1.a. How much pure antifreeze is in 72 liters of a 22% antifreeze solution?

For mixture problems, we use the equation base $\times$ rate (%) = percentage. The percent is always written as a decimal or a fraction.

$$72 \quad \cdot \quad 0.22 \quad = \quad 15.84 \text{ L}$$

72	0.22	15.84 L
↑	↑	↑
Amount of solution	Rate of concentration	Amount of pure antifreeze

Now Try:
1.a. How much pure acid is in 16 liters of a 15% acid solution?

_____

b. How much interest is earned if $7800 is invested at 1.2% simple interest for one year?

For annual interest problems, we use the equation principal $\times$ rate (%) = interest.

$$\$7800 \quad \cdot \quad 0.012 \quad = \quad \$93.60$$
$$\uparrow \qquad\qquad \uparrow \qquad\qquad\qquad \uparrow$$

Principal   Interest rate   Interest earned

b. How much interest is earned if $14,400 is invested at 2.3% simple interest for one year?

_____

Review this example for Objective 2:

2. How many liters of water must be added to 2 liters of pure alcohol to obtain a 10% alcohol solution?

Step 1 **Read** the problem carefully. Note the percent of each solution and of the mixture.

Step 2 **Assign a variable.**

Let x = the number of liters of water to be added. Note that water is 0% alcohol, so the amount of pure alcohol being added to the original solution is $0x$ L.

The amount of pure alcohol in the original solution is $1.00(2) = 2$ L.

The new solution will contain $(x + 2)$ liters of 10% solution. The amount of pure acid in this solution is $0.10(x + 2)$.

We summarize this information in a table.

Liters of Solution	*Percent alcohol*	*Liters of Pure Alcohol*
x	0	0
2	1.00	$2(1.00) = 2$
$(x + 2)$	0.10	$0.10(x + 2)$

Step 3 **Write an equation.**

Pure alcohol in water	plus	Pure alcohol in alcohol	is	Pure alcohol in alcohol
$\downarrow$	$\downarrow$	$\downarrow$	$\downarrow$	$\downarrow$
0	+	2	=	$0.10(x+2)$

Step 4 **Solve** the equation.

$2 = 0.10x + 0.10(2)$	Distributive property
$200 = 10x + 10(2)$	Multiply by 100.
$200 = 10x + 20$	Multiply.
$180 = 10x$	Subtract 20.
$18 = x$	Divide by 10.

Now Try:

2. A pharmacist has 2 liters of a solution containing 30% alcohol. If he wants to have a solution containing 44% alcohol, how much pure alcohol must he add?

_____

Step 5 **State the answer.** We must add 18 L of water to obtain a 10% alcohol solution.

Step 6 **Check.**

The answer checks since $0.10(18 + 2) = 2$.

Review this example for Objective 3:

3. A total of $2000 is invested for one year, part at $7\frac{1}{2}\%$ and the remainder at $8\frac{1}{2}\%$. If $156 interest is earned, how much is invested at each rate?

Step 1 **Read** the problem carefully. There will be two answers.

Step 2 **Assign a variable.**

Let x = the amount invested at $7\frac{1}{2}\%$ (in dollars).

Then, $2000 - x$ = the amount invested at $8\frac{1}{2}\%$ (in dollars).

We summarize the information in a table.

Amount Invested ($)	Interest Rate	Interest for One Year
x	0.075	$0.075x$
$2000 - x$	0.085	$0.085(2000 - x)$

Step 3 **Write an equation.**

Interest at $7\frac{1}{2}\%$	plus	Interest at $8\frac{1}{2}\%$	is	Total interest
$\downarrow$	$\downarrow$	$\downarrow$	$\downarrow$	$\downarrow$
$0.075x$	$+$	$0.085(2000 - x)$	$=$	156

Step 4 **Solve** the equation.

$0.075x + 0.085(2000) + 0.085(-x) = 156$

Distributive property

$$75x + 85(2000) + 85(-x) = 156$$

Multiply by 1000.

$$75x + 170,000 - 85x = 156,000$$

Multiply.

$$-10x + 170,000 = 156,000$$

Combine like terms.

Now Try:

3. Adam received an inheritance of $13,500. He wishes to divide the amount between investments at 4% and 7% to receive an average return of 6% on the two investments. How much should he invest at each rate?

_____

$-10x = -14,000$ Subtract 170,000.

$\quad x = 1400$ Divide by -10.

Step 5 **State the answer.** $1400 should be

invested at $7\frac{1}{2}\%$ and $2000 - \$1400 = \$600$

should be invested at $8\frac{1}{2}\%$.

Step 6 **Check.**

The answer checks since

$0.075(\$1400) + 0.085(\$600) = \$156.$

Review this example for Objective 4:

4. A collection of coins consisting of nickels and dimes has a value of $5.80. Find the number of nickels and dimes in the collection if there are 22 more dimes than nickels.

    Step 1 **Read** the problem carefully. There will be two answers.

    Step 2 **Assign a variable.**
    Let x = the number of nickels.
    Then, $22 + x$ = the number of dimes.
    We summarize the information in a table.

Number of Coins	Value of Each Coin	Total Value
x	0.05	$0.05x$
$22 + x$	0.10	$0.10(22 + x)$

    Step 3 **Write an equation.**

Value of nickels	plus	Value of dimes	is	Total Value
↓	↓	↓	↓	↓
$0.05x$	$+$	$0.10(22 + x)$	$=$	5.80

    Step 4 **Solve** the equation.

    $0.05x + 0.10(22) + 0.10(x) = 5.80$

    Distributive property

    $5x + 10(22) + 10(x) = 100(5.80)$

    Multiply by 100.

    $5x + 220 + 10x = 580$

    Multiply.

    $15x + 220 = 580$

    Combine like terms.

Now Try:

4. A cashier has $645 in ten-dollar bills and five-dollar bills. There are 90 bills in all. How many of each bill does the cashier have?

    _____

$15x = 360$ Subtract 220.

$x = 24$ Divide by 15.

Step 5 **State the answer.** There are 24 nickels and $22 + 24 = 46$ dimes.

Step 6 **Check.**

The answer checks since

$(24)(0.05) + 46(0.10) = \$5.80$.

Review these examples for Objective 5:

5. Marie drove from Cherry Hill, NJ to Yankee Stadium in Bronx, NY, a distance of 97.2 miles, in 1 hour, 45 minutes. What was her average rate in miles per hour? (Round your answer to the nearest tenth.)

   We use the formula $\text{rate} \times \text{time} = \text{distance}$ or $\text{rate} = \dfrac{\text{distance}}{\text{time}}$. Since we want the answer in miles per hour, we must convert 1 hour, 45 minutes to hours: $1 \text{ hr } 45 \text{ min} = 1\dfrac{45}{60} = 1.75 \text{ hr}$

   $r = \dfrac{97.2}{1.75} \approx 55.5 \text{ mph}$

   Marie drove at an average rate of 55.5 miles per hour.

6. Joy and Elizabeth started bicycling from the same point at the same time, traveling in opposite directions. Joy biked at 12 mph, while Elizabeth rode at 10 mph. After how many hours will they be 55 miles apart?

   Step 1 **Read** the problem carefully. We must find the time for the distance between the two to be 55 miles.

   Step 2 **Assign a variable.** We are looking for time.

   Let t = the number of hours until the distance between the two women is 55 miles.

   We summarize the information in a table.

	Rate	*Time*	*Distance*
Joy	12	t	$12t$
Elizabeth	10	t	$10t$

Now Try:

5. An Amtrak train traveled from Seattle to San Francisco, averaging 54 miles per hour. The distance between the two cities is 810 miles. How long did the trip take?

   _____

6. Sherri drove from her home on a trip, driving at 50 mph. Two hours later, Karen drove the same route at 60 mph. How many hours will it take for Karen to catch up to Sherri?

   _____

Step 3 **Write an equation.**

$12t + 10t = 55$

Step 4 **Solve** the equation.

$22t = 55$ Combine like terms.

$t = 2.5$ Divide by 22.

Step 5 **State the answer.** It will take 2.5 hours for the women to be 55 miles apart.

Step 6 **Check.**

After 2.5 hours, Joy will travel 12(2.5) = 30 mi and Elizabeth will travel 10(2.5) = 25 mi. The total distance is 30 + 25 = 55 mi, as required.

7. Two planes leave an airport at the same time, one traveling north and the other traveling south. The northbound plane travels 50 mph faster than the southbound plane. In 2.5 hr, they are 3000 mi apart. What are their rates?

Step 1 **Read** the problem carefully. We must find the rate for each plane.

Step 2 **Assign a variable.**

Let r = the rate of the southbound (slower) plane. Then $r + 50$ = the rate of the northbound (faster) plane.

We summarize the information in a table.

	Rate	*Time*	*Distance*
Southbound plane	r	2.5	$2.5r$
Northbound plane	$r + 50$	2.5	$2.5(r + 50)$

Step 3 **Write an equation.**

$2.5r + 2.5(r + 50) = 3000$

Step 4 **Solve** the equation.

$2.5r + 2.5r + 2.5(50) = 3000$ Distributive property

$5r + 125 = 3000$ Combine like terms; multiply.

$5r = 2875$ Subtract 125.

$r = 575$ Divide by 5.

Step 5 **State the answer.** The southbound plane travels at 575 mph and the northbound plane travels at 575 + 50 = 625 mph.

7. Two cars leave a parking garage, traveling in opposite directions. One car had an average speed of 8 mph more than the other. Fifteen minutes later, the cars were 22 miles apart. What are their rates?

_____

Step 6 **Check.**
After 2.5 hours, the southbound plane will
travel 2.5(575) = 1437.5 mi, and the northbound
plane will travel 1562.5 mi.
1437.5 + 1562.5 = 3000, as required.

Objective 1 Use percent in solving problems involving rates.
For extra help, see Example 1 on page 140 of your text, the Section Lecture video for
Section 2.7, and Exercise Solution Clip 1.

Solve the problem.

1. How much pure alcohol is in 50 liters of a 45% alcohol   1. _____
    solution?

2. If $10,000 is invested for one year at 8% simple interest,   2. _____
    how much interest is earned?

Objective 2 Solve problems involving mixtures.
For extra help, see Example 2 on pages 140–141 of your text, the Section Lecture video for
Section 2.7, and Exercise Solution Clip 13.

Solve the problem.

3. A car radiator contains 4 gallons of a coolant which is a   3. _____
    mixture of antifreeze and water. If the coolant in the
    radiator is 30% antifreeze, how much coolant must be
    added with 80% antifreeze to have a 50% solution?

4. A saline solution, weighing 48 oz, is 6.25% salt. How   4. _____
    many ounces of salt must be added to obtain a solution
    that is 10% salt?

5. A chemist has two acid solutions. One is a 60% solution   5. _____
    and the other a 30% solution. How many liters of each
    should she mix to obtain 10 liters of 51% acid solution?

Objective 3 Solve problems involving simple interest.
For extra help, see Example 3 on page 141–142 of your text, the Section Lecture video for Section 2.7, and Exercise Solution Clip 23.

Solve the problem. Assume that simple interest is being paid.

6. August has an annual interest income of $3390 from two investments. He has $10,000 more invested at 8% than he has invested at 6%. Find the amount invested at each rate.

6. _____

7. Georgia has 3 times as much money invested in 8% bonds as she has in stocks paying $9\frac{1}{2}$%. How much does she have invested in each if her yearly income from the investments is $5695?

7. _____

8. Noah has $1000 more invested at 9% than he has invested at 11%. If the annual income for the two investments is $1290, find how much he has invested at each rate.

8. _____

Objective 4 Solve problems involving denominations of money.
For extra help, see Example 4 on page 142–143 of your text, the Section Lecture video for Section 2.7, and Exercise Solution Clip 27.

Solve the problem.

9. Total receipts from the sale of 300 tickets to a school musical were $1130. If student tickets cost $3 each and adult tickets $5 each, how many student tickets were sold?

9. _____

10. A stamp collector buys some 20¢ stamps and some 35¢ stamps, paying $11.35 for them. The number of 35¢ stamps is one more than the number of 20¢ stamps. Find the number of 35¢ stamps she buys.

10. _____

11. Meredith has $10.35 in nickels, dimes, and quarters. If she has six more dimes than nickels and twice as many quarters as nickels, how many of each kind of coin does she have?

11. _____

Objective 5 Solve problems involving distance, rate, and time.
For extra help, see Example 5–7 on pages 143–145 of your text, the Section Lecture video for Section 2.7, and Exercise Solution Clips 35, 47, and 49.

Solve the problem, using $d = rt$, $r = \dfrac{d}{t}$, *or* $t = \dfrac{d}{r}$, *as necessary.*

12. A driver averaged 48 miles per hour and took 6 hours to travel from Chicago to St. Louis. What is the distance between Chicago and St. Louis?

12. _____

13. In the 2008 Olympics, Michael Phelps won the men's 200 m butterfly in a world-record time of 1 minute 52.03 seconds. Find his average speed in meters per second. (Round to the nearest hundredth.)

13. _____

Solve the problem.

14. Ron and Doug leave the same point at the same time traveling in cars going in opposite directions. Ron travels at 40 miles per hour and Doug travels at 60 miles per hour. In how many hours will they be 350 miles apart?

14. _____

15. Elly and Sam are jogging. Elly runs from point A to
point B in one hour at 2 miles per hour faster than Sam
does. If Sam takes $\frac{1}{2}$ hour more time than Elly to go
the same distance, find the distance between points A
and B.

15. _____

Chapter 2 LINEAR EQUATIONS AND INEQUALITIES IN ONE VARIABLE

2.8 Solving Linear Inequalities

Learning Objectives
1   Graph intervals on a number line.
2   Use the addition property of inequality.
3   Use the multiplication property of inequality.
4   Solve linear inequalities by using both properties in inequality.
5   Solve applied problems by using inequalities.
6   Solve linear inequalities with three parts.

Key Terms

Use the vocabulary terms listed below to complete each statement in exercises 1–5.

  inequality   interval   interval notation

  linear inequality in one variable   three part inequality

1.   A(n) _____ can be written in the form $Ax + B < C$, $Ax + B \leq C$, $Ax + B > C$, or $Ax + B \geq C$, where A, B, and C are real numbers, with $A \neq 0$.

2.   _____ is a simplified notation that uses parentheses ( ) and/or brackets [ ] to describe an interval on a number line.

3.   An inequality that says that one number is between two other numbers is called a _____.

4.   A(n) _____ is a portion of a number line.

5.   A(n) _____ is a statement that two expressions are not equal.

Guided Examples

Review this example for Objective 1:
1.   Write each inequality in interval notation, and graph the interval.

  a. $x > -3$

  In interval notation, $x > -3$ is written $(-3, \infty)$.

Now Try:
1.   Write each inequality in interval notation, and graph the interval.

  a. $x < -2$ _____

b. $x \le 4$

In interval notation, $x \le 4$ is written $(-\infty, 4]$.

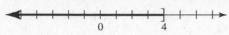

b. $x \ge -1$ _____

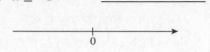

Review this example for Objective 2:

2. Solve the inequality $9 + 8b > 9b + 11$, and graph the solution set.

$$9 + 8b > 9b + 11$$

$9 + 8b - 8b > 9b + 11 - 8b$ Subtract $8b$.

           $9 > b + 11$ Combine like terms.

$9 - 11 > b + 11 - 11$ Subtract 11.

      $-2 > b$, or $b < -2$ Combine like terms.

The solution set is $(-\infty, -2)$.

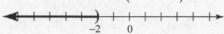

Now Try:

2. Solve the inequality, and graph the solution set.

$$6 + 3x < 4x + 4$$

_____

Review this example for Objective 3:

3. Solve the inequality $-5t \le -35$, and graph the solution set.

$$-5t \le -35$$

   $t \ge 7$ Divide by -5; reverse the symbol.

The solution set is $[7, \infty)$.

Now Try:

3. Solve the inequality, and graph the solution set.

$$-2s > 4$$

_____

Review these examples for Objective 4:

4. Solve the inequality $7m - 8 > 5m + 4$, and graph the solution set.

$7m - 8 > 5m + 4$

$2m - 8 > 4$ Subtract $5m$.

   $2m > 12$ Add 8.

    $m > 6$ Divide by 2.

The solution set is $(6, \infty)$.

Now Try:

4. Solve the inequality, and graph the solution set.

$$3x - \frac{3}{4} \ge 2x + \frac{1}{4}$$

_____

5. Solve the inequality $4(y-3)+16 < 6(y+2)$, and graph the solution set.

$$4y-12+16 < 6y+12 \quad \text{Distributive property}$$
$$4y+4 < 6y+12 \quad \text{Combine like terms.}$$
$$-2y+4 < 12 \quad \text{Subtract } 6y.$$
$$-2y < 8 \quad \text{Subtract 4.}$$
$$y > -4 \quad \begin{array}{l}\text{Divide by } -2;\\ \text{reverse inequality symbol.}\end{array}$$

The solution set is $(-4, \infty)$.

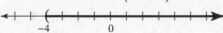

5. Solve the inequality, and graph the solution set.

$$-3(m+2)+1 \le -4(m-1)$$

Review this example for Objective 5:

6. Lauren has grades of 98 and 86 on her first two chemistry quizzes. What must she score on her third quiz to have an average of at least 91 on the three quizzes?

Step 1 **Read** the problem carefully.

Step 2 **Assign a variable.**
Let x = Lauren's score on the third quiz.

Step 3 **Write an inequality.**

$$\underset{\downarrow}{\text{average}} \quad \underset{\downarrow}{\underset{\text{least}}{\text{is at}}} \quad \underset{\downarrow}{91}$$

$$\frac{98+86+x}{3} \quad \ge \quad 91$$

Step 4 **Solve the inequality.**
$$\frac{184+x}{3} \ge 91 \quad \text{Add in the numerator.}$$
$$184 + x \ge 273 \quad \text{Multiply by 3.}$$
$$x \ge 89 \quad \text{Subtract 184.}$$

Step 5 **State the answer.** Lauren must score 89 or more on the third quiz to have an average of at least 91.

Step 6 **Check.**
$$\frac{98+86+89}{3} = \frac{273}{3} = 91$$

Now Try:

6. Jonathan sold two antique desks for $280 and $305. How much should he charge for the third in order to average at least $300 per desk?

_____

Name: _____   Date: _____
Instructor: _____   Section: _____

Review these examples for Objective 6:

7. Write the inequality $-3 < a \leq 2$ in interval form, and graph the interval.

    In interval notation, $-3 < a \leq 2$ is written $(-3, 2]$.

    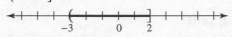

8. Solve the inequality $1 < 3z + 4 < 19,$ and graph the solution set.

    $1 - 4 < 3z + 4 - 4 < 19 - 4$ Subtract 4 from each part.

    $-3 < \quad 3z \quad < 15$

    $\dfrac{-3}{3} < \dfrac{3z}{3} < \dfrac{15}{3}$ Divide each part by 3.

    $-1 < \quad z \quad < 5$

    The solution set is $(-1, 5)$.

    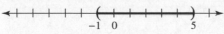

Now Try:

7. Write the inequality $-3 \leq y \leq 0$ in interval form, and graph the interval.

    _____

    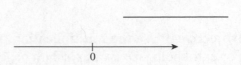

8. Solve the inequality $-5 < -2 - x \leq 4$, and graph the interval.

    _____

    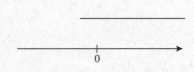

Objective 1 Graph intervals on a number line.
For extra help, see Example 1 on page 152 of your text, the Section Lecture video for Section 2.8, and Exercise Solution Clip 9.

Write each inequality in interval notation, and graph the interval.

1. $x \geq 3$

    1. _____

    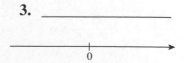

2. $7 < a$

    2. _____

Objective 2 Use the addition property of inequality.
For extra help, see Example 2 on page 153 of your text, the Section Lecture video for Section 2.8, and Exercise Solution Clip 21.

Solve each inequality. Write the solution in interval notation, and graph the interval.

3. $y - 7 > -12$

    3. _____

4. $5a + 3 \le 6a$

4. _____

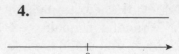

Objective 3 Use the multiplication property of inequality.

For extra help, see Example 3 on page 155 of your text, the Section Lecture video for Section 2.8, and Exercise Solution Clip 31.

Solve each inequality. Write the solution in interval notation, and graph the interval.

5. $-\frac{1}{2}r > 5$

5. _____

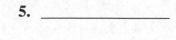

6. $4k \ge -16$

6. _____

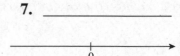

Objective 4 Solve linear inequalities by using both properties in inequality.

For extra help, see Examples 4 and 5 on page 156 of your text, the Section Lecture video for Section 2.8, and Exercise Solution Clips 45 and 51.

Solve each inequality. Write the solution in interval notation, and graph the interval.

7. $4(y - 3) + 2 > 3(y - 2)$

7. _____

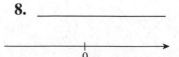

8. $7(2 - x) - 3 \le -2(x - 4) - x$

8. _____

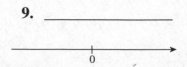

9. $3 - \frac{1}{4}z \ge 2 + \frac{3}{8}z$

9. _____

Name: Date:
Instructor: Section:

Objective 5 Solve applied problems by using inequalities.
For extra help, see Example 6 on page 157 of your text, the Section Lecture video for
Section 2.8, and Exercise Solution Clip 69.

Solve each problem.

10. Find every number such that one third the sum of that   10. _____
    number and 24 is less than or equal to 10.

11. Nina has a budget of $230 for gifts for this year. So far   11. _____
    she has bought gifts costing $47.52, $38.98, and $26.98.
    If she has three more gifts to buy, find the average
    amount she can spend on each gift and still stay within
    her budget.

12. Ruth tutors mathematics in the evenings in an office for   12. _____
    which she pays $600 per month rent. If rent is her only
    expense and she charges each student $40 per month,
    how many students must she teach to make a profit of at
    least $1600 per month?

Objective 6 Solve linear inequalities with three parts.
For extra help, see Examples 7 and 8 on pages 158–159 of your text, the Section Lecture
video for Section 2.8, and Exercise Solution Clip 93.

Solve each inequality Write the solution set in interval notation, and graph it.

13. $1 < 2r - 3 \le 5$ 13. _____

                                                               $\xrightarrow{\hspace{3cm}}$
                                                                        0

14. $-10 \le 4t - 2 \le 6$

14. _____

$\longrightarrow$
         0

15. $-5 < 3x - 8 < 6$

15. _____

$\longrightarrow$
         0

Chapter 3 LINEAR EQUATIONS AND INEQUALITIES IN TWO VARIABLES; FUNCTIONS

3.1 Linear Equations in Two Variables; The Rectangular Coordinate System

---

Learning Objectives
1   Interpret graphs.
2   Write a solution as an ordered pair.
3   Decide whether a given ordered pair is a solution of a given equation.
4   Complete ordered pairs for a given equation.
5   Complete a table of values.
6   Plot ordered pairs.

Key Terms
Use the vocabulary terms listed below to complete each statement in exercises 1–13.

line graph	linear equation in two variables	ordered pair
table of values	*x*-axis	*y*-axis
rectangular (Cartesian) coordinate system		quadrant
origin	plane	coordinates
plot	scatter diagram	

1. A(n) _____ is a pair of numbers written with parentheses in which the order of the numbers is important.

2. A(n) _____ is one of the four regions in the plane determined by a rectangular coordinate system.

3. The vertical line in a rectangular coordinate system is called the _____.

4. A(n) _____ is a series of line segments that connect points representing data.

5. The numbers in an ordered pair are called the _____ of the corresponding point.

6. The horizontal line in a rectangular coordinate system is called the _____.

7. A point at which the *x*-axis and *y*-axis of a rectangular coordinate system intersect is called the _____.

8. A(n) _____ is a graph of ordered pairs of data.

9. The *x*-axis and *y*-axis placed at a right angle at their zero points form a

    _____.

10. A(n) _____ is an equation that can be written in the form
    $Ax + By = C$, where *A*, *B*, and *C* are real numbers and *A* and *B* are not both zero.

11. To _____ an ordered pair is to locate it on a rectangular coordinate
    system.

12. The *x*-axis and *y*-axis determine a(n) _____.

13. A(n) _____ is an organized way of displaying ordered pairs.

Guided Examples

Review this example for Objective 1:

1. The line graph gives the value of one share
    of stock of Microchip Computer
    Corporation on the first trading day of the
    month for six consecutive months.

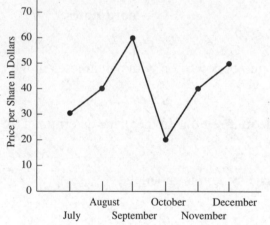

a. Between which consecutive months did
    the price of one share of stock decrease?

    September and October

b. Did the price of one share increase or
    decrease from July to December? By
    how much?

    The price of one share increased by $20
    from July to December.

Now Try:

1. Refer to the graph at the left to
    answer the questions.

a. In which months was the price
    of the stock double the stock
    price of other months?

    _____

b. About how much did the price
    of one share drop from
    September to October?

    _____

Review this example for Objective 3:	**Now Try:**

Review this example for Objective 3:

2. Decide whether each ordered pair is a solution of the equation $4x - 3y = 10$.

   a. $(1, 2)$ **b.** $(1, -2)$

     no   yes

Now Try:

2. Decide whether each ordered pair is a solution of the equation $-2x + y = 3$.

   a. $(-1, 2)$ _____

   b. $(-2, -1)$ _____

Review this example for Objective 4:

3. Complete each ordered pair for the equation $5x + 4y = 10$.

   a. $(2, \underline{})$

$$5(2) + 4y = 10 \quad \text{Let } x = 2.$$
$$10 + 4y = 10 \quad \text{Multiply.}$$
$$4y = 0 \quad \text{Subtract 10.}$$
$$y = 0 \quad \text{Divide by 4.}$$

The ordered pair is $(2, 0)$.

   b. $(\underline{}, -3)$

$$5x + 4(-3) = 10 \quad \text{Let } y = -3.$$
$$5x - 12 = 10 \quad \text{Multiply.}$$
$$5x = 22 \quad \text{Add 12.}$$
$$x = \frac{22}{5} \quad \text{Divide by 5.}$$

The ordered pair is $\left(\dfrac{22}{5}, -3 \right)$.

Now Try:

3. Complete each ordered pair for the equation $5x + 4y = 10$.

   a. $(-4, \underline{})$ _____

   b. $(\underline{}, 0)$ _____

Review this example for Objective 5:

4. Complete the table of values for the equation. Write the results as ordered pairs.

$$4x + 3y = 12$$

x	y
0	
	0
	-1
1	

If $x = 0$, then

$$4(0) + 3y = 12 \quad \text{Let } x = 0.$$
$$3y = 12 \quad 4 \cdot 0 = 0$$
$$y = 4 \quad \text{Divide by 3.}$$

Now Try:

4. Complete the table of values for the equation. Write the results as ordered pairs.

$$3x - 4y = -6$$

x	y
0	
	0
2	
	-2

   _____

   _____

   _____

   _____

If $y = 0$, then

$4x + 3(0) = 12$ Let $y = 0$.

$4x = 12$ $3 \cdot 0 = 0$

$x = 3$ Divide by 4.

If $y = -1$, then

$4x + 3(-1) = 12$ Let $y = -1$.

$4x - 3 = 12$ Multiply.

$4x = 15$ Add 3.

$x = \dfrac{15}{4}$ Divide by 4.

If $x = 1$, then

$4(1) + 3y = 12$ Let $x = 1$.

$4 + 3y = 12$ Multiply.

$3y = 8$ Subtract 4.

$y = \dfrac{8}{3}$ Divide by 3.

The ordered pairs are

$(0, 4), (3, 0), \left(\dfrac{15}{4}, -1\right),$ and $\left(1, \dfrac{8}{3}\right)$.

Review these examples for Objective 6:

5. Plot the given points in a coordinate system.
   $(0, -2), (2, 5), (-2, -7), (-3, 4), (4, -4),$
   $(-5, 0)$

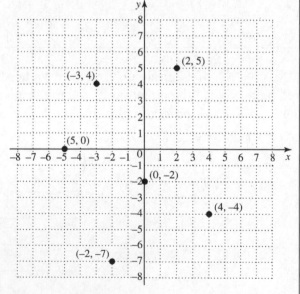

Now Try:

5. Plot the given points in a
   coordinate system.
   $(-2, 0), (5, 2), (-6, -2), (4, -3),$
   $(-4, 4), (0, -5)$

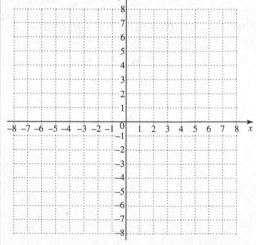

6. The cost to rent a truck for a week is $200 plus $0.52 per mile. Therefore, the cost to rent the truck for one week is given by $y = 0.52x + 200$, where x represents the number of miles driven and y represents the total cost for the week. How much did it cost to rent the truck if it was driven 400 miles?

We want to find y when $x = 400$.
$y = 0.52(400) + 200$ Let $x = 400$.
$y = 408$
It will cost $408 to rent the truck.

6. John paid $278 to rent the truck described at the left. How many miles did he drive the truck?

_____

Objective 1 Interpret graphs.
For extra help, see Example 1 on page 176 of your text and the Section Lecture video for Section 3.1.

The graph below shows the total number of degrees awarded by Jefferson University for the years 1990–1995. Use the graph to answer the questions in Exercises 1–3.

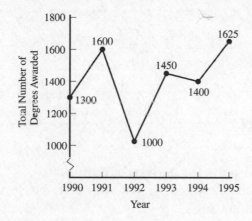

1. Between which two years did the total number of degrees awarded show the greatest decline?

1. _____

2. About how many more students received degrees in 1995 than 1994?

2. _____

3. Between which two years did the total number of degrees awarded show the smallest change?

3. _____

Objective 2 Write a solution as an ordered pair.

For extra help, see pages 177–178 of your text and the Section Lecture video for Section 3.1.

Write each solution as an ordered pair.

4. $x = -2$ and $y = -3$

4. _____

5. $y = \frac{1}{3}$ and $x = 0$

5. _____

Objective 3 Decide whether a given ordered pair is a solution of a given equation.

For extra help, see Example 2 on page 178 of your text, the Section Lecture video for Section 3.1, and Exercise Solution Clip 11.

Decide whether the given ordered pair is a solution of the given equation.

6. $5x - 2y = 6;\ (2, -2)$

6. _____

7. $2x - 3y = 1;\ \left(0, \frac{1}{3}\right)$

7. _____

8. $x = 1 - 2y;\ \left(0, \frac{1}{2}\right)$

8. _____

Objective 4 Complete ordered pairs for a given equation.

For extra help, see Example 3 on pages 178–179 of your text, the Section Lecture video for Section 3.1, and Exercise Solution Clip 23.

For each of the given equations, complete the ordered pairs beneath it.

9. $y = 2x - 5$

   (a) $(2,\ \ )$

   (b) $(0,\ \ )$

   (c) $(\ \ , 3)$

   (d) $(\ \ , -7)$

   (e) $(\ \ , 9)$

9.

   (a) _____

   (b) _____

   (c) _____

   (d) _____

   (e) _____

10. $y = 3 + 2x$

    (a) $(-4,\ \ )$

    (b) $(2,\ \ )$

    (c) $(\ \ , 0)$

    (d) $(-2,\ \ )$

    (e) $(\ \ , -7)$

10.

    (a) _____

    (b) _____

    (c) _____

    (d) _____

    (e) _____

11. $x = -2$

    (a) $(\ \ , -2)$

    (b) $(\ \ , 0)$

    (c) $(\ \ , 19)$

    (d) $(\ \ , 3)$

    (e) $\left(\ \ , -\frac{2}{3}\right)$

11.

    (a) _____

    (b) _____

    (c) _____

    (d) _____

    (e) _____

Objective 5 Complete a table of values.
For extra help, see Example 4 on pages 179–180 of your text, the Section Lecture video for Section 3.1, and Exercise Solution Clip 31.

Complete the table of ordered pairs for each equation.

12. $4x + y = 6$

x	2		1
y		4	

12. _____

13. $3x + 2y = 4$

x	0		4
y		0	

13. _____

14. $y - 4 = 0$

x	-6	0	6
y			

14. _____

Name: Date:
Instructor: Section:

Objective 6 Plot ordered pairs.

For extra help, see Examples 5 and 6 on pages 181–182 of your text and the Section Lecture video for Section 3.1.

Plot each point on the rectangular coordinate system below. Label each point with its coordinates.

15. (6, 1), (−2, 4), (−2, −6), (4, −2),   **15.**
     (0, −4), (−5, 0), (0, 0)

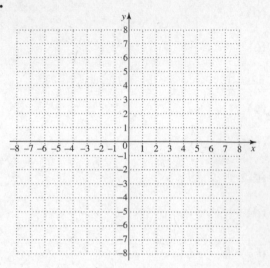

Copyright © 2012 Pearson Education, Inc. Publishing as Addison-Wesley.

Chapter 3 LINEAR EQUATIONS AND INEQUALITIES IN TWO VARIABLES; FUNCTIONS

3.2 Graphing Linear Equations in Two Variables

Learning Objectives
1   Graph linear equations by plotting ordered pairs.
2   Find intercepts.
3   Graph linear equations of the form $Ax + By = 0$.
4   Graph linear equations of the form $y = k$ or $x = k$.
5   Use a linear equation to model data.

Key Terms
Use the vocabulary terms listed below to complete each statement in exercises 1–3.

graph   **x-intercept**   **y-intercept**

1.   The _____ of any linear equation in two variables is a straight line.

2.   A point where a graph intersects the x-axis is called a(n) _____.

3.   A point where a graph intersects the y-axis is called a(n) _____.

Guided Examples

Review these examples for Objective 1:
1.   Graph $x - y = -1$.

   At least two different points are needed to draw the graph. First let $x = 0$ and then let $y = 0$ to find two ordered pairs.

   $0 - y = -1$ Let $x = 0$.
   $\quad -y = -1$ Additive identity
   $\quad\quad y = 1$ Multiply by -1.

   $x - 0 = -1$ Let $y = 0$.
   $\quad\quad x = -1$ Additive identity

   The ordered pairs are $(0, 1)$ and $(-1, 0)$.

   We find a third ordered pair (as a check) by choosing some other number for x or y.

   $4 - y = -1$ Let $x = 4$.
   $\quad -y = -5$ Subtract 4.
   $\quad\quad y = 5$ Multiply by -1.

   The third ordered pair is $(4, 5)$. Now plot the three ordered pairs.

Now Try:
1.   Graph $x + y = 3$.

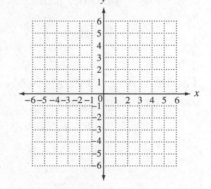

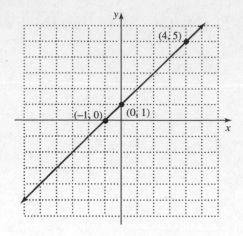

2. Graph $y = \dfrac{4}{3}x + 4$.

First let $x = 0$ and then let $y = 0$ to find two ordered pairs.

$y = \dfrac{4}{3}(0) + 4$ Let $x = 0$.

$y = 4$ Simplify.

$0 = \dfrac{4}{3}x + 4$ Let $y = 0$.

$-4 = \dfrac{4}{3}x$ Subtract 4.

$-3 = x$ Multiply by $\dfrac{3}{4}$.

The ordered pairs are $(0, 4)$ and $(-3, 0)$.

We find a third ordered pair (as a check) by choosing some other number for x or y.

$y = \dfrac{4}{3}(-6) + 4$ Let $x = -6$.

$y = -8 + 4 = -4$

The third ordered pair is $(-6, -4)$. Now plot the three ordered pairs.

2. Graph $y = -\dfrac{1}{2}x - 2$.

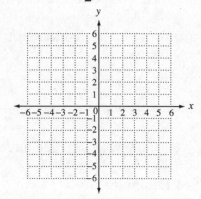

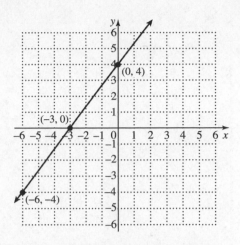

Review this example for Objective 2:

3. Find the intercepts for the graph $3x + 2y = 12$. Then draw the graph.

To find the y-intercept, let $x = 0$.

$3(0) + 2y = 12$ Let $x = 0$.

$\quad 0 + 2y = 12$

$\qquad 2y = 12$

$\qquad\quad y = 6$ Divide by 2.

The y-intercept is $(0, 6)$.

To find the x-intercept, let $y = 0$.

$3x + 2(0) = 12$ Let $y = 0$.

$\quad 3x + 0 = 12$

$\qquad 3x = 12$

$\qquad\quad x = 4$ Divide by 3.

The x-intercept is $(4, 0)$.

To find a third point, let $x = 2$.

$3(2) + 2y = 12$ Let $x = 2$.

$\quad 6 + 2y = 12$ Multiply.

$\qquad 2y = 6$ Subtract 6.

$\qquad\quad y = 3$ Divide by 2.

The third point is $(2, 3)$.

Now Try:

3. Find the intercepts for the graph $5x - 4y = 20$. Then draw the graph.

_____

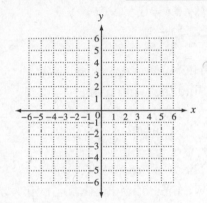

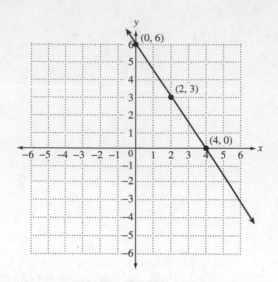

Review this example for Objective 3:

4. Graph $x + 5y = 0$.

    Find the y-intercept:
    $0 + 5y = 0$ Let $x = 0$.
    $\quad 5y = 0$
    $\quad\; y = 0$ Divide by 5.
    The y-intercept is $(0, 0)$.
    Find the x-intercept:
    $x + 5(0) = 0$ Let $y = 0$.
    $\quad x + 0 = 0$
    $\quad\quad x = 0$
    The x-intercept is $(0, 0)$.
    We must find two other points on the graph.
    Let $x = -5$.
    $-5 + 5y = 0$ Let $x = -5$.
    $\quad\; 5y = 5$ Add 5.
    $\quad\;\; y = 1$ Divide by 5.
    The ordered pair is $(-5, 1)$.
    Let $y = -1$.
    $x + 5(-1) = 0$ Let $y = -1$.
    $\quad x - 5 = 0$ Multiply.
    $\quad\quad x = 5$ Add 5.
    The ordered pair is $(5, -1)$.

Now Try:

4. Graph $2x - y = 0$.

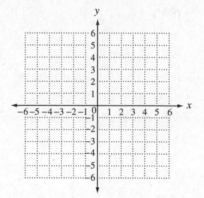

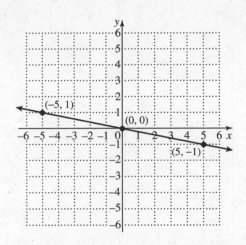

Review these examples for Objective 4:

5. Graph $y = -2$.

For any value of x, y is always -2. Three ordered pairs that satisfy the equation are $(-3, -2)$, $(0, -2)$, and $(4, -2)$.

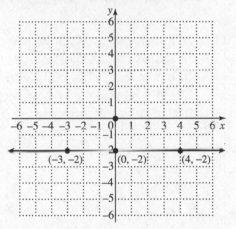

6. Graph $x + 2 = 0$

First subtract 2 from each side to obtain $x = -2$. For any value of y, x is always -2. Three ordered pairs that satisfy the equation are $(-2, -3)$, $(-2, 0)$, $(-2, 4)$.

Now Try:

5. Graph $y = 4$.

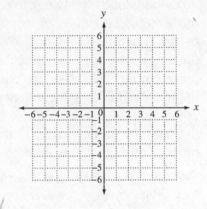

6. Graph $x - 4 = 0$

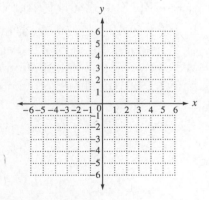

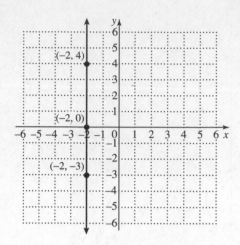

Review this example for Objective 5:

7. Every year sea turtles return to a certain group of islands to lay eggs. The number of turtle eggs that hatch can be approximated by the equation $y = -70x + 3260$, where y is the number of eggs that hatch and $x = 0$ representing 1990. Use this equation to find the number of eggs that hatched in 1995, 2000, and 2005. Use the three ordered pairs to graph the model. Then, estimate the number of eggs that will hatch in 2015.

For 1995, $x = 1995 - 1990 = 5$. Then,
$y = -70(5) + 3260 = 2910$.

For 2000, $x = 2000 - 1990 = 10$. Then,
$y = -70(10) + 3260 = 2560$.

For 2005, $x = 2005 - 1990 = 15$. Then,
$y = -70(15) + 3260 = 2210$.

For 2015, $x = 2015 - 1990 = 25$. Then,
$y = -70(25) + 3260 = 1510$.

In 1995, 2910 eggs hatched. In 2000, 2560 eggs hatched. In 2005, 2210 eggs hatched. We can write this information as the ordered pairs (5, 2910), (10, 2560), and (15, 2210).

Now Try:

7. The profit y in millions of dollars earned by a small computer company can be approximated by the linear equation
$y = 0.63x + 4.9$, where $x = 0$
corresponds to 2004, $x = 1$
corresponds to 2005, and so on. Use this equation to approximate the profit in 2005, 2007, and 2008. Use the answers to draw the graph of the model.

_____

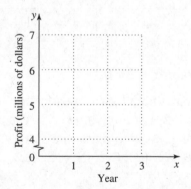

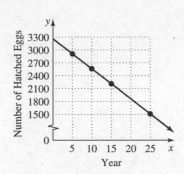

According to the model, 1510 eggs will hatch in 2015.

Objective 1 Graph linear equations by plotting ordered pairs.
For extra help, see Examples 1 and 2 on pages 188–190 of your text, the Section Lecture video for Section 3.2, and Exercise Solution Clips 1 and 3.

Complete the ordered pairs for each equation. Then graph the equation by plotting the points and drawing a line through them.

1. $2y - 4 = x$

    $(0, \quad)$

    $(\quad , 0)$

    $(-2, \quad)$

1.

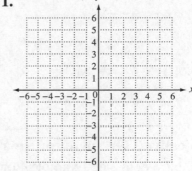

2. $x - y = 4$

    $(0, \quad)$

    $(\quad , 0)$

    $(-2, \quad)$

2.

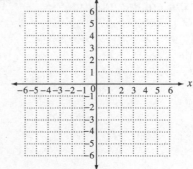

3. $2x + 3y = 6$

   $(0, \ )$

   $(\ , 0)$

   $(-3, \ )$

3.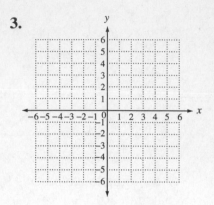

Objective 2 Find intercepts.

For extra help, see Example 3 on pages 190–191 of your text and the Section Lecture video for Section 3.2.

Find the intercepts for the graph of each equation.

4. $5x - 2y = 10$ **4.** _____

5. $4x + 3y = 9$ **5.** _____

6. $3x + 2y = -2$ **6.** _____

Objective 3 Graph linear equations of the form $Ax + By = 0$.

For extra help, see Example 4 on page 191 of your text, the Section Lecture video for Section 3.2, and Exercise Solution Clip 43.

Graph each equation.

7. $-3x - 2y = 0$ **7.**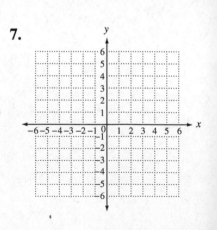

8. $4x = 3y$

8.

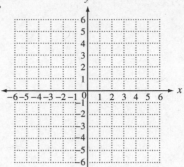

9. $3x - y = 0$

9.

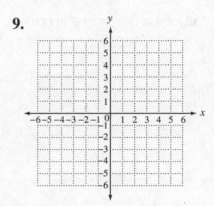

Objective 4 Graph linear equations of the form $y = k$ or $x = k$.
For extra help, see Examples 5 and 6 on page 192 of your text, the Section Lecture video for Section 3.2, and Exercise Solution Clips 47 and 49.

Graph each equation.

10. $y + 3 = 0$

10.

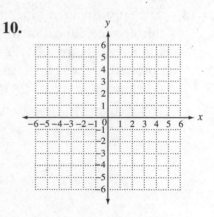

11. $x - 1 = 0$

11.
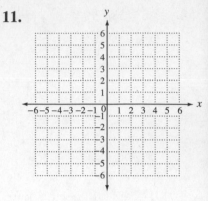

12. $x = 0$

12.
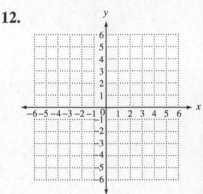

Objective 5 Use a linear equation to model data.
For extra help, see Example 7 on page 194 of your text, the Section Lecture video for
Section 3.2, and Exercise Solution Clip 67.

Solve each problem. Then graph the equation.

13. The enrollment at Lincolnwood High School decreased
during the years 2000 to 2005. If $x = 0$ represents 2000,
$x = 1$ represents 2001, and so on, the number of students
enrolled in the school can be approximated by the
equation $y = -85x + 2435$. Use this equation to
approximate the number of students in each year from
2000 through 2005.

13. _____

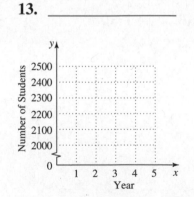

14. Suppose that the demand and price for a certain model of calculator are related by the equation $y = 45 - \frac{3}{5}x$, where y is the price (in dollars) and x is the demand (in thousands of calculators). Assuming that this model is valid for a demand up to 50,000 calculators, find the price at each of the following levels of demand.

a. 0 calculators

b. 5000 calculators

c. 20,000 calculators

d. 45,000 calculators

14.a. _____

b. _____

c. _____

d. _____

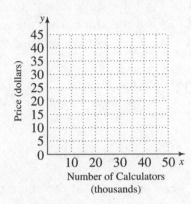

15. The number of band instruments sold by Elmer's Music Shop can be approximated by the equation $y = 325 + 42x$, where y is the number of instruments sold and x is the time in years, with $x = 0$ representing 2003. Use this equation to approximate the number of instruments sold in each year from 2003 through 2006.

15. _____

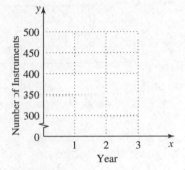

Chapter 3   LINEAR EQUATIONS AND INEQUALITIES IN TWO VARIABLES; FUNCTIONS

3.3 The Slope of a Line

Learning Objectives
1   Find the slope of a line, given two points.
2   Find the slope from the equation of a line.
3   Use slopes to determine whether two lines are parallel, perpendicular, or neither.

Key Terms
Use the vocabulary terms listed below to complete each statement in exercises 1–6.

rise	run	slope
subscript notation	**parallel**	**perpendicular**

1.   The ratio of the change in y to the change in x along a line is called the
   _____ of the line.

2.   The _____ is the horizontal change between two points on a line –
   that is, the change in x-values.

3.   _____ is a way of indicating nonspecific values, such as x_1 and x_2.

4.   Two lines that are _____ have the same slope.

5.   The _____ is the vertical change between two points on a line – that
   is, the change in y-values.

6.   Two lines that are _____ have slopes whose product is −1.

Name:

Instructor:

Date:

Section:

Guided Examples

Review these examples for Objective 1:

1. Find the slope of the line.

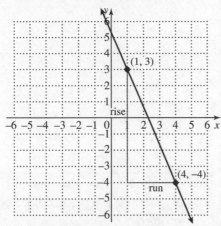

$$\text{slope} = \frac{\text{change in } y \text{ (rise)}}{\text{change in } x \text{ (run)}}$$

$$= \frac{3-(-4)}{1-4} = \frac{7}{-3} \text{ or } -\frac{7}{3}$$

2. Find the slope of the line through $(5, -2)$ and $(2, 7)$.

Let $(5, -2) = (x_1, y_1)$ and let $(2, 7) = (x_2, y_2)$.

$$\text{slope} = \frac{y_2 - y_1}{x_2 - x_1} = \frac{7-(-2)}{2-5} = \frac{9}{-3} = -3$$

3. Find the slope of the line through $(-7, 7)$ and $(2, 7)$.

$$m = \frac{7-7}{2-(-7)} = \frac{0}{9} = 0$$

All horizontal lines have slope 0.

4. Find the slope of the line through $(-7, 7)$ and $(-7, 2)$.

$$m = \frac{2-(-7)}{-7-(-7)} = \frac{9}{0}, \text{ which is undefined}$$

All vertical lines have undefined slope.

Now Try:

1. Find the slope of the line.

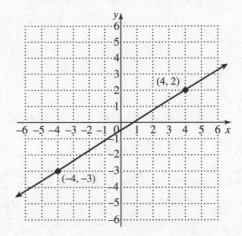

2. Find the slope of the line through $(2, -4)$ and $(-3, -1)$.

_____

3. Find the slope of the line through $(-4, -2)$ and $(2, -2)$.

_____

4. Find the slope of the line through $(-4, -2)$ and $(-4, 2)$.

_____

Review this example for Objective 2:

5. Find the slope of the line $2x + 8y = 9$.

   First, solve the equation for y.
   $$2x + 8y = 9$$
   $$8y = -2x + 9 \qquad \text{Subtract } 2x.$$
   $$y = -\frac{2}{8}x + \frac{9}{8} \qquad \text{Divide by 8.}$$
   $$= -\frac{1}{4}x + \frac{9}{8}$$

   The slope is given by the coefficient of x, so the slope is $-\frac{1}{4}$.

Now Try:

5. Find the slope of the line $4x - 3y = 0$.

   _____

Review this example for Objective 3:

6. Decide whether each pair of lines is *parallel*, *perpendicular*, or *neither*.

   a. $-4x + y = 4$
       $8x - 2y = 12$

   First, find the slope of each line by solving for y.
   $-4x + y = 4 \Rightarrow y = 4x + 4$
   Slope is 4.
   $8x - 2y = 12 \Rightarrow y = 4x - 6$
   Slope is 4
   Since the slopes are equal, the lines are parallel.

   b. $-4x + y = 4$
       $x + 4y = 12$

   First, find the slope of each line by solving for y.
   $-4x + y = 4 \Rightarrow y = 4x + 4$
   Slope is 4.
   $x + 4y = 12 \Rightarrow y = -\frac{1}{4}x + 3$

   Slope is $-\frac{1}{4}$.

   Since $4\left(-\frac{1}{4}\right) = -1$, the slopes are negative reciprocals, and the lines are perpendicular.

6. Decide whether the pair of lines is *parallel*, *perpendicular*, or *neither*.
   $9x + 3y = 2$
   $x - 3y = 5$

   _____

c. $8x + 2y = 7$

$x = 3 - y$

First, find the slope of each line by solving for y.

$8x + 2y = 7 \Rightarrow y = -4x + \frac{7}{2}$

Slope is -4.

$x = 3 - y \Rightarrow y = -x + 3$

Slope is -1.

Since the slopes are neither equal nor negative reciprocals, the lines are neither parallel nor perpendicular.

Objective 1 Find the slope of a line, given two points.
For extra help, see Examples 1–4 on pages 200–203 of your text, the Section Lecture video for Section 3.3, and Exercise Solution Clips 1, 27, 31, and 35.

Find the slope of each line.

1. Through $(4,3)$ and $(3,5)$ 1. _____

2. Through $(7,2)$ and $(-7,3)$ 2. _____

3. Through $(-3,2)$ and $(7,4)$ 3. _____

4. Through $(-7,-7)$ and $(2,-7)$ 4. _____

5. Through $(-4,6)$ and $(-4,-1)$ 5. _____

Objective 2 Find the slope from the equation of a line.
For extra help, see Example 5 on page 204 of your text, the Section Lecture video for Section 3.3, and Exercise Solution Clip 45.

Find the slope of each line.

6. $y = -5x$ 6. _____

7. $4y = 3x + 7$ 7. _____

8. $2x + 7y = 7$ 8. _____

9. $y = -4$ 9. _____

10. $x = 0$ 10. _____

Objective 3 Use slopes to determine whether two lines are parallel, perpendicular, or neither.
For extra help, see Example 6 on pages 205–206 of your text, the Section Lecture video for Section 3.3, and Exercise Solution Clip 57.

In each pair of equations, give the slope of each line, and then determine whether the two lines are parallel, perpendicular, *or* neither.

11. $y = 4x + 4$ 11. _____
    $y = 3 - \frac{1}{4}x$

12. $-x + y = -7$ 12. _____
    $x - y = -3$

13. $4x + 2y = 8$ 13. _____
    $x + 4y = -3$

14. $y + 4 = 0$

$y - 7 = 0$

14. _____

15. $y = 9$

$x = 0$

15. _____

Chapter 3 LINEAR EQUATIONS AND INEQUALITIES IN TWO VARIABLES; FUNCTIONS

3.4 Writing and Graphing Equations of Lines

Learning Objectives
1   Use the slope –intercept form of the equation of a line.
2   Graph a line by using its slope and a point on the line.
3   Write an equation of a line by using its slope and any point on the line.
4   Write an equation of a line by using two points on the line.
5   Write an equation of a line that fits a data set.

Key Terms
Use the vocabulary terms listed below to complete each statement in exercises 1–3.

slope-intercept form   **point-slope form**   **standard form**

1.   A linear equation is written in _Slope intercept_ if it is in the form $y = mx + b$, where m is the slope and $(0, b)$ is the y-intercept.

2.   A linear equation is written in _point slope_ if it is in the form $y - y_1 = m(x - x_1)$, where m is the slope and (x_1, y_1) is a point on the line.

3.   A linear equation in two variables written in the form $Ax + By = C$, with A and B both not 0, is in _Standard form_.

Guided Examples

Review these examples for Objective 1:
1.   Identify the slope and y-intercept of the line with each equation.

   a. $y = -\dfrac{1}{2}x + 4$

   slope: $-\dfrac{1}{2}$; y-intercept: $(0, 4)$

   b. $y = -6x - \dfrac{4}{3}$

   slope: -6; y-intercept: $\left(0, -\dfrac{4}{3}\right)$

Now Try:
1.   Identify the slope and y-intercept of the line with each equation.

   a. $y = \dfrac{2}{3}x + \dfrac{5}{3}$ _____

   b. $y = -\dfrac{x}{4} - \dfrac{5}{8}$ _____

Name: Date:

Instructor: Section:

2. Write an equation of the line with slope $-\frac{1}{2}$ and y-intercept $(0,-3)$

$$y = -\frac{1}{2}x - 3$$

2. Write an equation of the line with slope $\frac{3}{2}$ and y-intercept $\left(0, -\frac{2}{3}\right)$.

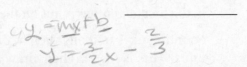

$$y = mx + b$$
$$y = \frac{3}{2}x - \frac{2}{3}$$

Review these examples for Objective 2:

3. Graph $x + 2y = 8$ by using the slope and y-intercept.

Step 1 Solve for y to write the equation in slope-intercept form.

$$x + 2y = 8$$
$$2y = -x + 8 \qquad \text{Subtract } x.$$
$$y = -\frac{1}{2}x + 4 \quad \text{Divide by 2.}$$

Step 2 The y-intercept is $(0, 4)$. Graph this point.

Step 3 The slope is $-\frac{1}{2}$, which can be written as either $\frac{-1}{2}$ or $\frac{1}{-2}$. We will use $\frac{-1}{2}$.

$$m = \frac{\text{change in } y \text{ (rise)}}{\text{change in } x \text{ (run)}} = \frac{-1}{2}$$

From the y-intercept, count down 1 unit and to the right 2 units, to obtain the point $(3, 2)$.

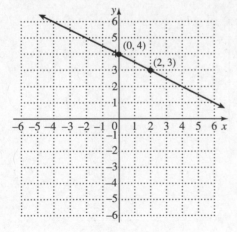

Now Try:

3. Graph $5x - 3y = 9$ using the slope and y-intercept.

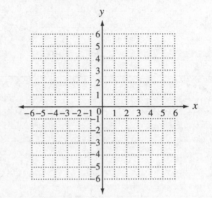

4. Graph the line through (2, 2) with slope $\frac{1}{3}$.

First, locate the point (2, 2). Since the slope is $\frac{1}{3}$, locate another point on the line by counting up 1 unit and then to the right 3 units. This is (3, 5). Then, draw the line through (2, 2) and (3, 5).

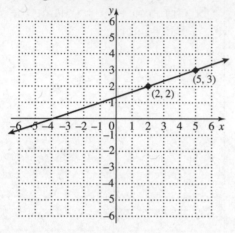

4. Graph the line through (3, −1) with slope −2.

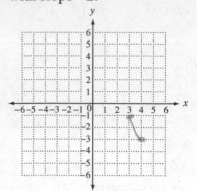

Review these examples for Objective 3:

5. Write an equation in slope-intercept form of the line having slope −5 and passing through the point (−1, 6).

Since the line passes through the point (−1, 6), substitute $x = -1$, $y = 6$, and slope $m = -5$ into $y = mx + b$, and solve for b.

$y = mx + b$ Slope-intercept form
$6 = -5(-1) + b$ Let $x = -1$, $y = 6$, $m = -5$.
$6 = 5 + b$ Multiply.
$1 = b$ Subtract 1.

Now substitute the values of m and b into slope-intercept form.
$y = -5x + 1$.

Now Try:

5. Write an equation in slope-intercept form of the line having slope $\frac{3}{4}$ and passing through the point (−4, −7).

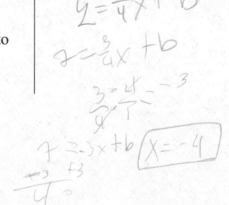

6. Write an equation of the line through $(-8, 2)$ with slope $\frac{3}{4}$. Give the final answer in slope-intercept form.

$$y - y_1 = m(x - x_1) \quad \text{Point-slope form}$$

$$y - 2 = \frac{3}{4}\left[x - (-8)\right] \quad y_1 = 2,\ x_1 = -8,\ m = \frac{3}{4}$$

$$y - 2 = \frac{3}{4}(x + 8)$$

$$y - 2 = \frac{3}{4}x + 6 \quad \text{Distributive property}$$

$$y = \frac{3}{4}x + 8 \quad \text{Add 2.}$$

6. Write an equation of the line through $(6, -2)$ with slope $\frac{3}{2}$. Give the final answer in slope-intercept form.

_____

$$y = mx + b$$

$$-2 = \frac{3}{2} \times (6) + b$$

$$-2 =$$

$$\frac{3}{2} \cdot \frac{-1}{1} = \frac{-3}{-1} = 3$$

Review this example for Objective 4:

7. Write an equation of the line through the points $(-2, 1)$ and $(3, 11)$. Give the final answer in slope-intercept form and then in standard form.

First, find the slope of the line.

$$m = \frac{y_2 - y_1}{x_2 - x_1} = \frac{11 - 1}{3 - (-2)} = \frac{10}{5} = 2$$

Now use either point and point-slope form.

$$y - y_1 = m(x - x_1) \quad \text{Point-slope form}$$

$$y - 1 = 2\left[x - (-2)\right] \quad y_1 = 1,\ x_1 = -2,\ m = 2$$

$$y - 1 = 2(x + 2)$$

$$y - 1 = 2x + 4 \quad \text{Distributive property}$$

$$y = 2x + 5 \quad \text{Add 1.}$$

The slope-intercept form of the equation is $y = 2x + 5$.

$$y = 2x + 5 \quad \text{Slope-intercept form}$$

$$-2x + y = 5 \quad \text{Subtract } 2x.$$

$$2x - y = -5 \quad \text{Multiply by } -1.$$

The standard from of the equation is $2x - y = -5$.

Now Try:

7. Write an equation of the line through the points $(1, -2)$ and $(-2, 8)$. Give the final answer in slope-intercept form and then in standard form.

_____

_____

$$\frac{10}{3} \times \frac{23}{3} = \frac{15}{3} = 3$$

$$\frac{16}{3} = \frac{6}{3} = \frac{4}{3}$$

Name: _____   Date: _____

Instructor: _____   Section: _____

Review this example for Objective 5:

8. The table shows the average annual telephone expenditures for cell phones from 2001 through 2007, where year 0 represents 2001. (Source: Bureau of Labor Statistics)

Year	Annual Cell Phone Expenditures
0	$210
1	$294
2	$316
3	$378
4	$455
5	$524
6	$608

Use the points (0, 210) and (4, 455) to write an equation in slope-intercept form to approximate the data. How well does this equation approximate the annual expenditure for 2007?

First, find the slope.

$$m = \frac{y_2 - y_1}{x_2 - x_1} = \frac{455 - 210}{4 - 0} = \frac{245}{4}$$

The y-intercept is (0, 210), so the equation is

$$y = \frac{245}{4}x + 210.$$

The year 2007 is represented by $x = 6$.

$$y = \frac{245}{4}(6) + 210 = 577.5$$

The model underestimates the expenditure in 2007. This is illustrated in the graph below.

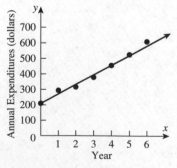

Now Try:

8. Using the table at the left, write an equation in slope-intercept form using the data for 2003 and 2006. Use this equation to approximate the annual expenditure for 2007.

_____

_____

Objective 1 Use the slope-intercept form of the equation of a line.
For extra help, see Examples 1 and 2 on pages 211–212 of your text, the Section Lecture video for Section 3.4, and Exercise Solution Clip 21.

Write an equation in slope-intercept form for each of the following lines.

1. $m = \frac{2}{3}$; $b = -4$ 1. _____

2. $m = -2$; $b = 0$ 2. _____

3. Slope 0; y-intercept $(0, -4)$ 3. _____

Objective 2 Graph a line by using its slope and a point on the line.
For extra help, see Examples 3 and 4 on pages 212–213 of your text, the Section Lecture video for Section 3.4, and Exercise Solution Clips 33 and 39.

Graph the line passing through the given point and having the given slope.

4. $(4, -2)$; $m = -1$ 4.

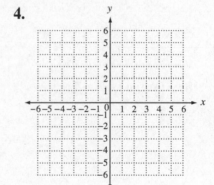

5. $(-3, -2)$; $m = \frac{2}{3}$ 5.

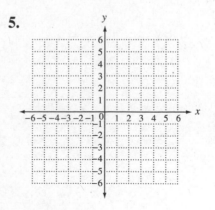

6. $(2,4)$; undefined slope

6.

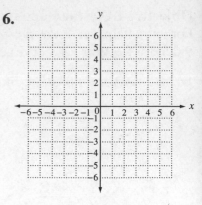

Objective 3 Write an equation of a line by using its slope and any point on the line.
For extra help, see Examples 5 and 6 on pages 214–215 of your text, the Section Lecture
video for Section 3.4, and Exercise Solution Clip 53.

*Write an equation for the line passing through the given point and having the given
slope. Write the equations in the slope-intercept form.*

7. $(5,4)$; $m = \frac{1}{3}$

7. _____

8. $(-2, 4)$; $m = 2$

8. _____

9. $(-3, -1)$; $m = -\frac{2}{3}$

9. _____

$y = mx + -3$

$-3 \left(-1 = -\frac{2}{3} x - 3 \right)$

Objective 4 Write an equation of a line by using two points on the line.
For extra help, see Example 7 on page 215 of your text, the Section Lecture video for
Section 3.4, and Exercise Solution Clip 65.

*Write an equation for the line passing through each pair of points. Give the final
answer in (a) slope-intercept form and (b) standard form.*

10. $(2,3)$ and $(7,5)$

10. a._____

b. _____

11. $(3,-4)$ and $(2,7)$

11. a._____

b. _____

12. $\left(\frac{1}{2},\frac{2}{3}\right)$ and $\left(-\frac{3}{2},2\right)$

12. a._____

b. _____

Objective 5 Write an equation of a line that fits a data set.
For extra help, see Example 8 on pages 216–217 of your text and the Section Lecture video
for Section 3.4.

*The total expenditures (in millions of dollars) for the purchase of raw materials for
Smith Manufacturing is given below. Use the information in the chart to answer
questions 13–15.*

Year	x	Millions of dollars (y)
2003	0	84
2004	1	101
2005	2	123
2006	3	136
2007	4	160
2008	5	181
2009	6	196

13. Use the data from 2004 and 2009 to find the slope of the   **13.**_____
line that approximates this information. Then use the
slope to find the equation of the line in slope-intercept
form.

14. To see how well the equation in exercise 27   **14.**_____
approximates the ordered pairs (x, y) in the table of data,
let $x = 4$ (for 2007) and find y.

15. Use the data from 2005 and 2008 to find the slope of the   **15.** _____
line that approximates this information. Then use the
slope to find the equation of the line in slope-intercept
form.

Chapter 3 LINEAR EQUATIONS AND INEQUALITIES IN TWO VARIABLES; FUNCTIONS

3.5   Graphing Linear Inequalities in Two Variables

Learning Objectives
1   Graph linear inequalities in two variables.
2   Graph an inequality with a boundary line through the origin.

Key Terms
Use the vocabulary terms listed below to complete each statement in exercises 1–3.

linear inequality in two variables   boundary line   test point

1. A _____ is used to determine which region of the graph of an inequality to shade.

2. A _____ divides a plane into two regions.

3. A _____ can be written in the form $Ax + By < C$ or $Ax + By > C$ (or with $\leq$ or $\geq$), where A, B, and C are real numbers and A and B are both not 0.

Guided Examples

Review these examples for Objective 1:
1. Graph $3x - 2y \leq 6$.

Begin by graphing the line $3x - 2y = 6$ with intercepts (2, 0) and (0, −3). This boundary line divides the plane into two regions, one of which satisfies the inequality. Choose a test point not on the boundary line to see whether the resulting statement is true or false. We will choose (0, 0).

$$3x - 2y \leq 6 \quad \text{Original inequality}$$
$$3(0) - 2(0) \overset{?}{\leq} 6 \quad \text{Let } x = 0 \text{ and } y = 0.$$
$$0 - 0 \overset{?}{\leq} 6$$
$$0 \leq 6 \quad \text{True}$$

Shade the region that includes the test point (0, 0).

Now Try:
1. Graph $3x + 2y \leq -6$.

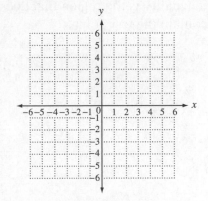

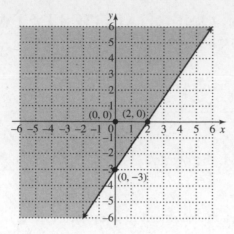

2. Graph $3x - 5y < -15$

To graph the inequality, first graph the equation $3x - 5y = -15$. Use a dashed line to show that the point on the line are not solution of the inequality. Then use $(0, 0)$ as a test point to see which side of the line satisfies the inequality.

$3x - 5y < -15$ Original inequality

$\overset{?}{}$

$3(0) - 5(0) \overset{?}{<} -15$ Let $x = 0$ and $y = 0$.

$\overset{?}{}$

$0 - 0 \overset{?}{<} -15$

$0 < -15$ False

Since $0 < -15$ is false, the graph of the inequality is the region that does not contain the test point.

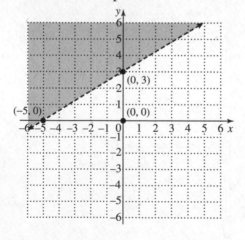

2. Graph $2x - 5y < 10$.

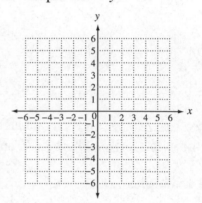

3. Graph $x < -3$.

First graph the vertical line $x = -3$. Use a dashed line and choose $(0, 0)$ as a test point.

$x < -3$ Original inequality

$\overset{?}{0 < -3}$ Let $x = 0$.

$0 < -3$ False

Since $0 < -3$ is false, the graph of the inequality is the region that does not contain the test point.

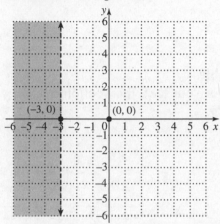

Now Try:

3. Graph $y > 3$.

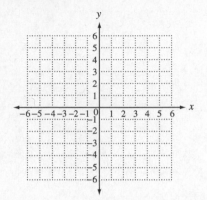

Review this example for Objective 2:

4. Graph $y < 3x$.

First graph the line $y = 3x$. Use a dashed line. We cannot use $(0, 0)$ as a test point since it is on the line, so choose another point, say $(2, 1)$.

$y < 3x$ Original inequality

$\overset{?}{1 < 3(2)}$ Let $x = 2$ and $y = 1$.

$1 < 6$ True

Since $1 < 6$ is true, the graph of the inequality is the region that contains the test point $(2, 1)$.

Now Try:

4. Graph $y \geq 2x$.

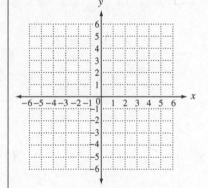

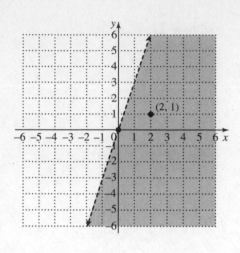

Objective 1 Graph linear inequalities in two variables.
For extra help, see Examples 1–3 on pages 224–226 of your text, the Section Lecture video for Section 3.5, and Exercise Solution Clips 21 and 25.

Graph each linear inequality.

1. $y \geq x - 1$

1.

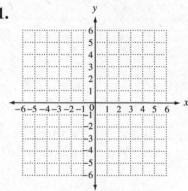

2. $x + y \geq 2$

2.

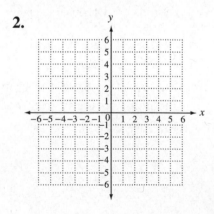

3. $y \geq -1$

3.

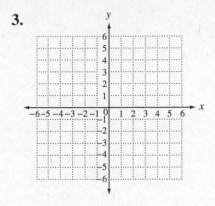

4. $x + 3y < 3$

4.

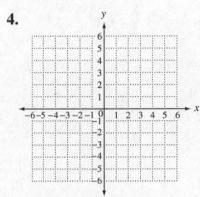

5. $2x + 5y > -10$

5.

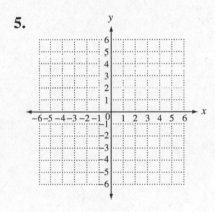

6. $y < x - 3$

6.

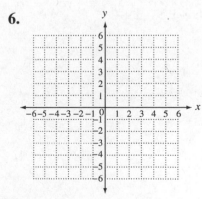

7. $x - 4 \le -1$

7.

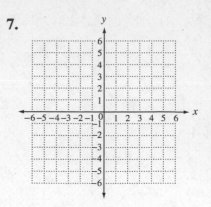

8. $y \le -\dfrac{2}{5}x + 2$

8.

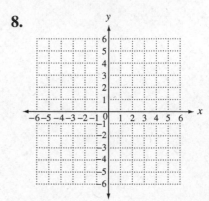

9. $y \ge 3x$

9.

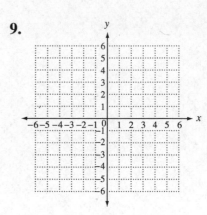

10. $y > -x + 2$

10.

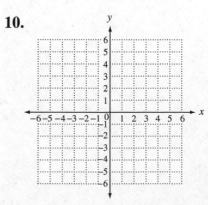

Objective 2 Graph an inequality with a boundary line through the origin.

For extra help, see Example 4 on pages 226–227 of your text, the Section Lecture video for Section 3.5, and Exercise Solution Clip 29.

Graph each linear inequality.

11. $y \geq \frac{1}{3}x$

11.

12. $x \geq -4y$

12.

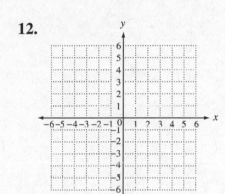

13. $3x - 4y \geq 0$

13.

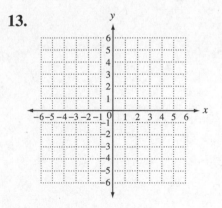

    155

14. $x > -2y$

14.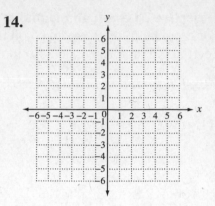

15. $3x - 2y < 0$

15.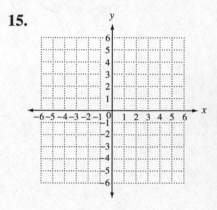

Chapter 3 LINEAR EQUATIONS AND INEQUALITIES IN TWO VARIABLES; FUNCTIONS

3.6   Introduction to Functions

Learning Objectives
1   Understand the definition of a relation.
2   Understand the definition of a function.
3   Decide whether an equation defines a function.
4   Find domains and ranges.
5   Use function notation.
6   Apply the function concept in an application.

Key Terms

Use the vocabulary terms listed below to complete each statement in exercises 1–6.

components	**relation**	**domain**
range	**function**	**function notation**

1.   A _____ is a set of ordered pairs in which each value of the first component x corresponds to exactly one value of the second component y.

2.   A _____ is a set of ordered pairs.

3.   _____ represents the value of the function at x – that is, the y-value which corresponds to x.

4.   The set of all second components (y-values) in the ordered pairs of a relation is the _____.

5.   In an ordered pair (x, y), x and y are called the _____ of the ordered pair.

6.   The set of all first components (x-values) in the ordered pairs of a relation is the _____.

Name: _____   Date: _____

Instructor: _____   Section: _____

Guided Examples

Review this example for Objective 1:

1. Identify the domain and range of the relation.

   $\{(5, 7), (6, 8), (7, 9), (8, 10)\}$

   Domain: $\{5, 6, 7, 8\}$
   Range: $\{7, 8, 9, 10\}$

Now Try:

1. Identify the domain and range of the relation.

   $\{(1, 4), (3, 4), (7, 4), (-2, 4)\}$

   _____

   _____

Review this example for Objective 2:

2. Determine whether each relation is a function.

   a. $\{(1, 3), (11, 9), (8, -2), (6, -7), (5, 7), (-4, -3)\}$
   Each first component appears once and only once. The relation is a function.

   b. $\{(3, 4), (5, 2), (4, 3), (5, 3), (-2, 2)\}$
   The first component 5 appears in two ordered pairs and corresponds to two different second components. Therefore, this relation is not a function.

Now Try:

2. Determine whether each relation is a function.

   a. $\{(1, 3), (2, -1), (-1, 4), (1, 4), (-2, -1)\}$

   _____

   b. $\{(4, 2), (3, 2), (2, 2), (1, 2), (0, 2)\}$

   _____

Review this example for Objective 3:

3. Determine whether each relation is a function.

   a.

   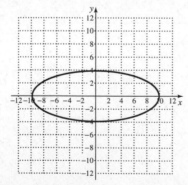

   The vertical line test shows that this graph is not the graph of a function.

Now Try:

3. Determine whether each relation is a function.

   a.

   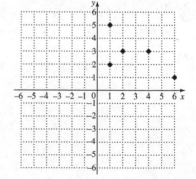

   _____

---

b. $y = 6x + 4$

The graph of $y = 6x + 4$ is a line that is not vertical, so the equation defines a function.

b. $x = -y^2$

_____

Review this example for Objective 4:

4. Find the domain and range of the function $y = -2x^2$.

Any number may be used for x, so the domain is the set of all real numbers or $(-\infty, \infty)$.

The square of any number is nonnegative. When a nonnegative number is multiplied by -2, the answer cannot be positive, so the range is the set of all nonpositive numbers or $(-\infty, 0]$.

Now Try:

4. Find the domain and range of the function $y = x^2 + 4$.

_____

Review this example for Objective 5:

5. Find $f(2)$ for the function $f(x) = 3x^2 - 7$.

$f(x) = 3x^2 - 7$

$f(2) = 3(2)^2 - 7$ Let $x = 2$.

$\quad = 3(4) - 7$ Apply the exponent.

$\quad = 12 - 7$ Multiply.

$\quad = 5$ Subtract.

Now Try:

5. Find $f(-2)$ for the function $f(x) = 8 - 5x$

_____

Review this example for Objective 6:

6. The total expenditures (in millions of dollars) for the purchase of raw materials for Smith Manufacturing is given below.

Year	Millions of dollars
2003	84
2004	101
2005	123
2006	136
2007	160
2008	181
2009	196

Now Try:

6. The table shows the average annual telephone expenditures for cell phones. (Source: Bureau of Labor Statistics)

Year	Annual Cell Phone Expenditures
2001	$210
2002	$294
2003	$316
2004	$378
2005	$455
2006	$524
2007	$608

a. Use the table to write a set of ordered pairs that defines a function *f*.

$f = \{(2003, 84), (2004, 101),$
$(2005, 123), (2006, 136),$
$(2007, 160), (2008, 181),$
$(2009, 196)\}$

b. What are the domain and range of *f*?

Domain: {2003, 2004, 2005, 2006, 2007, 2008, 2009}

Range: {84, 101, 123, 136, 160, 181, 196}

c. What is *f*(2005)?

f(2005) = 123

d. For what *x*-value does *f*(*x*) equal 181 million?

f(2008) = 181

a. Use the table to write a set of ordered pairs that defines a function *f*.

_____

b. What are the domain and range of *f*?

_____

c. What is *f*(2005)?

_____

d. For what *x*-value does *f*(*x*) equal $378?

_____

Objective 1 Understand the definition of a relation.
For extra help, see Example 1 on page 229 of your text, the Section Lecture video for Section 3.6, and Exercise Solution Clip 7.

Identify the domain and range of each relation.

1. $\{(1,3),(2,4),(4,7),(3,9)\}$

1. _____

2. $\{(1,4),(2,4),(3,4),(4,4)\}$

2. _____

Objective 2 Understand the definition of a function.
For extra help, see Example 2 on page 230 of your text, the Section Lecture video for Section 3.6, and Exercise Solution Clips 7 and 11.

Decide whether the relation is a function.

3. $\{(1,4),(2,4),(3,4),(4,4)\}$

3. _____

4. $\left\{\left(3,\dfrac{1}{2}\right),(4,7),(3,9),(2,6)\right\}$

4. _____

5.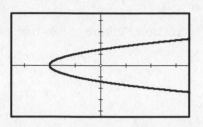

5. _____

Objective 3 Decide whether an equation defines a function.

For extra help, see Example 3 on pages 231–232 of your text, the Section Lecture video for Section 3.6, and Exercise Solution Clips 7 and 11.

Decide whether the equation or inequality defines y as a function of x.

6. $6x - y < 4$

6. _____

7. $y = \dfrac{-2}{3x-1}$

7. _____

8. $y = |x-1|$

8. _____

Objective 4 Find domains and ranges.

For extra help, see Example 4 on page 232 of your text, the Section Lecture video for Section 3.6, and Exercise Solution Clip 23.

Find the domain and the range for the function.

9. $y = 9 - x$

9. _____

10. $y = 2$

10. _____

Objective 5 Use function notation.
For extra help, see Example 5 on page 233 of your text, the Section Lecture video for
Section 3.6, and Exercise Solution Clip 35.

For the function, f, find (a) f(2) and (b) f(−1).

11. $f(x) = -x^2 + 8x - 2$ **11. a.** _____

                                                         b. _____

12. $f(x) = |x - 5|$ **12. a.** _____

                                                         b. _____

Objective 6 Apply the function concept in an application.
For extra help, see Example 6 on page 234 of your text and the Section Lecture video for
Section 3.6.

*The following table gives the weekend box office results for five random movies for
February 19–21, 2010. Use the table to answer questions 13–15.*

Movie	Number of Theaters (x)	Weekend Gross Receipts ($ millions) (y)
Shutter Island	2991	41.1
Valentine's Day	3665	16.7
Avatar	2581	16.2
The Wolfman	3223	9.9
Crazy Heart	1089	3.0

(*Source*: www.the-movie-times.com/thrsdir/TopTen.mv)

13.   What are the domain and range of *f*?   **13.** _____

14.   Find $f(2581)$ and $f(1089)$.   **14.** _____

15.   For what *x*-value does $f(x)$ equal 16.7?   **15.** _____

Chapter 4 SYSTEMS OF LINEAR EQUATIONS AND INEQUALITIES

4.1 Solving Systems of Linear Equations by Graphing

Learning Objectives
1   Decide whether a given ordered pair is a solution of a system.
2   Solve linear systems by graphing.
3   Solve special systems by graphing.
4   Identify special systems without graphing.

Key Terms
Use the vocabulary terms listed below to complete each statement in exercises 1–8.

system of linear equations (linear system)   **solution of a system**

solution set of a system   **set-builder notation**   **consistent system**

inconsistent system   **independent equations**   **dependent equations**

1.   _____ is used to describe a set of numbers without actually having to list all of the elements.

2.   The _____ includes all ordered pairs that satisfy all equations at the same time.

3.   Equations of a system that have the same graph are called _____.

4.   A(n) _____ of equations is a system with no solutions.

5.   A _____ is an ordered pair that makes all equations of the system true at the same time.

6.   Equations of a system that have different graphs are called _____.

7.   A system of equations with a solution is called a _____.

8.   A ∨_____ consists of two or more linear equations with the same variables.

Name: Date:
Instructor: Section:

Guided Examples

Review this example for Objective 1:

1. Decide whether the ordered pair (4, 1) is a solution of each system.

   a. $2x + 3y = 11$ **b.** $4x + 3y = 16$
       $3x - 2y = 10$ $x - 4y = -4$

   Substitute 4 for x and 1 for y in each equation.

   a. $2(4) + 3(1) \overset{?}{=} 11$ $3(4) - 2(1) \overset{?}{=} 10$

            $8 + 3 \overset{?}{=} 11$ $12 - 2 \overset{?}{=} 10$

              $11 = 11$ ✓ $10 = 10$ ✓

   Because (4, 1) satisfies both equations, it is a solution of the system.

   b. $4(4) + 3(1) \overset{?}{=} 16$

            $16 + 3 \overset{?}{=} 16$

             $19 \neq 16$

   Because (4, 1) does not satisfy the first equation, it is not a solution of the system.

Now Try:

1. Decide whether the ordered pair (−3, −1) is a solution of each system.

   a. $5x - 3y = -12$
       $2x + 3y = -9$

   _____

   b. $2x + 3y = 5$
       $3x - \ y = -8$

   _____

Review this example for Objective 2:

2. Solve the system by graphing
   $x - 2y = 6$
   $2x + y = 2$

   Graph each line by plotting three points for each line.

   $x - 2y = 6$ $2x + y = 2$

x	y
0	−3
6	0
4	−1

x	y
0	2
1	0
−1	4

Now Try:

2. Solve the system by graphing.
   $6x - 5y = 4$
   $2x - 5y = 8$

   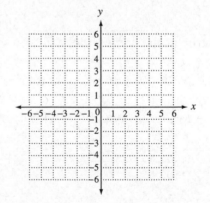

   _____

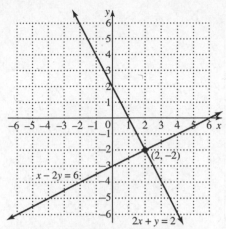

The lines intersect at $(2, -2)$.
Check by substituting 2 for x and -2 for y
in both equations.

$$2 - 2(-2) \overset{?}{=} 6 \qquad 2(2) + (-2) \overset{?}{=} 2$$
$$2 + 4 \overset{?}{=} 6 \qquad 4 - 2 \overset{?}{=} 2$$
$$6 = 6 \ \checkmark \qquad 2 = 2 \ \checkmark$$

Because $(2, -2)$ satisfies both equations, it
is a solution of the system.

Review this example for Objective 3:

3. Solve each system by graphing.

   a. $8x + 4y = -16$
   $\qquad 4x + 2y = 4$

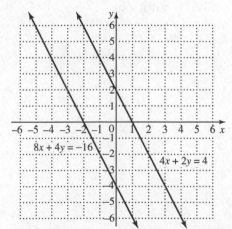

The two lines are parallel and have no
points in common. There is no solution,
so the solution set is $\varnothing$.

Now Try:

3. Solve each system by graphing.

   a. $3x + 3y = 8$
   $\qquad x = 4 - y$

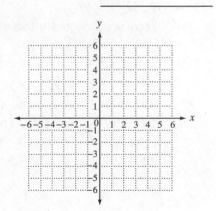

b. $-3x + 2y = 6$
   $-6x + 4y = 12$

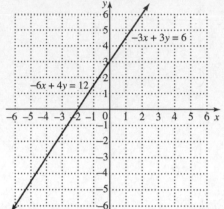

The graphs of these two equations are the same line since we can obtain the second equation by multiplying each side of the first equation by 2. Every point on the line is a solution of the system, so the solution set contains an infinite number of ordered pairs. The solution set is $\{(x, y) \mid -3x + 2y = 6\}$.

b. $x + 2y = 4$
   $8y = -4x + 16$

_____

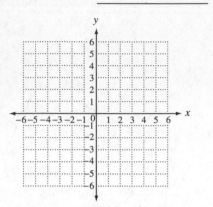

Review this example for Objective 4:

4. Describe each system without graphing. State the number of solutions.

   a. $x + y = 2$
   $x + y = 5$

   Write each equation in slope-intercept form.

   $\begin{array}{c|c} x + y = 2 & x + y = 5 \\ y = -x + 2 & y = -x + 5 \end{array}$

   Both equation have the same slope, but different y-intercepts. Thus, the equations have graphs that are parallel lines, and the system has no solution.

Now Try:

4. Describe each system without graphing. State the number of solutions.

   a. $x - 2y = 5$
   $2x - 4y = 10$

   _____

b. $2x + y = 5$
$3x - 2y = 4$

Write each equation in slope-intercept form.

$$\begin{array}{l|l} 2x + y = 5 & 3x - 2y = 4 \\ \quad y = -2x + 5 & \quad -2y = -3x + 4 \\ & \quad y = \dfrac{3}{2}x - 2 \end{array}$$

The slopes of the lines are different, so the graphs of the equations are neither parallel nor the same line. The system has exactly one solution.

c. $x - y = 1$
$-2x + 2y = -2$

Write each equation in slope-intercept form.

$$\begin{array}{l|l} x - y = 1 & -2x + 2y = -2 \\ \quad -y = -x + 1 & \quad 2y = 2x - 2 \\ \quad y = x - 1 & \quad y = x - 1 \end{array}$$

The equations are exactly the same, so the graphs are the same line. Thus, the system has an infinite number of solutions.

b. $3x + y = 7$
$6x + 2y = 9$

_____

c. $2x - y = -1$
$3x - y = 7$

_____

Objective 1 Decide whether a given ordered pair is a solution of a system.
For extra help, see Example 1 on page 248 of your text, the Section Lecture video for Section 4.1, and Exercise Solution Clip 3.

Decide whether the given ordered pair is a solution of the given system.

1. $(-5, -4)$
$x - y = -1$
$4x + y = -24$

1. _____

2. $(-1, -7)$
$x - y = -6$
$-2x + 3y = -19$

2. _____

3. $(-4, -1)$
$5x - 2y = 6$
$y = -3x - 11$

3. _____

Objective 2 Solve linear systems by graphing.
For extra help, see Example 2 on page 249 of your text, the Section Lecture video for Section 4.1, and Exercise Solution Clip 15.

Solve each system by graphing both equations on the same axes.

4. $3x - y = -7$
$2x + y = -3$

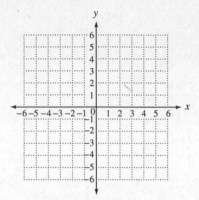

4. _____

5. $x - y = -7$
$x + 11 = 2y$

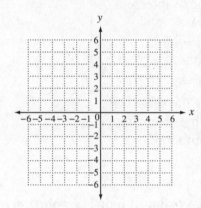

5. _____

6. $3x - 2y = 8$
$7x + 2y = 12$

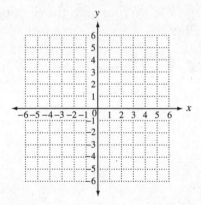

6. _____

7. $3x + 2 = y$
     $2x - y = 0$

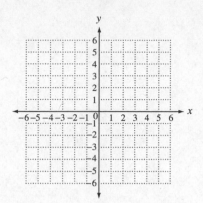

7. _____

Objective 3 Solve special systems by graphing.
For extra help, see Example 3 on page 250 of your text, the Section Lecture video for
Section 4.1, and Exercise Solution Clip 25.

*Solve each system of equations by graphing both equations on the same axes. State
whether the system is inconsistent or dependent and write the solution set using set
notation.*

8. $4x + 3y = 12$
     $6y + 8x = -24$

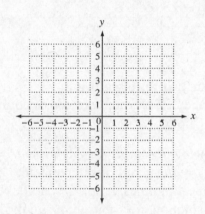

8. _____

9. $2x + 3y = 0$
     $6x = -9y$

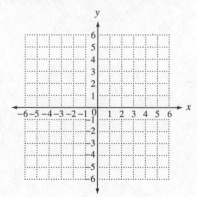

9. _____

10. $4x - 5y = 15$
        $10y - 8x = -30$

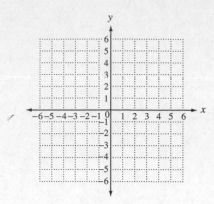

10. _____

11. $5x + 3y = 30$
        $10x + 6y = 50$

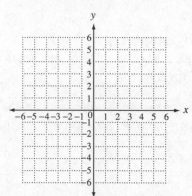

11. _____

Objective 4 Identify special systems without graphing.
For extra help, see Example 4 on page 252 of your text, the Section Lecture video for Section 4.1, and Exercise Solution Clip 35.

Without graphing, answer the following questions for each linear system.
  a.   Is the system inconsistent, *are the equations* dependent, *or* neither.
  b.   Is the graph a pair of intersecting lines, *a pair of* parallel lines, *or* one line?
  c.   Does the system have one solution, no solution, *or an* infinite number of solutions?

12. $2x - y = 4$
        $x + 3y = 2$

12. a. _____

      b. _____

      c. _____

13. $2x - y = 4$
        $2x = y + 3$

13. a. _____

      b. _____

      c. _____

Name: Date:
Instructor: Section:

14. $3x - 4y = 8$
$x - 2y = 2$

14. a. ______________
b. ______________
c. ______________

15. $3y = x + 6$
$y - \frac{1}{3}x = 2$

15. a. ______________
b. ______________
c. ______________

Chapter 4 SYSTEMS OF LINEAR EQUATIONS AND INEQUALITIES

4.2 Solving Systems of Linear Equations by Substitution

Learning Objectives
1   Solve linear systems by substitution.
2   Solve special systems by substitution.
3   Solve linear systems with fractions and decimals by substitution.

Key Terms

Use the vocabulary terms listed below to complete each statement in exercises 1–4.

substitution   ordered pair   inconsistent system

dependent system

1. The solution of a linear system of equations is written as a(n)
    ___Substitution___.

2. When one expression is replaced by another, _____ is being used.

3. A system of equations in which all solutions of the first equation are also solutions of
    the second equation is a(n) ___dependent system___.

4. A system of equations that has no common solution is called a(n)
    ___inconsistent___.

Guided Examples

Review these example for Objective 1:
1. Solve the system by the substitution method.

$$3x - 5y = 7 \quad (1)$$
$$y = 2x \quad (2)$$

Equation (2) is already solved for y, so substitute $2x$ for y in equation (1).

$3x - 5y = 7$ (1)
$3x - 5(2x) = 7$ Let $y = 2x$.
$3x - 10x = 7$ Multiply.
$-7x = 7$ Combine terms.
$x = -1$ Divide by -7.

Now Try:
1. Solve the system by the substitution method.

$$5x - 3y = -14$$
$$y = -3x$$

_____

Now find the value of y by substituting -1 for x in either equation. Using equation (2), we have $y = 2(-1) = -2$.

Check the solution $(-1, -2)$ in both equations.

$$
\begin{array}{c|c}
3x - 5y = 7 & y = 2x \\
\quad ? & \quad ? \\
3(-1) - 5(-2) = 7 & -2 = 2(-1) \\
\quad ? & -2 = -2 \ \checkmark \\
-3 + 10 = 7 & \\
7 = 7 \ \checkmark &
\end{array}
$$

Since $(-1, -2)$ satisfies both equations, the solution set is $\{(-1, -2)\}$.

2. Solve the system by the substitution method.

$$-8x + 5y = 11 \qquad (1)$$
$$x = y - 1 \qquad (2)$$

Equation (2) is already solved for x, so substitute $y - 1$ for x in equation (1).

$$-8x + 5y = 11 \quad (1)$$
$$-8(y - 1) + 5y = 11 \quad \text{Let } x = y - 1.$$
$$-8y + 8 + 5y = 11 \quad \text{Distributive property}$$
$$-3y + 8 = 11 \quad \text{Combine terms.}$$
$$-3y = 3 \quad \text{Subtract 8.}$$
$$y = -1 \quad \text{Divide by } -3.$$

Now find the value of x by substituting -1 for y in equation (2).

$$x = -1 - 1 = -2.$$

Check that the solution set of the equation is $\{(-2, -1)\}$.

3. Solve the system by the substitution method.

$$x - 4y = 17 \quad (1)$$
$$3x - 4y = 11 \quad (2)$$

Solve one of the equations for either x or y. Since the coefficient of x in equation (1) is 1, avoid fractions by solving this equation for x.

$$x - 4y = 17$$
$$x = 4y + 17$$

2. Solve the system by the substitution method.

$$x = y + 6$$
$$-2x + 3y = -19$$

_____

3. Solve the system by the substitution method.

$$x + 6y = -1$$
$$-2x - 9y = 0$$

_____

Now substitute $4y + 17$ for x in equation (2) and solve for y.

$$3x - 4y = 11 \quad (2)$$
$$3(4y + 17) - 4y = 11 \quad \text{Let } x = 4y - 17.$$
$$12y + 51 - 4y = 11 \quad \text{Distributive property}$$
$$8y + 51 = 11 \quad \text{Combine terms.}$$
$$8y = -40 \quad \text{Subtract 51.}$$
$$y = -5 \quad \text{Divide by 8.}$$

Now find the value of x by substituting -5 for y in equation (1).

$$x - 4y = 17 \quad (1)$$
$$x - 4(-5) = 17 \quad \text{Let } y = -5.$$
$$x + 20 = 17 \quad \text{Multiply.}$$
$$x = -3 \quad \text{Subtract 20.}$$

Check that the solution set of the system is $\{(-3, -5)\}$.

Review these examples for Objective 2:

4. Use substitution to solve the system.

$$y = -\frac{1}{3}x + 5 \quad (1)$$
$$3y + x = -9 \quad (2)$$

Equation (1) is already solved for y, so substitute $-\frac{1}{3}x + 5$ for y in equation (2).

$$3y + x = -9$$
$$3\left(-\frac{1}{3}x + 5\right) + x = -9 \quad \text{Let } y = -\frac{1}{3}x + 5.$$
$$-x + 15 + x = -9 \quad \text{Distributive property}$$
$$15 = -9 \quad \text{False}$$

The false result $15 = -9$ means that the equations in the system have graphs that are parallel lines. The system in inconsistent and has no solution. The solution set is $\varnothing$.

5. Use substitution to solve the system.

$$-x + 2y = -6 \quad (1)$$
$$2x - 4y = 12 \quad (2)$$

Solve equation (1) for x.

$$-x + 2y = -6 \quad (1)$$
$$-x = -2y - 6 \quad \text{Subtract } 2y.$$
$$x = 2y + 6 \quad \text{Multiply by } -1.$$

Now Try:

4. Use substitution to solve the system.

$$y = \frac{3}{2}x + 2$$
$$6x - 4y = 7$$

_____

5. Use substitution to solve the system.

$$2x - y = 4$$
$$8x - 4y = 16$$

_____

Now substitute $2y + 6$ for x in equation (2).

$$2x - 4y = 12 \quad (2)$$
$$2(2y + 6) - 4y = 12 \quad \text{Let } x = 2y + 6$$
$$4y + 12 - 4y = 12 \quad \text{Distributive property}$$
$$0 = 0 \quad \text{Subtract 12.}$$
$$\text{Combine like terms.}$$

The true result means that every solution of one equation is also a solution of the other. Therefore, the system has an infinite number of solutions, and the solution set is $\{(x, y) \mid -x + 2y = -6\}$.

Review these examples for Objective 3:	**Now Try:**
6. Solve the system by the substitution method.	**6.** Solve the system by the substitution method.

6. Solve the system by the substitution method.

$$\frac{5}{4}x - \frac{1}{2}y = -\frac{1}{4} \quad (1)$$
$$-\frac{5}{4}x + \frac{1}{8}y = 1 \quad (2)$$

Clear the fractions in equation (1) by multiplying each side by 4.

$$4\left(\frac{5}{4}x - \frac{1}{2}y\right) = 4\left(-\frac{1}{4}\right) \quad \text{Multiply by 4.}$$
$$5x - 2y = -1 \quad \text{Distributive property}$$

Clear the fractions in equation (2) by multiplying each side by 8.

$$8\left(-\frac{5}{4}x + \frac{1}{8}y\right) = 8(1) \quad (2)$$
$$-10x + y = 8 \quad \text{Distributive property}$$

The original system of equations has been simplified to an equivalent system.

$$5x - 2y = -1 \quad (3)$$
$$-10x + y = 8 \quad (4)$$

Now solve equation (4) for y.

$$y = 10x + 8 \quad \text{Add } 10x.$$

Substitute $10x + 8$ for y in equation (3) and solve for x.

$$5x - 2y = -1 \quad (3)$$
$$5x - 2(10x + 8) = -1 \quad \text{Let } y = 10x + 8.$$
$$5x - 20x - 16 = -1 \quad \text{Distributive property}$$
$$-15x - 16 = -1 \quad \text{Combine like terms.}$$
$$-15x = 15 \quad \text{Add 16.}$$
$$x = -1 \quad \text{Divide by } -15.$$

Now Try:

6. Solve the system by the substitution method.

$$\frac{1}{4}x + \frac{3}{8}y = -3$$
$$\frac{5}{6}x - \frac{3}{7}y = -10$$

_____

Substitute -1 for x in equation (4) to find y.
$y = 10(-1) + 8 = -10 + 8 = -2$

Check that the solution set of the original system is $\{(-1, -2)\}$.

7. Solve the system by the substitution method.
$0.2y + 0.8x = -0.2$　　(1)
$0.1x + 0.3y = -1.4$　　(2)

Clear the decimals in each equation by multiplying by 10.

$10(0.2y + 0.8x) = 10(-0.2)$　(1)
　　　　$2y + 8x = -2$　　　Distributive property

$10(0.1x + 0.3y) = 10(-1.4)$　(2)
　　　　$x + 3y = -14$　　　Distributive property

The original system of equations has been simplified to an equivalent system.
$2y + 8x = -2$　　(3)
$x + 3y = -14$　　(4)

Solve equation (4) for x.
$x = -3y - 14$

Substitute $-3y - 14$ for x in equation (3) and solve for y.

$$2y + 8x = -2 \quad (3)$$
$$2y + 8(-3y - 14) = -2 \quad \text{Let } x = -3y - 14.$$
$$2y - 24y - 112 = -2 \quad \text{Distributive property}$$
$$-22y - 112 = -2 \quad \text{Combine like terms.}$$
$$-22y = 110 \quad \text{Add 112.}$$
$$y = -5 \quad \text{Divide by } -22.$$

Now use equation (4) to find x.
$x = -3(-5) - 14 = 15 - 14 = 1.$

Check that the solution set of the original system is $\{(1, -5)\}$.

7. Solve the system by the substitution method.

$0.1x + 0.3y = 0.1$
$0.2x = -1 - 1.2y$

_____

Objective 1 Solve linear systems by substitution.
For extra help, see Examples 1–3 on pages 257–259 of your text, the Section Lecture video for Section 4.2, and Exercise Solution Clips 3, 5, and 11.

Solve each system by the substitution method. Check each solution.

1.　　$x + y = 7$
　　　　$y = 6x$

1. _____

2. $3x + 2y = 14$

     $y = x + 2$

2. _____

3. $y + 4x = -1$

     $x = -14 - 3y$

3. _____

4. $3x + 4y = 2$

    $2x + 3y = 2$

4. _____

5. $2x + 4y = -1$

    $-4x - 6y = 1$

5. _____

Objective 2 Solve special systems by substitution.

For extra help, see Examples 4 and 5 on pages 259–260 of your text, the Section Lecture video for Section 4.2, and Exercise Solution Clips 17 and 19.

Solve each system by the substitution method.

6. $x + y = 5$

    $-x - y = -5$

6. _____

7. $36x + 20y = 12$

    $-27x - 15y = -9$

7. _____

8. $12x - 18y = 25$
$\quad\;\; 4x - 6y = 5$

8. _____

9. $\quad\; 3x - 4y = 8$
$\quad 28y - 21x = -56$

9. _____

10. $3x - 5y = 12$
$\quad\; x - \dfrac{5}{3}y = 3$

10. _____

Objective 3 Solve linear systems with fractions and decimals by substitution.
For extra help, see Examples 6 and 7 on pages 261–262 of your text, the Section Lecture video for Section 4.2, and Exercise Solution Clip 23.

Solve each system by either the addition method or the substitution method. First clear all fractions. Check each solution.

11. $\dfrac{5}{3}x + y = 12$
$\quad\; x + \dfrac{1}{2}y = 7$

11. _____

12. $\dfrac{5x}{4} + \dfrac{2y}{3} = \dfrac{8}{3}$
$\quad \dfrac{2x}{3} - \dfrac{3y}{2} = -6$

12. _____

13. $\dfrac{x}{2} - \dfrac{y}{3} = -8$

$\dfrac{x}{4} - \dfrac{y}{8} = -4$

13. _____

14. $0.3x + 1.8y = -2.7$
$-0.2x - 0.6y = 0$

14. _____

15. $0.6x + 0.8y = 1$
$0.4y = 0.5 - 0.3x$

15. _____

Chapter 4 SYSTEMS OF LINEAR EQUATIONS AND INEQUALITIES

4.3 Solving Systems of Linear Equations by Elimination

Learning Objectives
1   Solve linear systems by elimination.
2   Multiply when using the elimination method.
3   Use an alternative method to find the second value in a solution.
4   Solve special systems by elimination.

Key Terms
Use the vocabulary terms listed below to complete each statement in exercises 1–3.

elimination method   opposites   true

1.   When using the elimination method, eliminate variables by identifying numerical coefficients of like terms that are _opposites_____.

2.   The _elimination method_____ is an algebraic method used to solve a system of equations in which the equations of the system are combined so that one or more variables is eliminated.

3.   When solving a system of linear equations using the elimination method, a _____true_____ statement indicates that the original equations are dependent.

Guided Examples

Review these examples for Objective 1:
1.   Use the elimination method to solve the system.

$$x + y = 5 \quad (1)$$
$$x - y = -3 \quad (2)$$

Add the two equations.

$$x + y = 5 \quad (1)$$
$$\underline{x - y = -3} \quad (2)$$
$$2x \quad = 2 \quad \text{Add the left sides; add right sides.}$$
$$x = 1 \quad \text{Divide by 2.}$$

Now find the value of y by substituting 1 for x in either equation.

$$x + y = 5 \quad (1)$$
$$1 + y = 5 \quad \text{Let } x = 1.$$
$$y = 4 \quad \text{Subtract 1.}$$

Now Try:
1.   Solve the system by the elimination method.

$$-x + 2y = -10$$
$$x + 3y = -10$$

_____

Check the solution $\{(1, 4)\}$ in both equations.

$$\begin{array}{c|c} x+y=5 & x-y=-3 \\ ? & ? \\ 1+4=5 & 1-4=-3 \\ 5=5 \;\checkmark & -3=-3 \;\checkmark \end{array}$$

Since $(1, 4)$ satisfies both equations, the solution set is $\{(1, 4)\}$.

2. Use the elimination method to solve the system.

$$y = 3x - 5 \quad (1)$$
$$2x = y + 4 \quad (2)$$

Write both equations in standard form.

$$-3x + y = -5 \quad (1)$$
$$2x - y = 4 \quad (2)$$

Now add the two equations.

$$\begin{array}{rl} -3x + y = -5 & (1) \\ \underline{2x - y = 4} & (2) \\ -x \quad\;\; = -1 & \text{Add.} \\ x = 1 & \text{Multiply by } -1. \end{array}$$

Now find the value of y by substituting 1 for x in equation (1).

$y = 3(1) - 5 = 3 - 5 = -2$.

Check the solution $(1, -2)$ in both equations.

$$\begin{array}{c|c} y = 3x - 5 & 2x = y + 4 \\ ? & ? \\ -2 = 3(1) - 5 & 2(1) = -2 + 4 \\ -2 = -2 \;\checkmark & 2 = 2 \;\checkmark \end{array}$$

Since $(1, -2)$ satisfies both equations, the solution set is $\{(1, -2)\}$.

2. Use the elimination method to solve the system.

$$4x + 3y = -4$$
$$2x - 3y = 16$$

_____

Review these examples for Objective 2:
3. Solve the system by the substitution method.

$$-4x + 5y = 22 \quad (1)$$
$$3x + 2y = -5 \quad (2)$$

To eliminate x, multiply each side of equation (1) by 3 and each side of equation (2) by 4.

Now Try:
3. Solve the system by the substitution method.

$$3x + 2y = 5$$
$$2x - 3y = 12$$

_____

$-12x + 15y = 66$ Multiply (1) by 3.

$\underline{12x + 8y = -20}$ Multiply (2) by 4.

$23y = 46$ Add.

$y = 2$ Divide by 23.

Find the value of x by substituting 2 for y in either equation. Using equation (1), we have

$-4x + 5(2) = 22$ Let $y = 2$.

$-4x + 10 = 22$ Multiply.

$-4x = 12$ Subtract 10.

$x = -3$ Divide by -4.

Check that the solution set of the system is $\{(-3, 2)\}$.

Review this example for Objective 3:

4. Solve the system.

$5x - 4y = 23$ (1)

$20 + 2y = 3x$ (2)

First, write equation (2) in standard form.

$5x - 4y = 23$ (1)

$-3x + 2y = -20$ (3)

Eliminate x by multiplying (1) by 3 and (3) by 5 and then adding.

$15x - 12y = 69$ $3 \times (1)$

$\underline{-15x + 10y = -100}$ $5 \times (3)$

$-2y = -31$ Add.

$y = \dfrac{31}{2}$ Divide by -2.

Instead of substituting $\dfrac{31}{2}$ for y in one of the equations, solve for x by starting again with the original equations and eliminating y. Multiply equation (3) by 2, then add.

$5x - 4y = 23$ (1)

$\underline{-6x + 4y = -40}$ $2 \times (3)$

$-x = -17$ Add.

$x = 17$ Multiply by -1.

Check that the solution set of the system is $\left\{\left(17, \frac{31}{2}\right)\right\}$.

Now Try:

4. Solve the system.

$5x + 3y = -4$

$3x + 5y = 2$

_____

Review this example for Objective 4:

5. Solve each system by the elimination method.

   a. $6x - 7y = 32$ (1)
   $-18x + 21y = -12$ (2)

   Multiply each side of equation (1) by 3, then add the equations.
   $18x - 21y = 96$ $3 \times (1)$
   $\underline{-18x + 21y = -12}$ (2)
   $\ 0 = 84$ False

   The false statement means that the solution set of the system is $\varnothing$.

   b. $2x + 4y = -6$ (1)
   $-x - 2y = 3$ (2)

   Multiply each side of equation (2) by 2, then add the equations.
   $2x + 4y = -6$ (1)
   $\underline{-2x - 4y = 6}$ $2 \times (2)$
   $\ 0 = 0$ True

   The true statement means that the equations are equivalent and that every solution of one equation is also a solution of the other. The solution set is $\{(x, y) \mid x + 2x = -3\}$.

Now Try:

5. Solve the system.

   a. $6y = -15x - 9$
   $10x + 4y = 6$

   _____

   b. $12x - 10y = -2$
   $25y - 30x = 5$

   _____

Objective 1 Solve linear systems by elimination.

For extra help, see Examples 1 and 2 on pages 264–266 of your text, the Section Lecture video for Section 4.3, and Exercise Solution Clips 3 and 7.

Solve each system by the elimination method. Check your answers.

1. $x - 3y = 5$
$-x + 4y = -5$

1. _____

2. $5x + 8y = 12$
$3x - 8y = 20$

2. _____

3. $15x - 3y = 8$
$21x + 3y = 10$

3. _____

Objective 2 Multiply when using the elimination method.
For extra help, see Example 3 on page 266 of your text, the Section Lecture video for
Section 4.3, and Exercise Solution Clip 17.

Solve each system by the elimination method. Check your answers.

4. $6x + 7y = 10$
$2x - 3y = 14$

4. _____

5. $x - 4y = 10$
$x + 6y = -10$

5. _____

6. $6x + y = 1$
$3x - 4y = 23$

6. _____

7. $3x - 4y = 16$

$4x + 5y = -20$

7. _____

Objective 3 Use an alternative method to find the second value in a solution.
For extra help, see Example 4 on page 267 of your text, the Section Lecture video for
Section 4.3, and Exercise Solution Clip 39.

Solve each system.

8. $6x = 16 - 7y$

$2x = 3y + 26$

8. _____

9. $2x = 15 + 4y$

$6y = -5x + 3$

9. _____

10. $3x + 1 = 2y$

$3y - 2 = 2x$

10. _____

11. $2x + 5y = -8$
       $5x - 2y = -1$

11. _____

Objective 4 Use the elimination method to solve special systems.
For extra help, see Example 5 on page 268 of your text, the Section Lecture video for Section 4.3, and Exercise Solution Clip 23.

Solve each system by the elimination method.

12. $12x - 8y = 3$
       $6x - 4y = 6$

12. _____

13. $6x - 12y = 3$
       $2x - 4y = 1$

13. _____

14. $4x - 2y = -8$
       $2x - y = 4$

14. _____

15. $18y = 12x - 6$
       $4x - 6y = 2$

15. _____

Chapter 4 SYSTEMS OF LINEAR EQUATIONS AND INEQUALITIES

4.4 Applications of Linear Systems

Learning Objectives
1   Solve problems about unknown numbers.
2   Solve problems about quantities and their costs.
3   Solve problems about mixtures.
4   Solve problems about distance, rate (or speed), and time.

Key Terms

Use the vocabulary terms listed below to complete each statement in exercises 1–3.

assign variables   **check**   **$d = rt$**

1.   When solving applications of linear systems you must read the problem and _____ for the unknown quantities.

2.   You must always _____ the solution to a linear system and verify that it makes sense in the original problem.

3.   The formula that relates distance, rate, and time is _____.

Guided Examples

Review this example for Objective 1:
1.   A rope 82 centimeters long is cut into two pieces with one piece four cm more than twice as long as the other. Find the length of each piece.

Step 1 Read the problem carefully. We must find the length of two pieces of rope. We know the total length of the rope and we also know that one piece is four cm more than twice as long as the other.

Step 2 Assign variables.
Let x = the length of the first piece and let y = the length of the second piece.

Step 3 Write two equations.
$$x + y = 82 \qquad (1)$$
$$x = 4 + 2y \qquad (2)$$

Now Try:
1.   There are a total of 49 students in the two second grade classes at Jefferson School. If there are 7 more students in Carla's class than in Linda's class, find the number of students in each class.

_____

Step 4 Solve the system.

$(4+2y)+y=82$ Substitute the expression for x from equation (2) into equation (1).

$4+3y=82$ Combine like terms.

$3y=78$ Subtract 4.

$y=26$ Divide by 3.

Now substitute 26 for y in equation (2) and solve for x.

$x=4+2(26)=56$

Step 5 State the answer.

One piece of rope is 56 cm long and the other piece of 26 cm long.

Step 6 Check the answer in the original problem.

26 + 56 = 82 and 56 = 4 + 2(26).

Review this example for Objective 2:

2. A total of 2250 general admission tickets and reserved seat tickets were sold for a basketball game. General admission tickets cost $10 and reserved seat tickets cost $15. If the total value of both kinds of tickets was $26,250, how many tickets of each kind were sold?

    Step 1 Read the problem several times.

    Step 2 Assign variables.

    Let x = the number of general admission tickets and let y = the number of reserved seat tickets.

    Summarize the information in a table.

	Number of tickets	Price per ticket	Total value
General admission	x	10	$10x$
Reserved seat	y	15	$15y$
Total	2250		26,250

    Step 3 Write two equations.

    $x+\ \ y=\ 2250$ (1)
    $10x+15y=26250$ (2)

Now Try:

2. The total receipts for a basketball game were $4690.50. There were 723 tickets sold, some for children and some for adults. If the adult tickets cost $9.50 and the children's tickets cost $4, how many of each type were there?

    _____

Step 4 Solve the system using the elimination method.

$$-15x - 15y = -33750 \quad \text{Multipy (1) by } -15.$$
$$10x + 15y = 26250 \quad (2)$$
$$-5x = -7500 \quad \text{Add.}$$
$$x = 1500 \quad \text{Divide by } -5.$$

Substitute 1500 for x in equation (1) and solve for y.

$$1500 + y = 2250$$
$$y = 750 \quad \text{Subtract 1500.}$$

Step 5 State the answer.

1500 general admission tickets and 750 reserved seat tickets were sold.

Step 6 Check.

The sum of 1500 and 750 is 2250, so the total number of tickets is correct. Since 1500 tickets were sold at $10 each and 750 at $15 each, the total of all the ticket prices is $10(1500) + $15(750) = $26,250, which agrees with the total amount stated in the problem.

Review this example for Objective 3:

3. Milton needs 45 liters of 20% solution. He has only 15% alcohol solution and 30% alcohol solution on hand to make the mixture. How many liters of each solution should he combine to make the mixture?

Step 1 Read the problem. Note the percentage of each solution and of the mixture.

Step 2 Assign variables.

Let x = the number of liters of 15% solution and let y = the number of liters of 30% solution.

Summarize the information in a table.

	Number of liters	Percent (as a decimal)	Liters of pure alcohol
15% solution	x	0.15	$0.15x$
30% solution	y	0.30	$0.30y$
20% solution	45	0.20	$(0.20)(45) = 9$

Now Try:

3. How many liters of 75% solution should be mixed with a 55% solution to get 70 liters of 63% solution?

_____

Step 3 Write two equations.

$$x + y = 45 \quad (1)$$
$$0.15x + 0.30y = 9 \quad (2)$$

Step 4 Solve the system using the elimination method.

$$-0.15x - 0.15y = -6.75 \quad \text{Multipy (1) by } -0.15.$$
$$0.15x + 0.30y = 9 \qquad (2)$$
$$0.15y = 2.25 \quad \text{Add.}$$
$$y = 15 \qquad \text{Divide by } 0.15.$$

Substitute 15 for y in equation (1) and solve for x.

$$x + 15 = 45$$
$$x = 30 \quad \text{Subtract 15.}$$

Step 5 State the answer.
Milton needs 30 liters of 15% alcohol solution and 15 liters of 30% alcohol solution.

Step 6 Check.
Since 30 and 15 is 45 and
$0.15(30) + 0.30(15) = 9$,
this mixture will give 45 liters of the 20% solution, as required.

Review this example for Objective 4:

4. Pablo left Somerset traveling to Akron 240 miles away at the same time as Shawn left Akron traveling to Somerset. They met after 2 hours. If Shawn was traveling twice as fast as Pablo, what were their speeds?

Step 1 Read the problem several times.

Step 2 Assign variables.
Let x = Pablo's rate and let y = Shawn's rate. Make a table and draw a sketch to summarize the information given in the problem.

	Rate	Time	Distance
Pablo	x	2	$2x$
Shawn	y	2	$2y$

Now Try:

4. Rick and Hilary drive from positions 378 miles apart toward each other. They meet after 3 hours. Find the average speed of each if Hilary travels 30 miles per hour faster than Rick.

_____

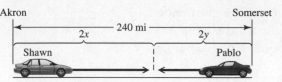

Akron   Somerset

Shawn and Pablo meet after 2 hours.

Step 3 Write two equations.

$$y = 2x \quad (1)$$
$$2x + 2y = 240 \quad (2)$$

Step 4 Solve the system using the substitution method.

$$2x + 2(2x) = 240 \quad \text{Substitute } 2x \text{ for } y \text{ in (2).}$$
$$6x = 240 \quad \text{Combine like terms}$$
$$x = 40 \quad \text{Divide by 6.}$$

Substitute 40 for x in equation (1) and solve for y.

$$y = 2(40) = 80$$

Step 5 State the answer.
Pablo traveled at 40 miles per hour and Shawn traveled at 80 miles per hour.

Step 6 Check.
Since each man travels for 2 hours, the total distance traveled is $2(40) + 2(80) = 240$ miles, as required.

Objective 1 Solve problems about unknown numbers.

For extra help, see Example 1 on pages 272–273 of your text and the Section Lecture video for Section 4.4.

Write a system of equations for each problem, and then solve the system.

1. The sum of two numbers is 20. Three times the smaller is equal to twice the larger. Find the numbers.

    1. _____

2. Two towns have a combined population of 9045. There are 2249 more people living in one than in the other. Find the population in each town.

    2. _____

3. The perimeter of a triangular pennant is 116 centimeters.   3. _____
    If two sides are of equal length, and the third side is 20
    centimeters longer than each of the equal sides, what are
    the lengths of the three sides

Objective 2 Solve problems about quantities and their costs.
For extra help, see Example 2 on pages 273–274 of your text and the Section Lecture video
for Section 4.4.

Write a system of equations for each problem, and then solve the system.

4. Admission prices at a football game were $12 for adults   4. _____
    and $9 for children. The total receipts for the game were
    $87,000. Tickets were sold to 8000 people. How many
    adults and how many children attended the game?

5. Charlie has only $5 bills and $20 bills and has a total of   5. _____
    $130. If there is a total of 11 bills, how many of each
    type are there?

6. A postal clerk has 1250 stamps in his drawer that are   6. _____
    worth a total of $287.50. If there are only 25-cent and
    15-cent stamps, how many of each kind are there?

7. A cashier has some $5 bills and some $10 bills. The total value of the money is $750. If the number of tens is equal to twice the number of fives, how many of each type are there?

7. _____

Objective 3 Solve problems about mixtures.
For extra help, see Example 3 on pages 274–275 of your text, the Section Lecture video for Section 4.4, and Solution Clip 25.

Write a system of equations for each problem, and then solve the system.

8. Steve wishes to mix coffee worth $6 a pound with coffee worth $9 a pound to get 45 pounds of a mixture worth $8 a pound. How many pounds of the $6 and the $9 coffee will be needed?

8. _____

9. Ben wishes to blend candy selling for $1.60 a pound with candy selling for $2.50 a pound to get a mixture that will be sold for $1.90 a pound. How many pounds of the $1.60 and the $2.50 candy should be used to get 30 pounds of the mixture?

9. _____

10. A 10% solution of antifreeze is mixed with a 50% solution of antifreeze to get 100 liters of a 22% solution. How many liters of each solution are used?

10. _____

11. A 20% alcohol solution is to be mixed with a 5%   **11.** _____
solution to get 15 liters of a 10% solution. How many
liters of each solution are used?

Objective 4 Solve problems about distance, rate (or speed), and time.
For extra help, see Example 4 on pages 275–276 of your text, the Section Lecture video for
Section 4.4, and Solution Clip 31.

Use a system of equations to solve each problem.

12. Two trains start from positions 1242 miles apart and   **12.** _____
travel toward each other. They meet after $4\frac{1}{2}$ hours.
Find the average speed of each train if one train travels
20 miles per hour faster than the other.

13. Two bicyclists leave from Washington DC and ride in   **13.** _____
opposite directions. One travels $1\frac{1}{2}$ times as fast as the
other. After 2 hours, they are 40 miles apart. Find the
speed of each bicyclist.

14. A plane can travel 300 miles per hour with the wind and   **14.** _____
230 miles per hour against the wind. Find the speed of
the wind and the speed of the plane in still air.

15. Two planes left Philadelphia traveling in opposite   **15.** _____
directions. Plane A left 15 minutes before plane B. After
plane B had been flying for 1 hour, the planes were 860
miles apart. What were the speeds of the two planes if
plane A was flying 40 miles per hour faster than plane
B?

Chapter 4 SYSTEMS OF LINEAR EQUATIONS AND INEQUALITIES

4.5 Solving Systems of Linear Inequalities

Learning Objectives
1   Solve systems of linear inequalities by graphing.

Key Terms
Use the vocabulary terms listed below to complete each statement in exercises 1–4.

  system of linear inequalities   **solution set of a system of linear inequalities**

  test point   **boundary line**

1.   The _____ includes all ordered pairs that make all inequalities of the system true at the same time.

2.   When solving a system of linear inequalities, use a _____ to determine which region is the correct solution.

3.   In the graph of a linear inequality, the _____ separates the region that satisfies the inequality from the region that does not satisfy the inequality.

4.   A _____ consists of two or more linear inequalities to be solved at the same time.

Guided Examples

Review these examples for Objective 1:
  1. Graph the solution set of the system.
$$x + y > -3$$
$$2x - 3y \le -2$$

Step 1 To graph $x + y > -3$, graph the dashed boundary line $x + y = -3$ and shade the region containing $(0, 0)$.

Now Try:
  1. Graph the solution set of the system.
$$4x - y > 2$$
$$y > -x - 2$$

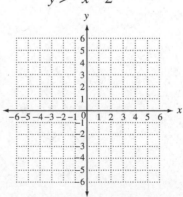

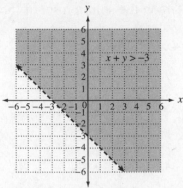

Then graph $2x - 3y \leq -2$ with the solid boundary line $2x - 3y = -2$. The test point $(0, 0)$ makes the inequality false, so shade the region on the other side of the boundary line.

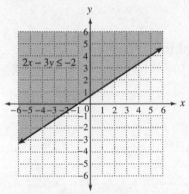

Step 2 The solution set of this system includes all points in the intersection (overlap) of the graphs of the two inequalities.

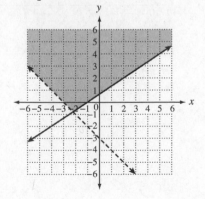

2. Graph the solution set of the system.

$$x + y \geq 4$$
$$x \geq 2$$
$$y < 5$$

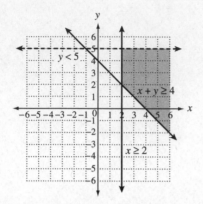

2. Graph the solution set of the system.

$$4x - y > 2$$
$$y > -x - 2$$
$$x \leq 3$$

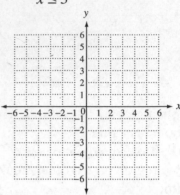

Objective 1 Solve systems of linear inequalities by graphing.

For extra help, see Examples 1–3 on pages 282–283 of your text, the Section Lecture video for section 4.5, and Solution Clips 5, 9, and 21.

Graph the solution of each system of linear inequalities.

1. $7x + 3y \geq 21$
   $x - y \leq 6$

1.

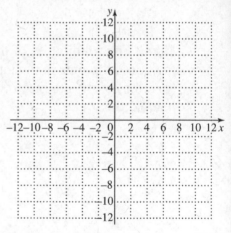

2. $3x - y \leq 6$

$3y - 6 \leq 2x$

2.

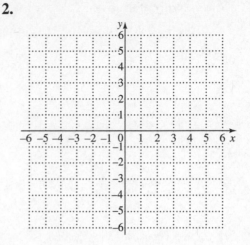

3. $x + y \leq 3$

$5x - y \geq 5$

3.

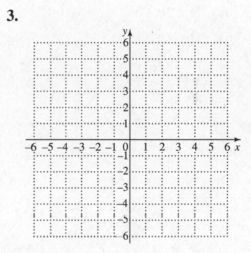

4. $3x - y > 3$

$4x + 3y < 12$

4.

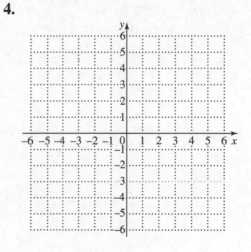

5. $2x - y \geq 4$

  $5y + 15 \geq -3x$

5.

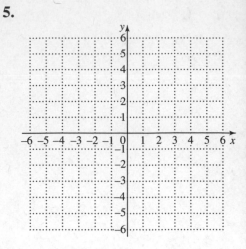

6. $x < 2y + 3$

  $0 < x + y$

6.

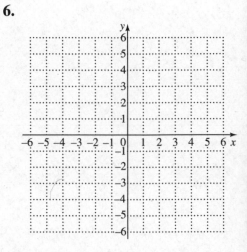

7. $2x - y > -4$

  $2x + y > 0$

7.

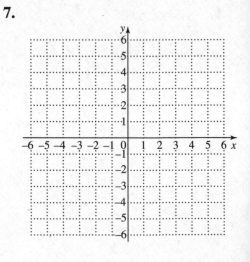

8. $3x - 4y < 12$

$y > -4$

8.

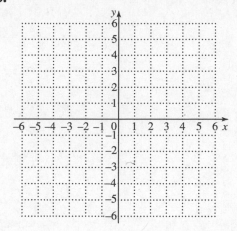

9. $3x + 5y \geq 15$

$y \geq x - 2$

9.

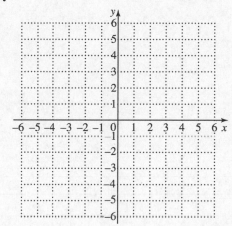

10. $4x - y \leq 4$

$7y + 14 \geq -2x$

10.

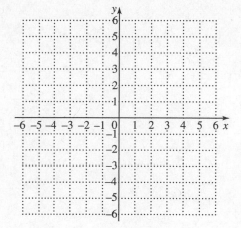

11. $x - 2y \le 3$
$\quad$ $2x + y \le -4$

11.

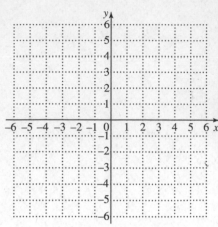

12. $x - 2y \ge -7$
$\quad$ $x - 2y < 2$

12.

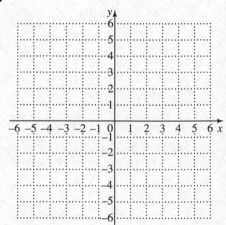

13. $x - y \le 2$
$\quad\quad$ $y \le 2$
$\quad\quad$ $x > 0$

13.

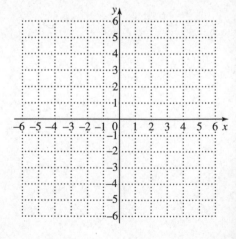

14. $3x - y \le 3$

$\quad\quad y < 3$

$\quad\quad x \ge -3$

14.

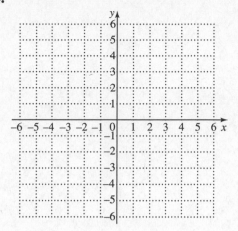

15. $x + 2y \ge -4$

$\quad\quad 5x \le 10 - 2y$

$\quad\quad x < 2$

15.

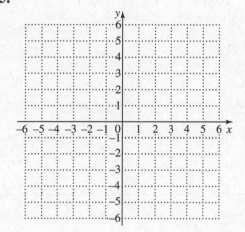

Chapter 5 EXPONENTS AND POLYNOMIALS

5.1 The Product Rule and Power Rules for Exponents

Learning Objectives

1   Use exponents.

2   Use the product rule for exponents.

3   Use the rule $\left(a^m\right)^n = a^{mn}$.

4   Use the rule $(ab)^m = a^m b^m$.

5   Use the rule $\left(\dfrac{a}{b}\right)^m = \dfrac{a^m}{b^m}$.

6   Use combinations of rules.

7   Use the rules for exponents in a geometry application.

Key Terms

Use the vocabulary terms listed below to complete each statement in exercises 1–4.

     base   **exponent**   **power**   **exponential expression**

1.   A number that indicates how many times a factor is repeated is a(n) _exponent_.

2.   A number or variable written with an exponent is a(n) _exponential expression_.

3.   The expression 2^5 is read "2 to the fifth _power_."

4.   In the expression 2^5, the number 2 is the _exponent_.

Guided Examples

Review these examples for Objective 1:

1. Write $2 \cdot 2 \cdot 2 \cdot 2 \cdot 2 \cdot 2$ in exponential form and evaluate.

   Since 2 occurs as a factor 6 times, the base is 2 and the exponent is 6.

   $2 \cdot 2 \cdot 2 \cdot 2 \cdot 2 \cdot 2 = 2^6 = 64$

Now Try:

1. Write $\left(-\dfrac{2}{5}\right)\left(-\dfrac{2}{5}\right)\left(-\dfrac{2}{5}\right)\left(-\dfrac{2}{5}\right)$ in exponential form and evaluate.

   _____

2. Evaluate $(-2)^7$. Name the base and the exponent.

$$(-2)^7 = (-2)(-2)(-2)(-2)(-2)(-2)(-2)$$
$$= -128$$

The base is -2 and the exponent is 7.

2. Evaluate -6^2. Name the base and the exponent.

_____

Review this example for Objective 2:

3. Use the product rule for exponents to find each product, if possible.

   a. $5^2 \cdot 5^3$ **b.** $(-2)^3(-2)^4$

   c. $y^4 \cdot y$ **d.** $4^3 \cdot 3^4$

   e. $x^3 x^7 x^2$ **f.** $4^3 + 4^2$

   g. $\left(-3m^6\right)\left(-4m^3\right)$

   a. $5^2 \cdot 5^3 = 5^{2+3} = 5^5 = 3125$

   b. $(-2)^3(-2)^4 = (-2)^{3+4} = (-2)^7 = -128$

   c. $y^4 \cdot y = y^4 \cdot y^1 = y^{4+1} = y^5$

   d. $4^3 \cdot 3^4$
      The product rule does not apply since the bases are different. Evaluate 4^3 and 3^4, then multiply.
      $4^3 \cdot 3^4 = 64 \cdot 81 = 5184$

   e. $x^3 x^7 x^2 = x^{3+7+2} = x^{12}$

   f. $4^3 + 4^2$
      The product rule does not apply since this is a sum, not a product. Evaluate 4^3 and 4^2, then add.
      $4^3 + 4^2 = 64 + 16 = 80$

   g. $\left(-3m^6\right)\left(-4m^3\right) = \left[(-3)(-4)\right] \cdot \left(m^6 \cdot m^3\right)$
         Commutative and associate properties
         $= 12m^{6+3} = 12m^9$
         Multiply; product rule

Now Try:

3. Use the product rule for exponents to find each product, if possible.

   a. $4^4 \cdot 4^3$

_____

   b. $(-3)^3(-3)^2$

_____

   c. $m^6 \cdot m^4$

_____

   d. $7^3 \cdot 3^2$

_____

   e. $a^4 a^5 a$

_____

   f. $5^2 - 5^3$

_____

   g. $10n^4 \cdot 6n^5$

_____

Review this example for Objective 3:

4. Simplify.

   a. $\left(9^5\right)^9$ **b.** $\left(a^3\right)^3$

   a. $\left(9^5\right)^9 = 9^{5 \cdot 9} = 9^{45}$

   b. $\left(a^3\right)^3 = a^{3 \cdot 3} = a^9$

Now Try:

4. Simplify.

   a. $\left[(-2)^7\right]^5$

   _____

   b. $\left(b^2\right)^5$

   _____

Review this example for Objective 4:

5. Simplify.

   a. $\left(yz^4\right)^3$ **b.** $\left(-2r^3s^2\right)^6$

   a. $\left(yz^4\right)^3 = y^3\left(z^4\right)^3 = y^3 z^{4 \cdot 3} = y^3 z^{12}$

   b. $\left(-2r^3s^2\right)^6 = (-2)^6\left(r^3\right)^6\left(s^2\right)^6 = 64 r^{3 \cdot 6} s^{2 \cdot 6}$

             $= 64 r^{18} s^{12}$

Now Try:

5. Simplify.

   a. $\left(2a^4b\right)^3$

   _____

   b. $\left(-3w^3z^7\right)^3$

   _____

Review this example for Objective 5:

6. Simplify.

   a. $\left(\dfrac{4}{g}\right)^3, g \neq 0$ **b.** $\left(\dfrac{2}{5}\right)^4$

   a. $\left(\dfrac{4}{g}\right)^3 = \dfrac{4^3}{g^3} = \dfrac{64}{g^3}$

   b. $\left(\dfrac{2}{5}\right)^4 = \dfrac{2^4}{5^4} = \dfrac{16}{625}$

Now Try:

6. Simplify.

   a. $\left(\dfrac{x}{y}\right)^8, y \neq 0$

   _____

   b. $\left(\dfrac{4}{3}\right)^4$

   _____

Review this example for Objective 6:

7. Simplify.

   a. $\left(\dfrac{3}{2}\right)^2 \cdot 4^3$ **b.** $\left(\dfrac{-2a}{b^2}\right)^7, b \neq 0$

   c. $\left(7a^2b^4c\right)^3\left(ab^3c^4\right)^4$

Now Try:

7. Simplify.

   a. $\left(\dfrac{2}{3}\right)^5 \cdot 3^4$

   _____

a. $\left(\dfrac{3}{2}\right)^2 \cdot 4^3 = \dfrac{3^2}{2^2} \cdot 4^3 = \dfrac{9}{4} \cdot 64 = 144$

b. $\left(\dfrac{-2a}{b^2}\right)^7 = \dfrac{(-2)^7 a^7}{\left(b^2\right)^7} = \dfrac{-128a^7}{b^{14}}$

c. $\left(7a^2 b^4 c\right)^3 \left(ab^3 c^4\right)^4$

$= 7^3 \left(a^2\right)^3 \left(b^4\right)^3 c^3 \cdot a^4 \left(b^3\right)^4 \left(c^4\right)^4$

$= 343 a^6 b^{12} c^3 \cdot a^4 b^{12} c^{16}$

$= 343 a^6 a^4 b^{12} b^{12} c^3 c^{16}$

$= 343 a^{10} b^{24} c^{19}$

b. $\left(\dfrac{3b^2}{11}\right)^4$

_____

c. $\left(2ab^2 c\right)^5 \left(a^2 b^4\right)^4$

_____

Review this example for Objective 7:

8. Write an expression that represents the area of the figure. Assume $m > 0$.

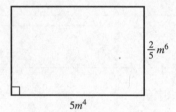

The formula for the area of a rectangle is $A = LW$.

$A = \left(5m^4\right)\left(\tfrac{2}{5} m^6\right)$ Area formula

$= 5 \cdot \dfrac{2}{5} \cdot m^4 \cdot m^6$ Commutative property

$= 2m^{4+6}$ Multiply; product rule

$= 2m^{10}$ Add exponents.

Now Try:

8. Write an expression that represents the area of the figure. Assume $x > 0$.

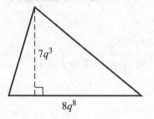

_____

Objective 1 Use exponents.

For extra help, see Examples 1 and 2 on page 296 of your text, the Section Lecture video for Section 5.1, and Exercise Solution Clips 7, 15, and 17.

1. Write the expression in exponential form.

$(-2y)(-2y)(-2y)(-2y)(-2y)$

1. _____

2. Identify the base and the exponent for the expression $(-4)^4$ and then evaluate.

2. _____

Objective 2 Use the product rule for exponents.
For extra help, see Example 3 on page 297 of your text, the Section Lecture video for Section 5.1, and Exercise Solution Clips 25, 31, 35, and 39.

Use the product rule to simplify each expression, if possible. Write each answer in exponential form.

3. $4^7 \cdot 4^4$

3. _____

4. $\left(-4x^3\right)\left(8x^{12}\right)$

4. _____

Objective 3 Use the rule $\left(a^m\right)^n = a^{mn}$.
For extra help, see Example 4 on page 298 of your text, the Section Lecture video for Section 5.1, and Exercise Solution Clips 43 and 45.

Simplify each expression. Write all answers in exponential form.

5. $\left(3^4\right)^3$

5. _____

6. $-\left(21^5\right)^4$

6. _____

Objective 4 Use the rule $(ab)^m = a^m b^m$.
For extra help, see Example 5 on pages 298–299 of your text, the Section Lecture video for Section 5.1, and Exercise Solution Clip 49.

Simplify each expression.

7. $5\left(ab^3\right)^3$

7. _____

8. $\left(4c^3 d^4\right)^3$

8. _____

Objective 5 Use the rule $\left(\dfrac{a}{b}\right)^m = \dfrac{a^m}{b^m}$.

For extra help, see Example 6 on page 299 of your text, the Section Lecture video for Section 5.1, and Exercise Solution Clip 59.

Simplify each expression. Assume all variables represent nonzero real numbers.

9. $\left(-\dfrac{2x}{5}\right)^3$

9. _____

10. $\left(\dfrac{x}{2y}\right)^4$

10. _____

Objective 6 Use combinations of rules.

For extra help, see Example 7 on page 300 of your text, the Section Lecture video for Section 5.1, and Exercise Solution Clips 67 and 79.

Simplify. Write all answers in exponential form.

11. $\left(5x^2 y^3\right)^7 \left(5xy^4\right)^4$

11. _____

12. $\left(\dfrac{7a^2 b^3}{2}\right)^7$

12. _____

13. $\left(\dfrac{km^4 p^2}{3n^4}\right)^7 \quad (n \neq 0)$

13. _____

Name: Date:
Instructor: Section:

Objective 7 Use the rules for exponents in a geometry application.
For extra help, see Example 8 on pages 300–301 of your text, the Section Lecture video for
Section 5.1, and Exercise Solution Clip 85.

Write an expression in simplified form that represents the area of the figure described.

14. A rectangle with length $5x^4$ and width $2x^3$ 14. _____

15. A circle with radius $4a^5$ 15. _____

Chapter 5 EXPONENTS AND POLYNOMIALS

5.2 Integer Exponents and the Quotient Rule

Learning Objectives
1   Use 0 as an exponent.
2   Use negative numbers as exponents.
3   Use the quotient rule for exponents.
4   Use combinations of rules.

Key Terms

Use the vocabulary terms listed below to complete each statement in exercises 1–3.

   zero exponent   negative exponent   quotient rule for exponents

1.   A number raised to a _negitive exponent_ can be rewritten as the reciprocal of the number raised to the opposite of the exponent.

2.   A nonzero number raised to a _zero exponent_ has a value of 1.

3.   The _Quotent rule exponent_ states that the quotient of two terms with the same base is equal the base raised to the difference of the exponents on each term.

Guided Examples

Review this example for Objective 1:
1. Evaluate.

   a. 10^0 **b.** $(-2)^0$

   c. -2^0 **d.** $4^0 - 3^0$

   a. $10^0 = 1$ **b.** $(-2)^0 = 1$

   c. $-2^0 = -1$ **d.** $4^0 - 3^0 = 1 - 1 = 0$

Now Try:
1.   Evaluate.

   a. 15^0

   _____

   b. -3^0

   _____

   c. $(-3)^0$

   _____

   d. $4^0 + 5^0$

   _____

Name: _____ Date: _____

Instructor: _____ Section: _____

Review these examples for Objective 2:

2. Simplify.

a. 5^{-2}

b. $\left(\dfrac{1}{5}\right)^{-3}$

c. $\left(\dfrac{2}{3}\right)^{-4}$

d. $2^{-3} - 4^{-2}$

e. $r^{-6}, r \neq 0$

a. $5^{-2} = \dfrac{1}{5^2} = \dfrac{1}{25}$

b. $\left(\dfrac{1}{5}\right)^{-3} = 5^3 = 125$

c. $\left(\dfrac{2}{3}\right)^{-4} = \left(\dfrac{3}{2}\right)^4 = \dfrac{81}{16}$

d. $2^{-3} - 4^{-2} = \dfrac{1}{2^3} - \dfrac{1}{4^2} = \dfrac{1}{8} - \dfrac{1}{16}$

$= \dfrac{2}{16} - \dfrac{1}{16} = \dfrac{1}{16}$

e. $r^{-6} = \dfrac{1}{r^6}, r \neq 0$

3. Simplify by writing with positive exponents. Assume that all variables represent nonzero real numbers.

a. $\dfrac{5^{-3}}{2^{-4}}$

b. $\dfrac{2x^{-4}}{3y^{-7}z}$

c. $\left(\dfrac{y^4}{3x^2}\right)^{-3}$

a. $\dfrac{5^{-3}}{2^{-4}} = \dfrac{2^4}{5^3} = \dfrac{16}{125}$

b. $\dfrac{2x^{-4}}{3y^{-7}z} = \dfrac{2y^7}{3x^4z}$

c. $\left(\dfrac{y^4}{3x^2}\right)^{-3} = \left(\dfrac{3x^2}{y^4}\right)^3 = \dfrac{3^3x^6}{y^{12}} = \dfrac{27x^6}{y^{12}}$

Now Try:

2. Simplify.

a. 6^{-3}

_____

b. $\left(-\dfrac{1}{5}\right)^{-4}$

_____

c. $\left(-\dfrac{5}{3}\right)^{-3}$

_____

d. $2^{-2} + 4^{-3}$

_____

e. $x^{-10}, x \neq 0$

_____

3. Simplify by writing with positive exponents. Assume that all variables represent nonzero real numbers.

a. $\dfrac{4^{-4}}{5^{-2}}$

_____

b. $\dfrac{3a^{-5}y}{4z^{-3}}$

_____

c. $\left(\dfrac{2y^2}{3x^3}\right)^{-5}$

_____

Review this example for Objective 3:

4. Simplify by writing with positive exponents. Assume that all variables represent nonzero real numbers.

a. $\dfrac{2^9}{2^5}$ **b.** $\dfrac{2^4 \cdot x^2}{2^5 \cdot x^8}$

c. $\dfrac{12x^9 y^5}{12^4 x^3 y^7}$ **d.** $\dfrac{8^4 x^{-3} y^{-5}}{8^6 x^{-2} y^2}$

a. $\dfrac{2^9}{2^5} = 2^{9-5} = 2^4 = 16$

b. $\dfrac{2^4 \cdot x^2}{2^5 \cdot x^8} = \dfrac{2^4}{2^5} \cdot \dfrac{x^2}{x^8} = 2^{-1} \cdot x^{-6} = \dfrac{1}{2} \cdot \dfrac{1}{x^6} = \dfrac{1}{2x^6}$

c. $\dfrac{12x^9 y^5}{12^4 x^3 y^7} = 12^{1-4} \cdot x^{9-3} \cdot y^{5-7} = 12^{-3} x^6 y^{-2}$

$= \dfrac{x^6}{12^3 y^2}$

d. $\dfrac{8^4 x^{-3} y^{-5}}{8^6 x^{-2} y^2} = 8^{4-6} \cdot x^{-3-(-2)} \cdot y^{-5-2}$

$= 8^{-2} x^{-1} y^{-7} = \dfrac{1}{64xy^7}$

Now Try:

4. Simplify by writing with positive exponents. Assume that all variables represent nonzero real numbers.

a. $\dfrac{12^{-7}}{12^{-6}}$

_____

b. $\dfrac{(3+a)^{-4}}{(3+a)^{-6}}$

_____

c. $\dfrac{15^{-4} x^4 yz^3}{15^2 x^6 z}$

_____

d. $\dfrac{5^{-4} x^{-6} yz^3}{5^{-2} x^6 y^{-2} z^5}$

_____

Review this example for Objective 4:

5. Simplify.

a. $\dfrac{6^8}{\left(6^3\right)^4}$ **b.** $(3y)^6 (3y)^{-4}$

c. $\left(\dfrac{2x^5}{3}\right)^{-4}$ **d.** $\dfrac{\left(3^{-2} x^{-5} y\right)^{-4}}{\left(2x^{-2} y^2\right)^{-2}}$

a. $\dfrac{6^8}{\left(6^3\right)^4} = \dfrac{6^8}{6^{12}} = 6^{8-12} = 6^{-4} = \dfrac{1}{6^4} = \dfrac{1}{1296}$

Now Try:

5. Simplify.

a. $\dfrac{\left(7^2\right)^6}{7^6}$

_____

b. $(3+a)^{-4}(3+a)^{-6}$

_____

b. $(3y)^6(3y)^{-4} = (3y)^{6+(-4)} = (3y)^2 = 9y^2$

c. $\left(\dfrac{5}{6x^6z}\right)^{-3}$

c. $\left(\dfrac{2x^5}{3}\right)^{-4} = \left(\dfrac{3}{2x^5}\right)^4 = \dfrac{3^4}{2^4 x^{20}} = \dfrac{243}{16x^{20}}$

_____

d. $\dfrac{\left(3^{-2}x^{-5}y\right)^{-4}}{\left(2x^{-2}y^2\right)^{-2}} = \dfrac{3^{-2(-4)} \cdot x^{-5(-4)} \cdot y^{-4}}{2^{-2} \cdot x^{-2(-2)} \cdot y^{2(-2)}}$

d. $\dfrac{\left(2^{-4}x^{-2}yz^3\right)^{-2}}{\left(5^{-2}x^6y^{-2}z^5\right)^{-1}}$

$= \dfrac{3^8 x^{20} y^{-4}}{2^{-2} x^4 y^{-4}} = 3^8 \cdot 2^2 x^{16}$

_____

Objective 1 Use 0 as an exponent.
For extra help, see Example 1 on page 304 of your text, the Section Lecture video for Section 5.2, and Exercise Solution Clip 1.

Evaluate each expression.

1. $-(-8)^0$

1. _____

2. $(-5)^0 - (-5)^0$

2. _____

3. $\dfrac{0^8}{8^0}$

3. _____

Objective 2 Use negative numbers as exponents.
For extra help, see Examples 2 and 3 on pages 305–306 of your text, the Section Lecture video for Section 5.2, and Exercise Solution Clips 21 and 37.

Evaluate each expression.

4. $\left(\frac{2}{7}\right)^{-1}$

4. _____

5. $10^{-2} + 5^{-2}$

5. _____

Write each expression with only positive exponents. Assume all variables represent nonzero real numbers

6. $\dfrac{2}{r^{-7}}$

6. _____

7. $\dfrac{2x^{-4}}{3y^{-7}}$

7. _____

Objective 3 Use the quotient rule for exponents.
For extra help, see Example 4 on page 307 of your text, the Section Lecture video for Section 5.2, and Exercise Solution Clip 35.

Use the quotient rule to simplify each expression. Write answers with only positive exponents. Assume that all variables represent nonzero real numbers.

8. $\dfrac{(-2)^8}{(-2)^3}$

8. _____

9. $\dfrac{a^4 b^3}{a^{-2} b^{-3}}$

9. _____

10. $\dfrac{3^{-1} m^{-4} p^6}{3^4 m^{-1} p^{-2}}$

10. _____

11. $\dfrac{8 b^{-3} c^4}{8^{-4} b^{-7} c^{-3}}$

11. _____

Name: Date:
Instructor: Section:

Objective 4 Use combinations of rules.
For extra help, see Example 5 on page 308 of your text, the Section Lecture video for Section 5.2, and Exercise Solution Clip 69.

Simplify each expression. Write answers with only positive exponents. Assume that all variables represent nonzero real numbers.

12. $\left(5w^2 y^2\right)^{-2}\left(4wy^{-3}\right)^2$ 12._____

13. $\dfrac{(2y)^{-4}}{(3y)^{-2}}$ 13._____

14. $\dfrac{c^{10}\left(c^2\right)^3}{\left(c^3\right)^3\left(c^2\right)^{-9}}$ 14._____

15. $\left(\dfrac{k^3 t^4}{k^2 t^{-1}}\right)^{-4}$ 15._____

Chapter 5 EXPONENTS AND POLYNOMIALS

5.3 An Application of Exponents: Scientific Notation

Learning Objectives
1 Express numbers in scientific notation.
2 Convert numbers in scientific notation to numbers without exponents.
3 Use scientific notation in calculations.

Key Terms

Use the vocabulary terms listed below to complete each statement in exercises 1–3.

  scientific notation   **left**   **right**

1. A number written in _____ has the form $a \times 10^n$, where $1 \le |a| < 10$ and n is an integer.

2. To write a number in the form $a \times 10^n$, where $1 \le |a| < 10$ and $n \ge 1$, without exponents, move the decimal to the _____ n places.

3. To write a number in the form $a \times 10^n$, where $1 \le |a| < 10$ and $n \le -1$, without exponents, move the decimal to the _____ n places.

Guided Examples

Review this example for Objective 1:
1. Write each number in scientific notation.
  a. 325,000   **b.** 0.0257
  c. −0.00768

  a. Move the decimal point to follow the first nonzero digit (the 3). Count the number of places the decimal point was moved. This is the exponent.

   $325{,}000 = 3.25 \times 10^5$
     5 places

  b. Move the decimal point to the right of the first nonzero digit (the 3). Count the number of places the decimal point was moved. Since the decimal was moved to the right, the exponent is negative.

   $0.0257 = 2.57 \times 10^{-2}$
     2 places

Now Try:
1. Write each number in scientific notation.
  a. 23,651,000,000

   _____

  b. −0.00047

   _____

  c. 0.0000503

   _____

c. $-0.00768 = -7.68 \times 10^{-3}$

Review this example for Objective 2:

2. Write each number without exponents.

   a. 2.3×10^4 **b.** 7.24×10^{-4}

   c. -8.301×10^{-6}

   a. Move the decimal four places to the right.

     $2.3 \times 10^4 = 23,000$

   b. Move the decimal four places to the left.

     $7.24 \times 10^{-4} = 0.000724$

   c. $-8.301 \times 10^{-6} = -0.000008301$

Now Try:

2. Write each number without exponents.

   a. 7.2×10^7

   _____

   b. 4.007×10^{-2}

   _____

   c. -4.5×10^{-5}

   _____

Review these examples for Objective 3:

3. Perform each calculation. Write answers in scientific notation and also without exponents.

   a. $\left(3 \times 10^6\right) \times \left(4 \times 10^{-2}\right)$ **b.** $\dfrac{4.6 \times 10^{-3}}{2.3 \times 10^{-1}}$

   a. $\left(3 \times 10^6\right) \times \left(4 \times 10^{-2}\right)$

     $= (3 \times 4)\left(10^6 \times 10^{-2}\right)$ Commutative and associative properties

     $= 12 \times 10^4$ Multiply; product rule

     $= \left(1.2 \times 10^1\right) \times 10^4$ Write 12 in scientific notation

     $= 1.2 \times \left(10^1 \times 10^4\right)$ Associative property

     $= 1.2 \times 10^5$ Product rule

     $= 120,000$

   b. $\dfrac{4.6 \times 10^{-3}}{2.3 \times 10^{-1}} = \dfrac{4.6}{2.3} \times \dfrac{10^{-3}}{10^{-1}} = 2 \times 10^{-2} = 0.02$

Now Try:

3. Perform each calculation. Write answers in scientific notation and also without exponents.

   a. $\left(2.3 \times 10^4\right) \times \left(1.1 \times 10^{-2}\right)$

   _____

   b. $\dfrac{8.5 \times 10^{-3}}{1.7 \times 10^{-7}}$

   _____

4. There are about 6×10^{23} atoms in a mole of atoms. About how many atoms are there in 81,000 moles?

First write 81,000 in scientific notation:
$81,000 = 8.1 \times 10^4$
Now multiply.
$\left(6 \times 10^{23}\right) \times \left(8.1 \times 10^4\right)$
$= \left(6 \times 8.1\right) \times \left(10^{23} \times 10^4\right)$
$= 48.6 \times 10^{26} = \left(4.86 \times 10^1\right) \times 10^{27}$
$= 4.86 \times 10^{28}$

There are about 4.86×10^{28} atoms in 81,000 moles.

4. Density is a physical property of matter that is defined as the ratio of an object's mass to its volume. Earth has a mass of 6×10^{24} kilograms and a volume of 1.08×10^{12} cubic kilometers. What is Earth's density in kilograms per cubic kilometer? Round your answer to two decimal places. (Source: NASA)

_____

Objective 1 Express numbers in scientific notation.
For extra help, see Example 1 on page 313 of your text, the Section Lecture video for Section 5.3, and Exercise Solution Clip 13.

Write each number in scientific notation.

1. 4,579,000   1. _____

2. 429,600,000,000   2. _____

3. 0.246   3. _____

4. 0.00000413   4. _____

5. 0.00426   5. _____

Objective 2 Convert numbers in scientific notation to numbers without exponents.
For extra help, see Example 2 on page 313 of your text, the Section Lecture video for Section 5.3, and Exercise Solution Clip 25.

Write each number without exponents.

6. 2.5×10^4 6. _____

7. -2.45×10^6 7. _____

8. 4.045×10^0 8. _____

9. 4.752×10^{-1} 9. _____

10. -9.11×10^{-4} 10. _____

Objective 3 Use scientific notation in calculations.
For extra help, see Examples 3–5 on pages 314–315 of your text, the Section Lecture video for Section 5.3, and Exercise Solution Clips 43 and 91.

Perform the indicated operations with the numbers in scientific notation, and then write your answers without exponents.

11. $\left(7 \times 10^7\right) \times \left(3 \times 10^0\right)$ 11. _____

12. $\left(2.3 \times 10^{-4}\right) \times \left(3.1 \times 10^{-2}\right)$ 12. _____

13. $\dfrac{9.39 \times 10^1}{3 \times 10^3}$ 13. _____

14. $\left(3 \times 10^4\right) \times \left(4 \times 10^2\right) \div \left(2 \times 10^3\right)$ 14. _____

15. The Milky Way galaxy is about 100,000 light-years in   **15.** _____

diameter. If one light-year is about 9.5×10^{15} meters,
then about how many meters is the diameter of the
Milky Way? (Source: NASA)

Chapter 5 EXPONENTS AND POLYNOMIALS

5.4 Adding and Subtracting Polynomials; Graphing Simple Polynomials

> **Learning Objectives**
> 1 Identify terms and coefficients.
> 2 Add like terms.
> 3 Know the vocabulary for polynomials.
> 4 Evaluate polynomials.
> 5 Add and subtract polynomials.
> 6 Graph equations defined by polynomials of degree 2.

Key Terms

Use the vocabulary terms listed below to complete each statement in exercises 1–14.

term	numerical coefficient	like terms	polynomial

descending powers   **degree of a term**

degree of a polynomial   **monomial**   **binomial**

trinomial   **parabola**   **vertex**   **axis**

line of symmetry

1. A _monomo_ is a polynomial with only one term.

2. The _degree_ is the sum of the exponents on the variables in the term.

3. The _degree polynomial_ is the greatest degree of any of the terms in the polynomial.

4. The _numerical co efficient_ is the numerical factor of a term.

5. The graph of a second-degree equation in two variables is called a _perraballa_ .

6. A _Binomial_ is a polynomial with exactly two terms.

7. A _trinomial_ is a number, variable, or the product or quotient of a number and one or more variables raised to powers.

8. The _access_ of a parabola is a line of symmetry for the graph.

9. A _trinomial_ is a polynomial with exactly three terms.

10. A polynomial in one variable is written in _decending order_ of the variable if the exponents on the terms of the polynomial decrease from left to right.

11. Terms with exactly the same variables (including the same exponents) are called _Like terms_ .

12. The point on a parabola that has the smallest y-value (if the parabola opens upward) or the largest y-value (if the parabola opens downward) is called the _Vertex_ of the parabola.

13. An algebraic expression made up of a term or a finite sum of terms in which all coefficients are real, all variables have whole number exponents, and no variables appear in denominators is called a _Polynomial_ .

14. The axis of a parabola is a _Line of symmetry_ for the graph.

Guided Examples

Review this example for Objective 1:	Now Try:
1. Name the coefficient of each term in the expression $3n^8 - n^2$. The coefficients are 8 and -1.	1. Name the coefficient of each term in the expression $-4x^3 + 10x^2 - 1$. _____
Review this example for Objective 2:	Now Try:
2. Simplify by adding like terms. $6v^3 - 9v^2 - 2v^2$ $6v^3 - 9v^2 - 2v^2 = 6v^3 + (-9-2)v^2$ Distributive property $= 6v^3 - 11v^2$	2. Simplify by adding like terms. $3c^2 - 6c + 9c^2$ _____
Review this example for Objective 3:	Now Try:
3. Simplify and write the resulting polynomial in descending powers of the variable. Then give the degree of this polynomial, and tell whether it is a *monomial*, a *binomial*, a *trinomial*, or *none of these*. $10y^4 + 6y^2 - 12y^3 - 5y^4$ $10y^4 + 6y^2 - 12y^3 - 5y^4$ $= 5y^4 - 12y^3 + 6y^2$ The degree of the polynomial is 4 and it is a trinomial.	3. Simplify and write the resulting polynomial in descending powers of the variable. Then give the degree of this polynomial, and tell whether it is a *monomial*, a *binomial*, a *trinomial*, or *none of these*. $3m^5 - m^2 + 3m^4 - 4m^2$ _____

Name: _____   Date: _____
Instructor: _____   Section: _____

Review this example for Objective 4:	Now Try:

4. Find the value of $3x^2 - 4x + 1$ for $x = -2$.

Substitute -2 for x.

$$3x^2 - 4x + 1 = 3(-2)^2 - 4(-2) + 1$$
$$= 3(4) - 4(-2) + 1$$
$$= 12 + 8 + 1$$
$$= 21$$

4. Find the value of $3x^2 - 4x + 1$ for $x = 3$.

_____

Review these examples for Objective 5:

5. Add.

  a. $5m^4 + 2m^3 - 4$
       $\underline{-3m^4 + 5m^3 - 3}$

  b. $\left(3x^2 + 2x^4 - 3\right) + \left(8x^3 - 5x^4 - 6x^2\right)$

  a. Add column by column.
       $5m^4 + 2m^3 - 4$
       $\underline{-3m^4 + 5m^3 - 3}$
       $2m^4 + 7m^3 - 7$

  b. Combine like terms.

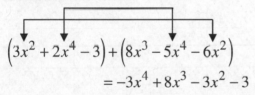

$$\left(3x^2 + 2x^4 - 3\right) + \left(8x^3 - 5x^4 - 6x^2\right)$$
$$= -3x^4 + 8x^3 - 3x^2 - 3$$

Now Try:

5. Add.

  a. $9m^3 + 4m^2 - 2m + 3$
       $\underline{-4m^3 - 6m^2 - 2m + 1}$

_____

  b. $\left(x^2 + 6x - 8\right) + \left(3x^2 - 10\right)$

_____

---

6. Subtract $8x^3 - 5x^4 - 6x^2$ from $3x^2 + 2x^4 - 3$.

Be careful to write the problem in the correct order.

$$\left(3x^2 + 2x^4 - 3\right) - \left(8x^3 - 5x^4 - 6x^2\right)$$
$$= \left(3x^2 + 2x^4 - 3\right) + \left(-8x^3 + 5x^4 + 6x^2\right)$$
$$= 7x^4 - 8x^3 + 9x^2 - 3$$

6. Subtract $6x^3 + 6x + 1$ from $9x^3 + 7x^2 - 6x + 3$.

_____

7. Subtract.

$$5m^4 + 2m^3 - 4$$
$$-3m^4 + 5m^3 - 3$$

Change all of the signs in the second row, then add.

$$5m^4 + 2m^3 - 4$$
$$3m^4 - 5m^3 + 3$$
$$\overline{8m^4 - 3m^3 - 1}$$

7. Subtract.

$$2m^2 - 5m + 1$$
$$-2m^2 - 5m + 3$$

_____

8. Subtract.

$$(4ab + 2bc - 9ac) - (3ca - 2cb - 9ba)$$

$$(4ab + 2bc - 9ac) - (3ca - 2cb - 9ba)$$
$$= (4ab + 2bc - 9ac) - (3ac - 2bc - 9ab)$$
$$= 4ab + 2bc - 9ac - 3ac + 2bc + 9ab$$
$$= 4ab + 9ab + 2bc + 2bc - 9ac - 3ac$$
$$= 13ab + 4bc - 12ac$$

8. Subtract.

$$(-4m^2n + 3mn - 6m) - (2n + 7mn)$$

_____

Review this example for Objective 6:

9. Graph $-x^2 + 1$.

Start by creating a table of values.

x	$y = -x^2 + 1$
-2	-3
-1	0
0	1
1	0
2	-3

Plot the ordered points and draw a smooth curve through them.

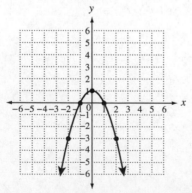

Now Try:

9. Graph $2x^2 - 1$.

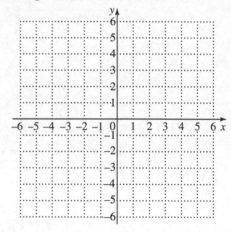

Name: Date:
Instructor: Section:

Objective 1 Identify terms and coefficients.

For extra help, see Example 1 on page 320 of your text, the Section Lecture video for Section 5.4, and Exercise Solution Clip 9.

For each of the following, determine the number of terms in the polynomial and name the coefficients of the terms.

1. $9x^3 + 3x^3 - 4x + 2$

                                                      1. _____

2. $-\frac{2}{5}y^2 + \frac{4}{3}y$

                                                      2. _____

Objective 2 Add like terms.

For extra help, see Example 2 on page 320 of your text, the Section Lecture video for Section 5.4, and Exercise Solution Clip 17.

In each polynomial, add like terms whenever possible. Write the result in descending powers of the variable.

3. $7z^3 - 4z^3 + 5z^3 - 11z^3$

                                                      3. _____

4. $4y^4 - 7y^2 + 4 - 7y^3 + 9y^4 - 2y^2 - 3y$

                                                      4. _____

5. $-\frac{1}{2}r^3 + \frac{1}{3}r + \frac{1}{4}r^3 - \frac{1}{3}r$

                                                      5. _____

Objective 3 Know the vocabulary for polynomials.
For extra help, see Example 3 on page 321 of your text, the Section Lecture video for Section 5.4, and Exercise Solution Clip 29.

For each polynomial, first simplify, if possible, and write the resulting polynomial in descending powers of the variable. Then give the degree of this polynomial, and tell whether it is a monomial, a binomial, a trinomial, or none of these.

6. $7m^2 + 3m + m^4$

6. _____

7. $3n^8 - n^2 - 2n^8$

7. _____

Objective 4 Evaluate polynomials.
For extra help, see Example 4 on page 322 of your text, the Section Lecture video for Section 5.4, and Exercise Solution Clip 37.

Find the value of the polynomial when (a) $x = 3$ and (b) $x = -2$.

8. $-x^2 + 5x - 9$

8. _____

9. $5x^4 - 2x^2 + 8x - 1$

9. _____

Name: Date:

Instructor: Section:

Objective 5 Add and subtract polynomials.

For extra help, see Examples 5–9 on pages 322–324 of your text, the Section Lecture video for Section 5.4, and Exercise Solution Clips 45, 55, 59, 61 and 73.

Add or subtract as indicated.

10. $\left(-4a^5 - 6a^3 + 5a + 2\right) + \left(4a^3 - 7a - 6\right)$ **10.** _____

11. $\left(2z^5 + z^4 - 2z^3 + 5\right) - \left(2z^4 - 7z^2 + 8z - 2\right)$ **11.** _____

12. $\left(2x^3 - 4x^2 + 3x + 10\right) - \left(6x^3 - 4x + 2\right)$ **12.** _____

13. $\left(8p^2 + 7p - 2\right) - \left(3p^2 - 3p + 7\right) + \left(2p^2 - 3\right)$ **13.** _____

Objective 6 Graph equations defined by polynomials of degree 2.

For extra help, see Example 10 on pages 324–325 of your text, the Section Lecture video for Section 5.4, and Exercise Solution Clip 89.

Select several values for x, then find the corresponding y-values, and graph.

14. $y = x^2 - 1$ **14.**

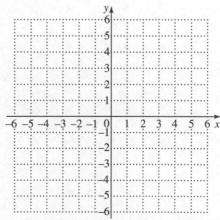

15. $y = 2 - x^2$

15.

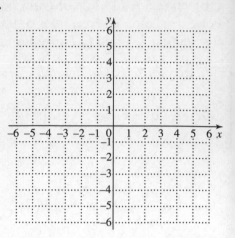

Chapter 5 EXPONENTS AND POLYNOMIALS

5.5 Multiplying Polynomials

Learning Objectives
1   Multiply a monomial and a polynomial.
2   Multiply two polynomials.
3   Multiply binomials by the FOIL method.

Key Terms

Use the vocabulary terms listed below to complete each statement in exercises 1–3.

outer product   **inner product**   **FOIL method**

1.   The _____ is a method used for multiplying two binomials.

2.   The _____ of $(2y - 5)(y + 8)$ is $-5y$.

3.   The _____ of $(2y - 5)(y + 8)$ is $16y$.

Guided Examples

Review this example for Objective 1:
1. Find the product.

$$2m\left(3m^3 + 7m^2 + 3\right)$$

$$2m\left(3m^3 + 7m^2 + 3\right)$$
$$= 2m\left(3m^3\right) + 2m\left(7m^2\right) + 2m(3)$$
$$\qquad\qquad\qquad\qquad \text{Distributive property}$$
$$= 6m^4 + 14m^3 + 6m \quad \text{Multiply monomials}$$

Now Try:
1. Find the product.

$$-7z^3\left(5z^3 - 4z^2 + 2\right)$$

_____

Name: Date:

Instructor: Section:

Review these examples for Objective 2:

2. Multiply.

$$(y+4)\left(y^2 - 4y + 16\right)$$

Multiply each term of the second polynomial by each term of the first. Then combine like terms.

$$(y+4)\left(y^2 - 4y + 16\right)$$
$$= y\left(y^2\right) + y(-4y) + y(16) + 4\left(y^2\right)$$
$$+ 4(-4y) + 4(16)$$
$$= y^3 - 4y^2 + 16y + 4y^2 - 16y + 64$$
$$= y^3 + 64$$

3. Multiply $(r-2)\left(2r^2 + r - 1\right)$ vertically.

$$
\begin{array}{r}
2r^2 + \ r - 1 \\
r - 3 \\
\hline
- 6r^2 - 3r + 3 \\
2r^3 + \ r^2 - r \\
\hline
2r^3 - 5r^2 - r + 3
\end{array}
$$

Begin by multiplying each of the terms in the top row by -3.

Then multiply each of the terms in the top row by r. Finally, add like terms.

Review these examples for Objective 3:

4. Use the FOIL method to find the product.

$$(4m+3)(m-7)$$

Step 1 **Multiply the <u>First</u> terms:**

$$4m(m) = 4m^2$$

Step 2 **Find the <u>Outer</u> product:**

$$4m(-7) = -28m$$

      Find the <u>Inner</u> product:

$$3(m) = 3m$$

      Add the outer and inner products:

$$-28m + 3m = -25m$$

Step 3 **Multiply the <u>Last</u> terms:**

$$3(-7) = -21$$

Now add the terms found in Steps 1–3.

$$(4m+3)(m-7) = 4m^2 - 25m - 21$$

Now Try:

2. Simplify by adding like terms.

$$(x-3)\left(x^2 + x - 9\right)$$

_____

3. Multiply

$$\left(3x^2 + \tfrac{1}{2}x\right)\left(2x^2 + 6x - 4\right)$$

vertically.

_____

Now Try:

4. Use the FOIL method to find the product.

$$(x-3)(2x-9)$$

_____

5. Find each product.

   a. $(3x+2y)(2x-3y)$

   b. $-y^2(3y+1)(2y-5)$

   a. $(3x+2y)(2x-3y)$

$$= 3x(2x)+3x(-3y)+2y(2x)+2y(-3y)$$

       F   **O**   **I**   **L**

$$= 6x^2 \quad -9xy \quad +4xy \quad -6y^2$$

$$= 6x^2 - 5xy - 6y^2$$

   b. First multiply the two binomials using the FOIL method. Then multiply that product by $-y^2$ using the distributive property.

$$(3y+1)(2y-4)$$

$$= 3y(2y)+3y(-4)+1(2y)+1(-4)$$

       F   **O**   **I**   **L**

$$= 6y^2 \quad -12y \quad +2y \quad -4$$

$$= 6y^2 - 10y - 4$$

$$-y^2(3y+1)(2y-5) = -y^2\left(6y^2 - 10y - 4\right)$$

$$= -6y^4 + 10y^3 + 4y^2$$

Alternatively, we could have multiplied the first binomial by $-y^2$ using the distributive property first, and then multiplied that product and the second binomial using the FOIL method.

5. Find each product.

   a. $(2m+3n)(-3m+4n)$

   _____

   b. $a^2(3-4a)(1-2a)$

   _____

Objective 1 Multiply a monomial and a polynomial.
For extra help, see Example 1 on page 329 of your text, the Section Lecture video for Section 5.5, and Exercise Solution Clip 15.

Find each product.

1. $-2x^4\left(3+6x+2x^2\right)$

1. _____

2. $-6z\left(z^5 + 3z^3 + 4z + 2\right)$

2. _____

3. $-3y^2\left(2y^3 + 3y^2 - 4y + 11\right)$ 3. _____

4. $7b^2\left(-5b^2 + 1 - 4b\right)$ 4. _____

5. $-4r^4\left(2r^2 - 3r + 2\right)$ 5. _____

Objective 2 Multiply two polynomials.
For extra help, see Examples 2–4 on pages 330–331 of your text, the Section Lecture video for Section 5.5, and Exercise Solution Clips 29, 33, and 39.

Find the product.

6. $(3m - 5)(2m + 4)$ 6. _____

7. $(r + 3)\left(2r^2 - 3r + 5\right)$ 7. _____

8. $\left(2m^2 + 1\right)\left(3m^3 + 2m^2 - 4m\right)$ 8. _____

Find each product, using the vertical method of multiplication.

9. $(2y - 3)\left(3y^3 + 2y^2 - y + 2\right)$ 9. _____

10. $\left(2x^2 + 3x + 2\right)\left(4x^3 + 2x + 3\right)$

10. _____

Objective 3 Multiply binomials by the FOIL method.

For extra help, see Examples 5–7 on page 332 of your text, the Section Lecture video for Section 5.5, and Exercise Solution Clips 41, 61, and 63.

Use the FOIL method to find each product.

11. $(5a - b)(4a + 3b)$

11. _____

12. $(11k - 4)(11k + 4)$

12. _____

13. $\left(3 + 4a^2\right)\left(1 + 2a^2\right)$

13. _____

14. $\left(2v^2 + w^2\right)\left(v^2 - 3w^2\right)$

14. _____

15. $\left(x - \frac{1}{3}\right)\left(x - \frac{4}{3}\right)$

15. _____

Chapter 5 EXPONENTS AND POLYNOMIALS

5.6 Special Products

Learning Objectives
1   Square binomials.
2   Find the product of the sum and difference of two terms.
3   Find greater powers of binomials.

Key Terms

Use the vocabulary terms listed below to complete each statement in exercises 1–2.

    square of a binomial   **product of the sum and difference of two terms**

1.   The _____ is a binomial consisting of the square of the first term minus the square of the second term.

2.   The _____ is a trinomial consisting of the square of the first term of the binomial, plus twice the product of the two terms, plus the square of the last term of the binomial.

Guided Examples

Review these examples for Objective 1:

1. Find $(m-2)^2$

$$(m-2)^2 = (m-2)(m-2)$$
$$= m^2 - 2m - 2m + 4 \quad \text{FOIL}$$
$$= m^2 - 4m + 4$$

2. Find each binomial square and simplify.

   a. $(2y+5)^2$ **b.** $(8k+5p)^2$

   c. $\left(4x-\frac{1}{4}y\right)^2$ **d.** $y(5y-3)^2$

   a. $(2y+5)^2 = (2y)^2 + 2(10y) + 5^2$
$$= 4y^2 + 20y + 25$$

   b. $(8k+5p)^2 = (8k)^2 + 2(40kp) + (5p)^2$
$$= 64k^2 + 80kp + 25p^2$$

Now Try:

1. Find $(z+6)^2$

_____

2. Find each binomial square and simplify.

   a. $(a+2b)^2$

_____

   b. $(2m-3p)^2$

_____

c. $\left(4x-\frac{1}{4}y\right)^2 = (4x)^2 - 2(4x)\left(\frac{1}{4}y\right) + \left(\frac{1}{4}y\right)^2$

$$= 16x^2 - 2xy + \frac{1}{16}y^2$$

d. $y(5y-3)^2 = y\left[(5y)^2 - 2(15y) + 3^2\right]$

$$= y\left(25y^2 - 30y + 9\right)$$

$$= 25y^3 - 30y^2 + 9y$$

c. $\left(3x+\frac{2}{3}y\right)^2$

_____

d. $x^2(2x+5)^2$

_____

Review this example for Objective 2:

3. Find each product.

a. $(y+5)(y-5)$

b. $(2k+5)(2k-5)$

c. $(2r+3x)(2r-3x)$

d. $\left(4x-\frac{1}{4}y\right)\left(4x+\frac{1}{4}y\right)$

e. $y(5y-3)(5y+3)$

a. $(y+5)(y-5) = y^2 - 5^2$

$$= y^2 - 25$$

b. $(2k+5)(2k-5) = (2k)^2 - 5^2$

$$= 4k^2 - 25$$

c. $(2r+3x)(2r-3x) = (2r)^2 - (3x)^2$

$$= 4r^2 - 9x^2$$

d. $\left(4x-\frac{1}{4}y\right)\left(4x+\frac{1}{4}y\right) = (4x)^2 - \left(\frac{1}{4}y\right)^2$

$$= 16x^2 - \frac{1}{16}y^2$$

e. $y(5y-3)(5y+3) = y\left[(5y)^2 - 3^2\right]$

$$= y\left(25y^2 - 9\right)$$

$$= 25y^3 - 9y$$

Now Try:

3. Find each product.

a. $(7-x)(7+x)$

_____

b. $(5m-4)(5m+4)$

_____

c. $(3m-7n)(3m+7n)$

_____

d. $\left(6x+\frac{2}{3}y\right)\left(6x-\frac{2}{3}y\right)$

_____

e. $x^2(8x-3)(8x+3)$

_____

Review this example for Objective 3:

4. Find each product.

    a. $(4y+5)^3$

    b. $(3x+y)^3$

    c. $(2r-3)^4$

    a. $(4y-5)^3 = (4y-5)^2(4y-5)$ $a^3 = a^2 \cdot a$

$$= \left(16y^2 - 40y + 25\right)(4y-5)$$

           Square the binomial.

$$= 64y^3 - 160y^2 + 100y - 80y^2$$
$$+ 200y - 125$$

           Multiply polynomials.

$$= 64y^3 - 240y^2 + 300y - 125$$

           Combine like terms.

    b. $(3x+y)^3 = (3x+y)^2(3x+y)$ $a^3 = a^2 \cdot a$

$$= \left(9x^2 + 6xy + y^2\right)(3x+y)$$

           Square the binomial.

$$= 27x^3 + 18x^2y + 3xy^2 + 9x^2y$$
$$+ 6xy^2 + y^3$$

           Multiply polynomials.

$$= 27x^3 + 27x^2y + 9xy^2 + y^3$$

           Combine like terms.

    c. $(2r-3)^4$

$$= (2r-3)^2(2r-3)^2 \qquad a^4 = a^2 \cdot a^2$$
$$= \left(4r^2 - 12r + 9\right)\left(4r^2 - 12r + 9\right)$$

           Square each binomial.

$$= 16r^4 - 48r^3 + 36r^2 - 48r^3$$
$$+ 144r^2 - 108r + 36r^2 - 108r + 81$$

           Multiply polynomials.

$$= 16r^4 - 96r^3 + 216r^2 - 216r + 81$$

           Combine like terms.

Now Try:

4. Find each product.

    a. $(7-x)^3$

    ————————

    b. $(5m-4n)^3$

    ————————

    c. $(3a-2)^4$

    ————————

Name: Date:
Instructor: Section:

Objective 1 Square binomials.
For extra help, see Examples 1 and 2 on pages 335–336 of your text, the Section Lecture video for Section 5.6, and Exercise Solution Clips 7 and 13.

Find each square by using the pattern for the square of a binomial.

1. $(2m+5)^2$

1. _____

2. $(5m+3n)^2$

2. _____

3. $(2p+3q)^2$

3. _____

4. $y^3(4y-7)^2$

4. _____

5. $\left(3a+\frac{1}{2}b\right)^2$

5. _____

Objective 2 Find the product of the sum and difference of two terms.
For extra help, see Examples 3 and 4 on pages 336–337 of your text, the Section Lecture video for Section 5.6, and Exercise Solution Clips 23, 31, and 39.

Find each product by using the pattern for the sum and difference of two terms.

6. $(12+x)(12-x)$

6. _____

7. $(4p+7q)(4p-7q)$

7. _____

8. $(9-4y)(9+4y)$

8. _____

9. $\left(7m - \frac{3}{4}\right)\left(7m + \frac{3}{4}\right)$

9. _____

10. $\left(y^2 + 2\right)\left(y^2 - 2\right)$

10. _____

Objective 3 Find greater powers of binomials.
For extra help, see Example 5 on page 337 of your text, the Section Lecture video for Section 5.6, and Exercise Solution Clips 43 and 51.

Find each product.

11. $(2x - 3)^3$

11. _____

12. $(2x + 1)^3$

12. _____

13. $-k(4k + 2)^3$

13. _____

14. $(3b - 2)^4$

14. _____

15. $(4s + 3t)^4$

15. _____

Chapter 5 EXPONENTS AND POLYNOMIALS

5.7   Dividing Polynomials

Learning Objectives
1   Divide a polynomial by a monomial.
2   Divide a polynomial by a polynomial.

Key Terms
Use the vocabulary terms listed below to complete each statement in exercises 1–3.

> **dividing a polynomial by a monomial   dividing a polynomial by a polynomial**
>
> **long division**

1.   When _____ divide each term of the polynomial by the monomial.

2.   We use the method of _____ to divide a polynomial by a
       polynomial.

3.   When _____ each polynomial must be written in descending
       powers.

Guided Examples

Review these examples for Objective 1:

1. Divide $6a^3 + 27a^2$ by $3a^2$.

$$\frac{6a^3 + 27a^2}{3a^2} = \frac{6a^3}{3a^2} + \frac{27a^2}{3a^2}$$
$$= 2a + 9$$

2. Divide $-54m^3 + 30m^2 + 6m$ by $3m^2$.

$$\frac{-54m^3 + 30m^2 + 6m}{3m^2} = -\frac{54m^3}{3m^2} + \frac{30m^2}{3m^2} + \frac{6m}{3m^2}$$
$$= -18m + 10 + \frac{2}{m}$$

3. Divide $6x^5 - 4x^3 + 12x$ by $-4x$.

$$\frac{6x^5 - 4x^3 + 12x}{-4x} = -\frac{3}{2}x^4 + x^2 - 3$$

Now Try:

1. Divide $72y^5 - 27y^3$ by $9y^2$.

   _____

2. Divide $63b^5 - 27b^3 - b$ by $9b^3$.

   _____

3. Divide $-24w^8 + 12w^6 - 18w^4$ by
     $-6w^5$.

   _____

4. Divide $30x^3y^3 - 45x^2y^2 + 15xy^2$ by $-15xy^2$.

$$\frac{30x^3y^3 - 45x^2y^2 + 15xy^2}{-15xy^2} = -2x^2y + 3x - 1$$

4. Divide
$5x^2y^4 - 30x^4y^3 + 30x^5y^2$ by $-5x^2y^2$.

_____

Review these examples for Objective 2:

5. Divide $\dfrac{3x^2 - 3x - 6}{x+1}$.

$$\begin{array}{r}
3x - 6 \\
x+1\overline{\smash{)}3x^2 - 3x - 6} \\
\underline{-(3x^2 + 3x)} \\
-6x - 6 \\
\underline{-(6x - 6)} \\
0
\end{array}$$

$$\frac{3x^2 - 3x - 6}{x+1} = 3x - 6$$

6. Divide $\dfrac{5b^2 + 32b + 3}{5b + 7}$.

$$\begin{array}{r}
b + 5 \\
5b+7\overline{\smash{)}5b^2 - 32b + 3} \\
\underline{-(5b^2 + \ 7b)} \\
-25b + \ 3 \\
\underline{-(-25b + 35)} \\
-32 \leftarrow \text{Remainder}
\end{array}$$

$$\frac{5b^2 + 32b + 3}{5b + 7} = b + 5 - \frac{32}{5b + 7}$$

7. Divide $\dfrac{y^3 - 1}{y + 1}$.

Note that the dividend is missing the y^2 and y terms. Use 0 as the coefficient for each missing term. Thus, $y^3 - 1 = y^3 + 0y^2 + 0y - 1$.

Now Try:

5. Divide $\dfrac{2r^2 + 3r - 20}{r + 4}$.

_____

6. Divide $\dfrac{9m^2 - 18m + 16}{3m - 4}$.

7. Divide $\dfrac{3x^4 - 6x^2 + 3x + 6}{x + 1}$.

_____

$$\begin{array}{r} y^2 - y + 1 \\ y+1\overline{\smash{\big)}\,y^3 + 0y^2 + 0y - 1} \\ \underline{y^3 + \ y^2} \\ -y^2 + 0y \\ \underline{-y^2 - \ y} \\ y - 1 \\ \underline{y + 1} \\ -2 \end{array}$$

$$\frac{y^3 - 1}{y + 1} = y^2 - y + 1 - \frac{2}{y + 1}$$

8. Divide $\dfrac{3x^4 - 2x^3 - 2x^2 - 2x - 1}{x^2 - 1}$.

Note that the divisor is missing the x term. Use 0 as the coefficient for each the term. Thus, $x^2 - 1 = x^2 + 0x - 1$.

$$\begin{array}{r} 3x^2 - 2x + 1 \\ x^2 + 0x - 1\overline{\smash{\big)}\,3x^4 - 2x^3 - 2x^2 - 2x - 1} \\ \underline{3x^4 + 0x^3 - 3x^2} \\ -2x^3 + \ x^2 - 2x \\ \underline{-2x^3 + 0x^2 + 2x} \\ x^2 - 4x - 1 \\ \underline{x^2 + 0x - 1} \\ -4x \end{array}$$

$$\frac{3x^4 - 2x^3 - 2x^2 - 2x - 1}{x^2 - 1}$$

$$= 3x^2 - 2x + 1 - \frac{4x}{x^2 - 1}$$

8. Divide
$$\frac{6x^4 - 12x^3 + 13x^2 - 5x - 1}{2x^2 + 3}.$$

_____

9. Divide $\dfrac{4x^3 - 5x^2 + x - 16}{2x - 4}$.

$$2x - 4 \overline{\smash{)}\ 4x^3 - 5x^2 + x - 17} \quad \begin{array}{c} 2x^2 + \frac{3}{2}x + \frac{7}{2} \end{array}$$

$$\begin{array}{r}
\underline{4x^3 - 8x^2} \\
3x^2 + x \\
\underline{3x^2 - 6x} \\
7x - 17 \\
\underline{7x - 14} \\
-3
\end{array}$$

$$\frac{4x^3 - 5x^2 + x - 16}{2x - 4} = 2x^2 + \frac{3}{2}x + \frac{7}{2} - \frac{3}{2x - 4}$$

9. Divide $\dfrac{24x^3 + 16x^2 - 6x - 9}{6x + 3}$.

_____

Objective 1 Divide a polynomial by a monomial.
For extra help, see Examples 1–4 on page 341–342 of your text, the Section Lecture video for Section 5.7, and Exercise Solution Clips 13, 19, 29 and 37.

Divide each polynomial by $3m^2$.

1. $12m^4 - 9m^3 + 6m^2$

1. _____

2. $27m^3 + 18m^2 - 6m$

2. _____

3. $3m^2 - 3$

3. _____

Perform each division.

4. $\dfrac{12x^6 + 18x^5 + 30x^3}{6x^2}$

4. _____

5. $\left(m^2 + 3m - 12\right) \div (2m)$

5. _____

6. $\dfrac{6y^5 - 3y^4 + 9y^2 + 27}{-3y}$

6. _____

7. $\dfrac{14y^2 - 14y + 70}{-7y^2}$

7. _____

Objective 2 Divide a polynomial by a polynomial.
For extra help, see Examples 5–9 on page 344–346 of your text, the Section Lecture video for Section 5.7, and Exercise Solution Clips 53, 59, 63, 75, and 79.

Perform each division.

8. $\dfrac{18a^2 - 9a - 5}{3a + 1}$

8. _____

9. $\left(2a^2 - 11a + 16\right) \div (2a + 3)$

9. _____

10. $\dfrac{5w^2 - 22w + 4}{w - 4}$

10. _____

11. $\dfrac{2z^3 - 7z^2 + 3z + 2}{2z + 3}$

11. _____

12. $\left(27p^4 - 36p^3 - 6p^2 + 26p - 24\right) \div \left(3p - 4\right)$

12. _____

13. $\left(3x^3 - 11x^2 + 25x - 25\right) \div \left(x^2 - 5\right)$

13. _____

14. $\dfrac{2a^4 + 5a^2 + 3}{2a^2 + 3}$

14. _____

15. $\dfrac{6x^5 + 7x^4 - 7x^3 + 7x + 4}{3x + 2}$

15. _____

Chapter 6 FACTORING AND APPLICATIONS

6.1 The Greatest Common Factor; Factoring by Grouping

Learning Objectives
1   Find the greatest common factor of a list of terms.
2   Factor out the greatest common factor.
3   Factor by grouping.

Key Terms

Use the vocabulary terms listed below to complete each statement in exercises 1–4.

factor   **factored form**   **common factor**

greatest common factor (GCF)

1.   An expression is in _____ when it is written as a product.

2.   The _____ of a polynomial is the largest term that is a factor of all terms in the polynomial.

3.   A _____ of a given number is any number that divides without remainder into the given number.

4.   An integer that is a factor of two or more integers is a _____ of those integers.

Guided Examples

Review these examples for Objective 1:
1.   Find the greatest common factor for each list of numbers.

   a. 60, 75, 120   **b.** 80, 280, 112

   Write the prime factored form of each number. Then use each prime the least number of times it appears in all the factored forms.

   a. $60 = 2^2 \cdot 3 \cdot 5$ **b.** $80 = 2^4 \cdot 5$
              $75 = 3 \cdot 5^2$ $280 = 2^3 \cdot 5 \cdot 7$
            $120 = 2^3 \cdot 3 \cdot 5$ $112 = 2^4 \cdot 7$
            $\text{GCF} = 3 \cdot 5 = 15$ $\text{GCF} = 2^3 = 8$

Now Try:
1.   Find the greatest common factor for each list of numbers.

   a. 108, 48, 84

   _____

   b. 70, 126, 42, 56

   _____

247

2. Find the greatest common factor for each list of terms.

   a. $27x^2, 54x^4, 81x^5$

   b. $72u^2v^3, 54uv^2, 63uv^4$

   a. $27x^2 = 3^3 \cdot x^2$
      $54x^4 = 2 \cdot 3^3 \cdot x^4$
      $81x^5 = 3^4 \cdot x^5$

      The GCF of the coefficients is $3^3 = 27$. The least exponent on x is 2. So the GCF of the terms is $27x^2$.

   b. $72u^2v^3 = 2^3 \cdot 3^2 \cdot u^2 \cdot v^3$
      $54uv^2 = 2 \cdot 3^3 \cdot u \cdot v^2$
      $63uv^4 = 3^2 \cdot 7 \cdot u \cdot v^4$

      The GCF of the coefficients is $3^2 = 9$. The least exponent on u is 1 and least exponent on v is 2. So the GCF of the terms is $9uv^2$.

2. Find the greatest common factor for each list of terms.

   a. $56y^4, 24y^3, 32y^2$

      _____

   b. $15ab^2, 45a^3b^4, 90ab^3$

      _____

Review these examples for Objective 2:

3. Write in factored form by factoring out the greatest common factor.

   a. $16x^2 + 40x^4$

   b. $26x^8 - 13x^{12} - 52x^{10}$

   c. $t^8 - t^5$

   d. $24ab^3 - 8a^2b + 40a^3b^2$

   a. $16x^2 + 40x^4$
      $= 8x^2(2) + 8x^2(5x^2)$ GCF $= 8x^2$
      $= 8x^2(2 + 5x^2)$

   b. $26x^8 - 13x^{12} - 52x^{10}$
      $= 13x^8(2) + 13x^8(-x^4) + 13x^8(-4x^2)$
      GCF $= 13x^8$
      $= 13x^8(2 - x^4 - 4x^2)$

Now Try:

3. Write in factored form by factoring out the greatest common factor.

   a. $-15y^2 + 18y$

      _____

   b. $96a^3 - 48a^2 + 60a$

      _____

   c. $x^4z^2 + x^6z^2$

      _____

   d. $14x^3y^2 + 7x^2y - 21x^5y^3$

      _____

c. $t^8 - t^5 = t^5\left(t^3\right) + t^5(-1)$ GCF $= t^5$

       $= t^5\left(t^3 - 1\right)$

d. $24ab^3 - 8a^2b + 40a^3b^2$

     $= 8ab\left(3b^2\right) + 8ab(-a) + 8ab\left(5a^2b\right)$

                    GCF $= 8ab$

     $= 8ab\left(3b^2 - a + 5a^2b\right)$

4. Write in factored form by factoring out the greatest common factor.

   a. $2a(x - 2y) + 9b(x - 2y)$

   b. $(x-1)(3x-1) + x(3x-1)$

   a. $2a(x-2y) + 9b(x-2y)$ GCF $= (x-2y)$

        $= (x-2y)(2a+9b)$

   b. $(x-1)(3x-1) + x(3x-1)$ GCF $= (3x-1)$

        $= (3x-1)(x-1+x)$

        $= (3x-1)(2x-1)$

4. Write in factored form by factoring out the greatest common factor.

   a. $x^2(r-4s) + z^2(r-4s)$

   _____

   b. $(b+5)(3x+1) - (3x+1)$

   _____

Review these examples for Objective 3:

5. Factor by grouping.

   a. $4ax + 4ay + 3bx + 3by$

   b. $5rs - 5rt - 2qs + 2qt$

   c. $1 + p - q - pq$

   d. $3r^3 - 2r^2s + 3s^2r - 2s^3$

   a. Group the first two terms and the last two terms since the first two terms have a common factor of $4a$ and the last two terms have a common factor of $3b$.

     $4ax + 4ay + 3bx + 3by$

     $= (4ax + 4ay) + (3bx + 3by)$

     $= 4a(x + y) + 3b(x + y)$

            Factor each group.

     $= (x + y)(4a + 3b)$

           Factor out $(x + 3)$.

Now Try:

5. Factor by grouping.

   a. $12bc - 4bd - 15xc + 5xd$

   _____

   b. $6v^3 - 16v^2 + 21v - 56$

   _____

   c. $4xy - x - 24y + 6$

   _____

   d. $28x^3y^2 + 16x^2y - 21xy - 12$

   _____

b. $5rs - 5rt - 2qs + 2qt$
$= (5rs - 5rt) + (-2qs + 2qt)$
$= 5r(s - t) - 2q(s - t)$

Factor each group.

$= (s - t)(5r - 2q)$ Factor out $(s - t)$.

c. $1 + p - q - pq$
$= (1 + p) + (-q - pq)$
$= 1(1 + p) - q(1 + p)$ Factor each group.
$= (1 + p)(1 - q)$ Factor out $(1 - q)$.

d. $3r^3 - 2r^2s + 3s^2r - 2s^3$
$= \left(3r^3 - 2r^2s\right) + \left(3s^2r - 2s^3\right)$
$= r^2(3r - 2s) + s^2(3r - 2s)$

Factor each group.

$= (3r - 2s)\left(r^2 + s^2\right)$

Factor out $(3r - 2s)$.

6. Factor by grouping.

a. $28xy + 25 + 35x + 20y$

b. $21xy - 12b^2 + 14xb - 18by$

a. We must rearrange the terms in order to find a common factor in each group of two terms.
$28xy + 25 + 35x + 20y$
$= 28xy + 35x + 20y + 25$

Commutative property

$= (28xy + 35x) + (20y + 25)$

Group the terms.

$= 7x(4y + 5) + 5(4y + 5)$

Factor each group.

$= (4y + 5)(7x + 5)$

Factor out $(4y + 5)$.

b. $21xy - 12b^2 + 14xb - 18by$
$= (21xy + 14xb) + \left(-12b^2 - 18by\right)$
$= 7x(3y + 2b) - 6b(2b + 3y)$

Factor each group.

$= 7x(2b + 3y) - 6b(2b + 3y)$

Commutative property

$= (2b + 3y)(7x - 6b)$

Factor out $(2b + 3y)$.

6. Factor by grouping.

a. $30uv + 30u + 36u^2 + 25v$

_____

b. $9mz - 4nc + 3mc - 12nz$

_____

Name:

Date:

Instructor:

Section:

Objective 1 Find the greatest common factor of a list of terms.

For extra help, see Examples 1 and 2 on pages 361–362 of your text, the Section Lecture video for Section 6.1, and Exercise Solution Clip 1.

Find the greatest common factor for each list of numbers.

1. 9, 18, 24, 48

1. _____

2. 42, 48, 72

2. _____

Find the greatest common factor for each list of terms.

3. $12ab^3$, $18a^2b^5$, $26a^2b^3$, $32a^3b^4$

3. _____

4. $6k^2m^4n^5$, $8k^3m^7n^4$, k^4m^8, n^7

4. _____

5. $45a^7y^4$, $75a^3y^2$, $-90a^2y$, $30a^4y^3$

5. _____

Objective 2 Factor out the greatest common factor.

For extra help, see Examples 3 and 4 on pages 362–363 of your text, the Section Lecture video for Section 6.1, and Exercise Solution Clips 39 and 55.

Write in factored form by factoring out the greatest common factor.

6. $12t^5 - 6t^3 + 8t^2$

6. _____

7. $2x^2y^8 + 5p^3q$

7. _____

8. $9x^3y^2 - 12x^5y + 15x^6y^3$

8. _____

9. $(x+2)(x+y)-(x+3)(x+y)$

9. _____

10. $(x-1)(x+1)+(x-1)$

10. _____

Objective 3 Factor by grouping.
For extra help, see Examples 5 and 6 on pages 363–365 of your text, the Section Lecture video for Section 6.1, and Exercise Solution Clips 71 and 85.

Factor by grouping.

11. $1-x+xy-y$

11. _____

12. $8-12p-6p^3+9p^4$

12. _____

13. $x^3+x^3y^2-3y^2-3$

13. _____

14. $-3x+2y-6+xy$

14. _____

15. $6wx-35wx+14w^2-15x^2$

15. _____

Chapter 6 FACTORING AND APPLICATIONS

6.2 Factoring Trinomials

Learning Objectives
1   Factor trinomials with a coefficient of 1 for the second-degree term.
2   Factor such trinomials after factoring out the greatest common factor.

Key Terms

Use the vocabulary terms listed below to complete each statement in exercises 1–3.

FOIL   **prime polynomial**   **common factor**

1.   A _____ is a polynomial that cannot be factored using only integer coefficients.

2.   After factoring a trinomial, you can use _____ to check your results.

3.   When factoring a trinomial, always look for a _____ first.

Guided Examples

Review these examples for Objective 1:

1.   Factor $x^2 + 8x + 15$.

We must find two integers whose product is 15 and whose sum is 8.

Factors of 15	Sum of Factors
1, 15	16
3, 5	8

From the list, 3 and 5 are the required integers since $3 \cdot 5 = 15$ and $3 + 5 = 8$.
Thus, $x^2 + 8x + 15 = (x + 3)(x + 5)$.

Now Try:

1.   Factor $x^2 + 14x + 24$.

_____

2. Factor $x^2 - 13x + 30$.

We must find two integers whose product is 30 and whose sum is -1. These integers have a product that is positive and a sum that is negative, so we consider only pairs of negative integers.

Factors of 15	Sum of Factors
$-1, -30$	-31
$-2, -15$	-17
$-3, -10$	-13
$-5, -6$	-11

From the list, -3 and -10 are the required integers since $(-3)(-10) = 30$ and $-3 + (-10) = -13$.

Thus, $x^2 - 13x + 30 = (x - 3)(x - 10)$.

3. Factor $x^2 + 5x - 24$.

We must find two integers whose product is -24 and whose sum is 5. These integers have a product that is negative, so the pairs of integers must have different signs.

Factors of -24	Sum of Factors
$-1, 24$	23
$1, -24$	-23
$-2, 12$	10
$2, -12$	-10
$-3, 8$	5
$3, -8$	-5
$-4, 6$	2
$4, -6$	-2

From the list, -3 and 8 are the required integers since $(-3)(8) = -24$ and $-3 + 8 = 5$.

Thus, $x^2 + 5x - 24 = (x + 3)(x - 8)$.

2. Factor $x^2 - 15x + 36$.

_____

3. Factor $x^2 + 3x - 28$.

_____

4. Factor $x^2 - 5x - 24$.

We must find two integers whose product is −24 and whose sum is −5. These integers have a product that is negative, so the pairs of integers must have different signs.

Factors of −24	Sum of Factors
−1, 24	23
1, −24	−23
−2, 12	10
2, −12	−10
−3, 8	5
3, −8	−5
−4, 6	2
4, −6	−2

From the list, 3 and −8 are the required integers since $(3)(-8) = -24$ and $3 + (-8) = -5$.

Thus, $x^2 - 5x - 24 = (x+3)(x-8)$.

5. Factor $r^2 - r - 5$.

We must find two integers whose product is −5 and whose sum is −1. These integers have a product that is negative, so the pairs of integers must have different signs.

Factors of −24	Sum of Factors
−1, 5	4
1, −5	−4

Neither of the pairs of integers has a sum of −1. Therefore, $r^2 - r - 5$ cannot be factored by using only integers. It is a *prime polynomial*.

4. Factor $t^2 - t - 20$.

_____

5. Factor $x^2 + 14x - 49$.

_____

6. Factor $r^2 + 4rs - 21s^2$.

We must find two expressions whose product is $-21s^2$ and whose sum is $4s$.

Factors of $-21s^2$	Sum of Factors
$-s, 21s$	$20s$
$s, -21s$	$-20s$
$-3s, 7s$	$4s$
$3s, -7s$	$-4s$

Thus, $r^2 + 4rs - 21s^2 = (r - 3s)(r + 7s)$.

6. Factor $p^2 - 7mp + 12m^2$.

_____

Review this example for Objective 2:

7. Factor $3p^6 - 18p^5 + 24p^4$

First factor out the greatest common factor, $3p^4$.

$$3p^6 - 18p^5 + 24p^4 = 3p^4\left(p^2 - 6p + 8\right)$$

Now factor $p^2 - 6p + 8$. The integers -2 and -4 have a product of 8 and a sum of -6.

$$3p^6 - 18p^5 + 24p^4 = 3p^4\left(p^2 - 6p + 8\right)$$
$$= 3p^4(p - 2)(p - 4)$$

Now Try:

7. Factor $2s^2t - 16st - 40t$.

_____

Objective 1 Factor trinomials with a coefficient of 1 for the second-degree term.
For extra help, see Examples 1–6 on pages 369–371 of your text, the Section Lecture video for Section 6.2, and Exercise Solution Clips 27, 31, 33, 37, and 47.

*Factor completely. If a polynomial cannot be factored, write **prime**.*

1. $z^2 + 5z + 6$

1. _____

2. $m^2 - m - 12$

2. _____

3. $r^2 + r + 3$ **3.** _____

4. $r^2 - 6r - 16$ **4.** _____

5. $k^2 + 9k + 20$ **5.** _____

6. $n^2 + 6n + 9$ **6.** _____

7. $a^2 - 4a - 21$ **7.** _____

8. $b^2 + 10bc + 25c^2$ **8.** _____

9. $a^2 - 6ab - 16b^2$ **9.** _____

10. $x^2 + 8xy + 15y^2$ **10.** _____

Name: Date:
Instructor: Section:

Objective 2 Factor such trinomials after factoring out the greatest common factor.
For extra help, see Example 7 on page 371 of your text, the Section Lecture video for
Section 6.2, and Exercise Solution Clip 55.

Factor completely. If a polynomial cannot be factored, write prime.

11. $2m^3 - 2m^2 - 4m$ 11. _____

12. $2n^4 - 16n^3 + 30n^2$ 12. _____

13. $3xy^2 - 24xy + 36x$ 13. _____

14. $2x^2y^2 - 2xy^3 - 12y^4$ 14. _____

15. $2x^3 - 14x^2y + 20xy^2$ 15. _____

Chapter 6 FACTORING AND APPLICATIONS

6.3 More on Factoring Trinomials

Learning Objectives
1   Factor trinomials by grouping when the coefficient of the squared term is not 1.
2   Factor trinomials by using the FOIL method.

Key Terms

Use the vocabulary terms listed below to complete each statement in exercises 1–3.

FOIL   **outer product**   **inner product**

1.   The _____ of $(2y-5)(y+8)$ is $-5y$.

2.   _____ is a shortcut method for finding the product of two binomials.

3.   The _____ of $(2y-5)(y+8)$ is $16y$.

Guided Examples

Review these examples for Objective 1:
1.   Factor each trinomial.

   a. $2x^2 + 5x - 3$

   b. $3x^2 - 2x - 5$

   c. $6x^2 + 5xy - 6y^2$

   a.   We must find two integers whose
        product is $2(-3) = -6$ and whose sum
        is 5.

Integers	Product	Sum
−1, 6	−6	5
1, −6	−6	−5
−2, 3	−6	1
2, −3	−6	−1

   The integers are −1 and 6. Write the
   middle term as $-x + 5x$, then factor by
   grouping.

Now Try:
1.   Factor each trinomial.

   a. $8b^2 + 18b + 9$

   _____

   b. $3m^2 - 5m - 12$

   _____

   c. $6n^2 + 11np + 4p^2$

   _____

$$2x^2 + 5x - 3 = 2x^2 - x + 6x - 3$$
$$= \left(2x^2 - x\right) + (6x - 3)$$
$$= x(2x - 1) + 3(2x - 1)$$
$$= (2x - 1)(x + 3)$$

b. $3x^2 - 2x - 5$

We must find two integers whose product is $3(-5) = -15$ and whose sum is -2.

Integers	Product	Sum
$-1, 15$	-15	14
$1, -15$	-15	-14
$-3, 5$	-15	2
$3, -5$	-15	-2

The integers are 3 and -5. Write the middle term as $3x - 5x$, then factor by grouping.

$$3x^2 - 2x - 5 = 3x^2 + 3x - 5x - 5$$
$$= \left(3x^2 + 3x\right) - (5x + 5)$$
$$= 3x(x + 1) - 5(x + 1)$$
$$= (x + 1)(3x - 5)$$

c. $6x^2 + 5xy - 6y^2$

We must find two integers whose product is $6(-6) = -36$ and whose sum is 5. The integers are 9 and -4. Write the middle term as $9xy - 4xy$, then factor by grouping.

$$6x^2 + 5xy - 6y^2$$
$$= 6x^2 + 9xy - 4xy - 6y^2$$
$$= \left(6x^2 + 9xy\right) + \left(-4xy - 6y^2\right)$$
$$= 3x(2x + 3y) - 2y(2x + 3y)$$
$$= (2x + 3y)(3x - 2y)$$

2. Factor $9c^2 + 24c + 12$.

Start by factoring out the common factor, 3, and then factor by grouping.
$$9c^2 + 24c + 12 = 3\left(3c^2 + 8c + 4\right)$$

To factor $3c^2 + 8c + 4$, find two integers whose product is $3(4) = 12$ and whose sum is 8. These integers are 2 and 6.
$$9c^2 + 24c + 12 = 3\left(3c^2 + 8c + 4\right)$$
$$= 3\left(3c^2 + 2c + 6c + 4\right)$$
$$= 3\left[\left(3c^2 + 2c\right) + (6c + 4)\right]$$
$$= 3\left[c(3c + 2) + 2(3c + 2)\right]$$
$$= 3(3c + 2)(c + 2)$$

2. Factor $16x^2 + 60x - 100$.

_____

Review these examples for Objective 2:

3. Factor $2x^2 + 11x + 5$.

We use the FOIL method in reverse. We want to write $2x^2 + 11x + 5$ as the product of two binomials.
$$2x^2 + 11x + 5 = (\_\_\_\_\_)(\_\_\_\_\_)$$
The product of the first two terms of the binomials is $2x^2$. Since all coefficients in the trinomial are positive, consider only positive factors. The factors of $2x^2$ are $2x$ and x.
$$2x^2 + 11x + 5 = (2x + \_\_\_)(x + \_\_\_)$$
The factors of the last term, 5, are 1 and 5 or 5 and 1. Try each pair to find the pair that gives the correct middle term, $11x$.

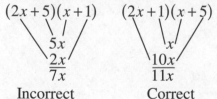

$(2x + 5)(x + 1)$	$(2x + 1)(x + 5)$
$5x$	x
$2x$	$10x$
$7x$	$11x$
Incorrect	Correct

Thus, $2x^2 + 11x + 5 = (2x + 1)(x + 5)$.

Now Try:

3. Factor $3y^2 + 13y + 4$.

_____

4. Factor $4x^2 - 12x + 5$.

The only factors of 5 are 1 and 5 or -1 and -5, so it is easier to begin by factoring 5. The middle term has a negative coefficient and the last term is positive, so consider only negative factors.

$$4x^2 - 12x + 5 = (\underline{\quad} - 1)(\underline{\quad} - 5)$$

The factors of the first term, $4x^2$, are x and $4x$ or $2x$ and $2x$. Try each pair to find the pair that gives the correct middle term, $11x$.

$$(x-1)(4x-5) \qquad (x-5)(4x-1)$$

$\begin{array}{c} -4x \\ -5x \\ \hline -9x \end{array}$ $\qquad$ $\begin{array}{c} -20x \\ -x \\ \hline -21x \end{array}$

Incorrect $\qquad\qquad$ Incorrect

$$(2x-1)(2x-5)$$

$\begin{array}{c} -2x \\ -10x \\ \hline -12x \end{array}$

Correct

Thus, $4x^2 - 12x + 5 = (2x-1)(2x-5)$.

5. Factor $2x^2 + 5x - 3$.

Since the constant term is negative, one positive factor and one negative factor of -3 are needed.

$$(2x-3)(x+1) \qquad (2x+3)(x-1)$$

$\begin{array}{c} -3x \\ 2x \\ \hline -x \end{array}$ $\qquad$ $\begin{array}{c} -2x \\ 3x \\ \hline x \end{array}$

Incorrect $\qquad\qquad$ Incorrect

$$(2x+1)(x-3) \qquad (2x-1)(x+3)$$

$\begin{array}{c} x \\ -6x \\ \hline -5x \end{array}$ $\qquad$ $\begin{array}{c} -x \\ 6x \\ \hline 5x \end{array}$

Incorrect $\qquad\qquad$ Correct

Thus, $2x^2 + 5x - 3 = (2x-1)(x+3)$.

4. Factor $5x^2 - 18x + 9$.

_____

5. Factor $3z^2 + 2z - 8$

_____

6. Factor $10x^2 - 29xy + 21y^2$.

The factors of $10x^2$ are x and $10x$ or $2x$ and $5x$. The coefficient of the middle term is negative and the coefficient of the third term is positive, so we need the factors of the third term are both negative. The factors of $21y^2$ are $-y$ and $-21y$ or $-3y$ and $-7y$.

$10x^2 - 29xy + 21y^2 = (5x - 7y)(2x - 3y)$

6. Factor $40z^2 + 18zy - 9y^2$.

_____

7. Factor $-9x^3 + 30x^2 - 24x$.

First, factor out the common factor, $-3x$.
$-9x^3 + 30x^2 - 24x = -3x(3x^2 - 10x + 8)$

Now factor $3x^2 - 10x + 8$.
$$-9x^3 + 30x^2 - 24x = -3x(3x^2 - 10x + 8)$$
$$= -3x(x - 2)(x - 4)$$

7. Factor $-30a^4b + 3a^3b + 6a^2b$.

_____

Objective 1 Factor trinomials by grouping when the coefficient of the squared term is not 1.

For extra help, see Examples 1 and 2 on pages 374–375 of your text and the Section Lecture video for Section 6.3.

Factor by grouping.

1. $6x^2 + 19x + 10$

1. _____

2. $16x^2 + 4x - 6$

2. _____

3. $24y^2 - 17y + 3$

3. _____

4. $3p^3 + 8p^2 + 4p$

4. _____

5. $2s + 5st - 3st^2$ 5. _____

Objective 2 Factor trinomials by using the FOIL method.
For extra help, see Examples 3–7 on pages 376–378 of your text, the Section Lecture video for Section 6.3, and Exercise Solution Clips 23, 25, 31, 35, 45, 47, and 51.

Factor completely using the FOIL method.

6. $20r^2 - 28r - 3$ 6. _____

7. $21s^2 + 13s + 2$ 7. _____

8. $20x^2 + 39x - 11$ 8. _____

9. $15q^2 - 7q - 4$ 9. _____

10. $8x^2 + 2x - 15$ 10. _____

11. $6p^2 - p - 15$ 11. _____

12. $6y^2 - 5yz - 6z^2$

12. _____

13. $6x^2 - 5xy - y^2$

13. _____

14. $8x^3 - 10x^2y + 3xy^2$

14. _____

15. $2y^5z^2 - 5y^4z^3 - 3y^3z^4$

15. _____

Chapter 6 FACTORING AND APPLICATIONS

6.4 Special Factoring Techniques

Learning Objectives
1   Factor a difference of squares.
2   Factor a perfect square trinomial.
3   Factor a difference of cubes.
4   Factor a sum of cubes.

Key Terms
Use the vocabulary terms listed below to complete each statement in exercises 1–4.

  difference of squares   **perfect square trinomial** **difference of cubes**

  sum of cubes

1. The _____, $x^2 - y^2$, can be factored as the product of the sum and
  difference of two terms, or $x^2 - y^2 = (x + y)(x - y)$.

2. The _____, $x^3 + y^3$, can be factored as
  $x^3 + y^3 = (x + y)\left(x^2 - xy + y^2\right)$.

3. A _____ is a trinomial that can be factored as the square of a
  binomial.

4. The _____, $x^3 - y^3$, can be factored as
  $x^3 - y^3 = (x - y)\left(x^2 + xy + y^2\right)$.

Guided Examples

Review these examples for Objective 1:
1. Factor each binomial if possible.

  a. $x^2 - 25$

  b. $y^2 + 81$

  The pattern to factor the difference of
  squares is $x^2 - y^2 = (x + y)(x - y)$.

  a. $x^2 - 25 = (x + 5)(x - 5)$

Now Try:
1. Factor each binomial if
  possible.

  a. $y^2 - 16$

  _____

  b. $p^2 + 25$

  _____

Name:

Instructor:

Date:

Section:

b. Since $y^2 + 81$ is the sum of squares, it is not equal to $(y + 9)(y - 9)$ and cannot be factored. Therefore, $y^2 + 81$ is a prime polynomial.

2. Factor each difference of squares.

  a. $9j^2 - \frac{16}{49}$

  b. $121m^2 - 9n^2$

  a. $9j^2 - \frac{16}{49} = \left(3j + \frac{4}{7}\right)\left(3j - \frac{4}{7}\right)$

  b. $121m^2 - 9n^2 = (11m + 3n)(11m - 3n)$

3. Factor completely.

  a. $98r^2 - 200$

  b. $a^4 - 81$

  c. $m^4n^2 - m^2$

  a. $98r^2 - 200 = 2\left(49r^2 - 100\right)$

                  Factor out the GCF, 2.

        $= 2(7r + 10)(7r - 10)$

  b. $a^4 - 81 = \left(a^2 + 9\right)\left(a^2 - 9\right)$

        Factor the difference of squares.

        $= \left(a^2 + 9\right)(a + 3)(a - 3)$

        Factor the difference of squares.

  c. $m^4n^2 - m^2 = m^2\left(m^2n^2 - 1\right)$

               Factor out the GCF, m^2.

        $= m^2(mn + 1)(mn - 1)$

        Factor the difference of squares.

2. Factor each difference of squares.

  a. $25a^2 - 36$

  _____

  b. $169m^2 - 225p^2$

  _____

3. Factor completely.

  a. $81x^4 - 900x^2$

  _____

  b. $m^4 - 625$

  _____

  c. $100x^4y^2 - 81z^2$

  _____

Name: Date:
Instructor: Section:

Review these examples for Objective 2:

4. Factor $a^2 + 14a + 49$.

   There are two patterns to factor perfect square trinomials:

   $x^2 + 2xy + y^2 = (x + y)^2$
   $x^2 - 2xy + y^2 = (x - y)^2$

   $a^2 + 14a + 49 = (a + 7)^2$

5. Factor each trinomial.

   a. $r^2 - 32r + 256$

   b. $100x^2 + 180x + 81$

   c. $9x^2 + 56x + 64$

   d. $200x^4 + 80x^3 + 8x^2$

   a. $r^2 - 32r + 256 = (r - 16)^2$

   b. $100x^2 + 180x + 81 = (10x + 9)^2$

   c. Although $9x^2$ and 64 are perfect squares, $9x^2 + 56x + 64$ is not a perfect square trinomial. Note that $(3x + 8)^2 = 9x^2 + 48x + 64$. Trying other methods of factoring shows that $9x^2 + 48x + 64$ cannot be factored and is prime.

   d. $200x^4 + 16x^3 + 8x^2$
      $= 8x^2(25x^2 + 2x + 1)$
      Factor out the GCF, $8x^2$.
      $= 8x^2(5x + 1)^2$

Review this example for Objective 3:

6. Factor each polynomial.

   a. $m^3 - 27$

   b. $8p^3 - 125$

   c. $2x^3 - 128$

   d. $2000x^6 - 54y^3$

Now Try:

4. Factor $x^2 + 4x + 4$.

   _____

5. Factor each trinomial.

   a. $x^2 - 8x + 16$

   _____

   b. $16x^2 - 40x + 25$

   _____

   c. $16y^2 - 24y - 81$

   _____

   d. $27x^4 - 18x^3 + 3x^2$

   _____

Now Try:

6. Factor each polynomial.

   a. $x^3 - 343$

   _____

   b. $27x^3 - 8$

   _____

The pattern to factor the difference of cubes is $x^3 - y^3 = (x - y)(x^2 + xy + y^2)$.

a. $m^3 - 27 = m^3 - 3^2$
$$= (m - 3)(m^2 + 3m + 9)$$

b. $8p^3 - 125 = (2p)^3 - 5^3$
$$= (2p - 5)(4p^2 + 10p + 25)$$

c. $2x^3 - 128 = 2(x^3 - 64)$

Factor out the GCF, 2.

$$= 2(x^3 - 4^3)$$
$$= 2(x - 4)(x^2 + 4x + 16)$$

d. $2000x^6 - 54y^3$
$$= 2(1000x^6 - 27y^3)$$

Factor out the GCF, 2.

$$= 2\left[(10x^2)^3 - (3y)^3\right]$$
$$= 2(10x^2 - 3y)(100x^4 + 30x^2y + 9y^2)$$

c. $375 - 81y^3$

_____

d. $250x^4 - 128xy^3$

_____

Review this example for Objective 4:

7. Factor each polynomial.

a. $q^3 + 8$

b. $250p^3 + 128r^3$

c. $1029x^4 + 648x$

The pattern to factor the sum of cubes is $x^3 + y^3 = (x + y)(x^2 - xy + y^2)$.

a. $q^3 + 8 = q^3 + 2^3$
$$= (q + 2)(q^2 - 2q + 4)$$

b. $250p^3 + 128r^3$
$$= 2(125p^3 + 64r^3)$$

Factor out the GCF, 2.

$$= 2\left[(5p)^3 + (4r)^3\right]$$
$$= 2(5p + 4r)(25p^2 - 20pr + 16r^2)$$

Now Try:

7. Factor each polynomial.

a. $27x^3 + 125$

_____

b. $8x^3 + 64b^3$

_____

c. $u^3v^6 + 216$

_____

c. $1029x^4 + 648x$

$= 3x\left(343x^3 + 216\right)$

Factor out the GCF, $3x$.

$= 3x\left[(7x)^3 + 6^3\right]$

$= 3x(7x + 6)\left(49x^2 - 42x + 36\right)$

Objective 1 Factor a difference of squares.
For extra help, see Examples 1–3 on pages 381–382 of your text, the Section Lecture video for Section 6.4, and Exercise Solution Clips 7, 15 and 17.

Factor.

1. $9x^2 - 1$

1. _____

2. $36z^2 - 121$

2. _____

3. $144x^2 - 25y^2$

3. _____

4. $s^4 - 16$

4. _____

Objective 2 Factor a perfect square trinomial.
For extra help, see Examples 4 and 5 on pages 383–384 of your text, the Section Lecture video for Section 6.4, and Exercise Solution Clips 37 and 39.

Factor.

5. $y^2 + 22y + 121$

5. _____

6. $16a^2 - 40ab + 25b^2$

6. _____

7. $16t^2 + 56t + 49$ 7. _____

Objective 3 Factor a difference of cubes.
For extra help, see Example 6 on page 385 of your text, the Section Lecture video for Section 6.4, and Exercise Solution Clip 55.

Factor.

8. $8r^3 - 27s^3$ 8. _____

9. $64a^3 - 343b^3$ 9. _____

10. $216m^3 - 125p^6$ 10. _____

11. $8a^3 - 125b^3$ 11. _____

Objective 4 Factor a sum of cubes.
For extra help, see Example 7 on page 386 of your text, the Section Lecture video for Section 6.4, and Exercise Solution Clip 57.

Factor.

12. $27r^3 + 8s^3$ 12. _____

13. $8a^3 + 64b^3$ 13. _____

14. $125p^3 + q^3$

14. _____

15. $192x^6 + 1029$

15. _____

Chapter 6 FACTORING AND APPLICATIONS

6.5 Solving Quadratic Equations by Factoring

Learning Objectives
1 Solve quadratic equations by factoring.
2 Solve other equations by factoring.

Key Terms

Use the vocabulary terms listed below to complete each statement in exercises 1–2.

quadratic equation   **standard form**

1. An equation written in the form $ax^2 + bx + c = 0$ is written in the
    _____ of a quadratic equation.

2. A _____ is an equation that can be written in the form
    $ax^2 + bx + c = 0$, for real numbers a, b, and c, with $a \neq 0$.

Guided Examples

Review these examples for Objective 1:

1. Solve each equation.

    a. $(3x + 7)(x - 4) = 0$

    b. $x(4x - 3) = 0$ /

    a. Since $(3x + 7)(x - 4) = 0$, the zero-factor property tells us that at least one of the factors equals 0. Therefore, either $3x + 7 = 0$ or $x - 4 = 0$.

    $3x + 7 = 0$ or $x - 4 = 0$
    $3x = -7$ $x = 4$
    $x = -\dfrac{7}{3}$

    Check each proposed solution in the original equation.

    $$(3x + 7)(x - 4) = 0$$
    $$\left[3\left(-\tfrac{7}{3}\right) + 7\right]\left(-\tfrac{7}{3} - 4\right) = 0$$
    $$0\left(-\tfrac{19}{3}\right) = 0$$
    $$0 = 0 \checkmark$$

Now Try:

1. Solve each equation.

    a. $(5x - 2)(3x + 1) = 0$

    _____

    b. $p(5p + 25) = 0$

    _____

$$(3x+7)(x-4)=0$$
$$[3(4)+7](4-4)=0$$
$$19(0)=0$$
$$0=0 \checkmark$$

The solution set is $\left\{-\frac{7}{3}, 0\right\}$.

b. $x(4x-3)=0$

$$x=0 \quad \text{or} \quad 4x-3=0$$
$$4x=3$$
$$x=\frac{3}{4}$$

Check these solutions by substituting each one into the original equation.

The solution set is $\left\{0, \frac{3}{4}\right\}$.

2. Solve each equation.

a. $z^2 = 7z - 12$

b. $x^2 - 2x = 63$

a. First, rewrite the equation in standard form.

$$z^2 - 7z + 12 = 0$$

Now factor $z^2 - 7z + 12$ and then apply the zero-factor property to solve the equation.

$$z^2 - 7z + 12 = 0$$
$$(z-3)(z-4)=0$$

$$z-3=0 \quad \text{or} \quad z-4=0$$
$$z=3 \qquad\qquad z=4$$

Be sure to check each proposed solution in the original equation.

The solution set is $\{3, 4\}$.

b. $$x^2 - 2x = 63$$
$$x^2 - 2x - 63 = 0$$
$$(x-9)(x+7)=0$$

$$x-9=0 \quad \text{or} \quad x+7=0$$
$$x=9 \qquad\qquad x=-7$$

Be sure to check each proposed solution in the original equation.

The solution set is $\{9, -7\}$.

2. Solve each equation.

a. $a^2 = -18 - 9a$

_____

b. $p^2 - 20 = -p$

_____

3. Solve $4x^2 - 24x = 64$.

$$4x^2 - 24x = 64$$

$4x^2 - 24x - 64 = 0$ Standard form

$4(x^2 - 6x - 16) = 0$ Factor out the GCF, 4.

$x^2 - 6x - 16 = 0$ Divide each side by 4.

$(x - 8)(x + 2) = 0$ Factor.

$x - 8 = 0$ or $x + 2 = 0$ Zero-factor property

$x = 8$ $x = -2$ Solve.

Be sure to check each proposed solution in the original equation.

The solution set is $\{8, -2\}$

4. Solve each equation.

    a. $36z^2 - 25 = 0$

    b. $25x^2 = 20x$

    c. $x(5x + 17) = 12$

    a. $36z^2 - 25 = 0$

       $(6z - 5)(6z + 5) = 0$ Factor the difference of squares.

       $6z - 5 = 0$ or $6z + 5 = 0$ Zero factor property

       $6z = 5$ $6z = -5$ Solve.

       $z = \dfrac{5}{6}$ $z = -\dfrac{5}{6}$

       Be sure to check each proposed solution in the original equation.

       The solution set is $\left\{\dfrac{5}{6}, -\dfrac{5}{6}\right\}$.

    b. $25x^2 = 20x$

       $25x^2 - 20x = 0$ Standard form

       $5x(5x - 4) = 0$ Factor.

       $5x = 0$ or $5x - 4 = 0$ Zero factor property

       $x = 0$ $5x = 4$ Solve.

                       $x = \dfrac{4}{5}$

       Be sure to check each proposed solution in the original equation.

       The solution set is $\left\{0, \dfrac{4}{5}\right\}$.

3. Solve $16x^2 = -4x + 6$.

_____

4. Solve each equation.

    a. $9a^2 = 49$

       _____

    b. $24r = -8r^2$

       _____

    c. $x(3x - 7) = 6$

       _____

c.
$$x(5x+17)=12$$
$$5x^2+17x=12 \quad \text{Distributive property}$$
$$5x^2+17x-12=0 \quad \text{Standard form}$$
$$(5x-3)(x+4)=0 \quad \text{Factor.}$$

$$5x-3=0 \quad \text{or} \quad x+4=0 \quad \begin{array}{l}\text{Zero factor} \\ \text{property}\end{array}$$
$$5x=3 \qquad\qquad x=-4 \quad \text{Solve.}$$
$$x=\frac{3}{5}$$

Be sure to check each proposed solution in the original equation.

The solution set is $\left\{\frac{3}{5}, 4\right\}$.

5. Solve $9x^2+12x+4=0$.

$$9x^2+12x+4=0$$
$$(3x+2)^2=0 \quad \text{Factor.}$$
$$(3x+2)(3x+2)=0 \quad a^2=a\cdot a$$

Because the two factors are identical, they both lead to the same solution.

$$3x+2=0 \quad \text{or} \quad 3x+2=0 \quad \begin{array}{l}\text{Zero factor} \\ \text{property}\end{array}$$
$$3x=-2 \quad \text{Solve.}$$
$$x=-\frac{2}{3}$$

Be sure to check the proposed solution in the original equation.

The solution set is $\left\{-\frac{2}{3}\right\}$.

5. Solve $36x^2-60x=-25$.

_____

Review these examples for Objective 2:

6. Solve each equation.

a. $4x^3-64x=0$

b. $(2x-3)\left(2x^2+11x+15\right)=0$

a.
$$4x^3-64x=0$$
$$4x\left(x^2-16\right)=0 \quad \text{Factor out } 4x.$$

$$4x(x+4)(x-4)=0 \quad \begin{array}{l}\text{Factor the difference} \\ \text{of squares.}\end{array}$$

Now apply the zero-factor property.
$$4x=0 \quad \text{or} \quad x+4=0 \quad \text{or} \quad x-4=0$$
$$x=0 \qquad\qquad x=-4 \qquad\qquad x=4$$

Check each solution to verify that the solution set is $\{0, -4, 4\}$.

Now Try:

6. Solve each equation.

a. $98a^3=162a$

b. $(x+4)\left(x^2+7x+10\right)=0$

_____

b. $(2x-3)\left(2x^2+11x+15\right)=0$

$(2x-3)(2x+5)(x+3)=0$ Factor.

Now apply the zero-factor property.

$2x-3=0$ or $2x+5=0$

$\quad 2x=3 \qquad\qquad 2x=-5$

$\quad\quad x=\dfrac{3}{2} \qquad\qquad x=-\dfrac{5}{2}$

$\qquad\qquad$ or $x-3=0$

$\qquad\qquad\qquad x=3$

Check each solution to verify that the solution set is $\left\{\dfrac{3}{2}, -\dfrac{5}{2}, 3\right\}$.

7. Solve $(3x+10)(x-1)=(x-4)^2-33$.

$(3x+10)(x-1)=(x-4)^2-33$

$3x^2+7x-10=x^2-8x+16-33$ Multiply.

$3x^2+7x-10=x^2-8x-17$

$\qquad\qquad\qquad\qquad$ Combine like terms.

$2x^2+15x+7=0$ Standard form

$(2x+1)(x+7)=0$ Factor.

$2x+1=0$ or $x+7=0$

$\quad 2x=-1 \qquad\qquad x=-7$

$\quad\quad x=-\dfrac{1}{2}$

Check each solution to verify that the solution set is $\left\{-\dfrac{1}{2}, -7\right\}$.

7. Solve

$$(4x+3)(2x+1)=(3x+2)^2-16.$$

_____

Objective 1 Solve quadratic equations by factoring.

For extra help, see Examples 1–5 on pages 392–396 of your text, the Section Lecture video for Section 6.5, and Exercise Solution Clips 13, 27, 37, and 51.

Solve each equation and check your solutions.

1. $9s^2-18s=0$

1. _____

2. $r^2+r=72$

2. _____

3. $6z^2 = -19z - 10$

3. _____

4. $8x^2 + 2x - 15 = 0$

4. _____

5. $24a^2 + 22a - 10 = 0$

5. _____

6. $5x^2 - 8x - 4 = 0$

6. _____

7. $r(6r + 11) = -3$

7. _____

8. $2k^2 + 3k - 9 = 0$

8. _____

9. $16x^2 = 25$

9. _____

10. $4q^2 - 5q - 6 = 0$

10. _____

Objective 2 Solve other equations by factoring.
For extra help, see Examples 6 and 7 on pages 396–397 of your text, the Section Lecture video for Section 6.5, and Exercise Solution Clips 57, and 71.

Solve each equation and check your solutions.

11. $z\left(4z^2 - 9\right) = 0$

11. _____

12. $2m^3 + m^2 - 6m = 0$

12. _____

13. $4x^3 - 9x = 8x^2 - 18$

13. _____

14. $(3x + 1)(4x + 1) = (7x + 1)(x - 1) - 4$

14. _____

15. $(y - 7)\left(2y^2 + 7y - 15\right) = 0$

15. _____

Chapter 6 FACTORING AND APPLICATIONS

6.6 Applications of Quadratic Equations

Learning Objectives
1 Solve problems involving geometric figures.
2 Solve problems involving consecutive integers.
3 Solve problems by applying the Pythagorean theorem.
4 Solve problems by using given quadratic models.

Key Terms

Use the vocabulary terms listed below to complete each statement in exercises 1–2.

hypotenuse   legs

1. In a right triangle, the sides that form the right angle are the _____.

2. The longest side of a right triangle is the _____.

Guided Examples

Review this example for Objective 1:
1. The top of a rectangular table has an area of 63 square feet. It has a length that is 2 feet more than the width. Find the width of the table top.

Step 1 Read the problem. We need to find the width of the table top.

Step 2 Assign a variable.
Let w = the width of the table top.
Then $w + 2$ = the length of the table top.

Step 3 Write an equation. The area of a rectangle is given by
A = length $\times$ width, so for this problem, we have $A = 63 = (w + 2)w$.

Step 4 Solve.
$$63 = (w + 2)w$$
$$63 = w^2 + 2w \quad \text{Multiply.}$$
$$w^2 + 2w - 63 = 0 \quad \text{Standard form}$$
$$(w + 9)(w - 7) = 0 \quad \text{Factor.}$$

Now Try:
1. The area of a triangle is 42 square centimeters. The base is 2 centimeters less than twice the height. Find the base and height of the triangle.

base _____

height _____

$w + 9 = 0$ or $w - 7 = 0$ Zero factor property

$w = -9$ $w = 7$ Solve.

Step 5 State the answer. The solutions are −9 and 7. Because a rectangle cannot have a side of negative length, we discard the solution −9. The width is 7 feet.

Step 6 Check. The width is 7 feet, so the length is 2 feet more or 9 feet. The area is 7(9) = 63 sq ft, as required.

Review this example for Objective 2:

2. Find two consecutive positive even integers whose product is six more than three times their sum.

Step 1 Read the problem. We are seeking two consecutive positive even integers.

Step 2 Assign a variable.
Let x = the first even integer.
Then $x + 2$ = the second even integer.

Step 3 Write an equation.

$$x(x + 2) = 6 + 3[x + (x + 2)]$$

Step 4 Solve.

$x(x + 2) = 6 + 3[x + (x + 2)]$

$x(x + 2) = 6 + 3(2x + 2)$ Combine terms.

$x^2 + 2x = 6 + 6x + 6$ Distributive property

$x^2 - 4x - 12 = 0$ Standard form

$(x - 6)(x + 2) = 0$ Factor.

$x - 6 = 0$ or $x + 2 = 0$ Zero factor property

$x = 6$ $x = -2$ Solve.

Step 5 State the answer. The solutions are −2 and 6. Because we are seeking a positive integer, we discard the solution −2. Thus, the first integer is 6 and the second integer is 8.

Step 6 Check. The product of the integers is 6(8) = 48. Three times their sum is 3(6 + 8) = 3(14) = 42. The product, 48, is six more than three times the sum, 42, as required.

Now Try:

2. Find three consecutive positive odd integers such that four times the sum of all three equals 13 more than the product of the smaller two.

Review this example for Objective 3:

3. Mindy and Lissa started biking from the same corner. Mindy biked east and Lissa biked south. When they were 26 miles apart, Lissa had biked 14 miles further than Mindy. Find the distance each biked.

Step 1 Read the problem. We are seeking the distance that each girl biked.

Step 2 Assign a variable.

Let x = the distance that Mindy biked. Then $x + 14$ = the distance that Lissa biked.

Now draw a labeled diagram.

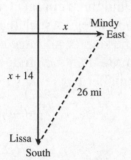

Step 3 Write an equation.

$$x^2 + (x+14)^2 = 26^2$$

Step 4 Solve.

$$x^2 + (x+14)^2 = 26^2$$

$x^2 + x^2 + 28x + 196 = 676$ Multiply.

$2x^2 + 28x - 480 = 0$ Combine terms; standard form

$2(x^2 + 14x - 240) = 0$ Factor out the GCF, 2.

$x^2 + 14x - 240 = 0$ Divide by 2.

$(x + 24)(x - 10) = 0$ Factor.

$x + 24 = 0$ or $x - 10 = 0$ Zero factor property

$x = -24$ $x = 10$ Solve.

Step 5 State the answer. The solutions are −24 and 10. Because distance cannot be negative, we discard the solution −24. Thus, Mindy biked 10 miles and Lissa biked 24 miles.

Step 6 Check. Since $10^2 + 24^2 = 26^2$, the answers are correct.

Now Try:

3. A ladder is leaning against a building. The distance from the bottom of the ladder to the building is 8 feet less than the length of the ladder. How high up the side of the building is the top of the ladder if that distance is 4 feet less than the length of the ladder?

_____

Review these examples for Objective 4:

4. A baseball is thrown with an upward velocity of 160 feet per second from the top of a building that is 500 feet tall. The height of the baseball after t seconds is modeled by the equation $h = -16t^2 + 160t + 500$. How long will it take for the projectile to reach the ground?

When the ball reaches the ground, $h = 0$, so substitute 0 for h in the equation and solve for t.

$$0 = -16t^2 + 160t + 500$$

$-16t^2 + 160t + 500 = 0$ Interchange sides.

$-4(4t^2 - 40t - 125) = 0$ Factor out -4.

$4t^2 - 40t - 125 = 0$ Divide by -4.

$(2t + 5)(2t - 25) = 0$ Factor.

$2t + 5 = 0$ or $2t - 25 = 0$ Zero factor property

$2t = -5$ $2t = 25$ Solve.

$t = -\frac{5}{2}$ $t = \frac{25}{2}$

Since time cannot be negative, we discard the solution $t = -\frac{5}{2}$. Thus, the baseball will reach the ground after $\frac{25}{2} = 12.5$ seconds.

5. Suppose that the population of a certain city t years after the end of 2000 is described (in thousands) by the quadratic model $p = 0.07t^2 + 4t + 110$, where $t = 0$ represents 2000, $t = 1$ represents 2001, etc., and p is the number of people in thousands. Use the model to estimate the population at the end of 2012.

Since $t = 0$ represents 2000, $t = 12$ represents 2012. Substitute 12 for t in the equation.

$p = 0.07(12)^2 + 4(12) + 110$

$\quad = 168.08$

According to the model, the population at the end of 2012 will be 168.08 thousand or 168,080.

Now Try:

4. Jeff threw a stone straight upward at 46 feet per second from a dock 6 feet above a lake. The height of the stone above the lake t seconds after it is thrown is given by $h = -16t^2 + 46t + 6$. How long will it take for the stone to reach a height of 39 feet?

_____

5. The unemployment rate in a certain community can be modeled by the equation $y = 0.025x^2 - 0.481x + 7.854$, where y is the unemployment rate (percent) and x is the month ($x = 1$ represents January, $x = 2$ represents February, etc.) Use the model to find the unemployment rate in August. Round your answer to the nearest tenth.

_____

Name: Date:
Instructor: Section:

Objective 1 Solve problems about geometric figures.
For extra help, see Example 1 on pages 400–401 of your text, the Section Lecture video for
Section 6.6, and Exercise Solution Clip 7.

Solve each equation. Check your answers to be sure they are reasonable.

1. A book is three times as long as it is wide. Find the
    length and width of the book in inches if its area is
    numerically 128 more than its perimeter.

    1. width _____

       length _____

2. Each side of one square is 1 meter less than twice the
    length of each side of a second square. If the difference
    between the areas of the two squares is 16 meters, find
    the lengths of the sides of the two rectangles.

    2. square 1 _____

       square 2 _____

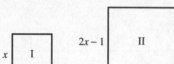

3. The length of a rectangle is three times its width. If the
    width were increased by 4 and the length remained the
    same, the resulting rectangle would have an area of 231
    square inches. Find the dimensions of the original
    rectangle.

    3. width _____

       length _____

4. The volume of a box is 192 cubic feet. If the length of
    the box is 8 feet and the width is 2 feet more than the
    height, find the height and width of the box.

    4. width _____

       height _____

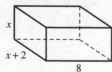

Objective 2 Solve problems involving consecutive integers.
For extra help, see Example 2 on page 402 of your text, the Section Lecture video for
Section 6.6, and Exercise Solution Clip 19.

Solve each problem.

5. If the square of the sum of two consecutive integers is
    reduced by twice their product, the result is 25. Find the
    integers.

    5. _____

6. Find all possible pairs of consecutive odd integers
    whose sum is equal to their product decreased by 47.

    6. _____

7. Find two consecutive positive even integers whose
    product is six more than three times their sum.

    7. _____

Objective 3 Solve problems by applying the Pythagorean theorem.
For extra help, see Example 3 on pages 402–403 of your text, the Section Lecture video for
Section 6.6, and Exercise Solution Clip 25.

Solve each problem.

8. A train and a car leave a station at
    the same time, the train traveling
    due north and the car traveling
    west. When they are 100 miles
    apart, the train has traveled 20
    miles farther than the car. Find the
    distance each has traveled.

    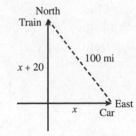

    8. car_____

       train _____

9. Two trains leave New York City at the same time. One train travels due north and the other travels due east. When they are 75 miles apart, the train going north has gone 30 miles less than twice the distance traveled by the train going east. Find the distance traveled by the train going north.

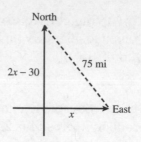

9. _____

10. Mark is standing directly beneath a kite attached to a string which Nina is holding, with her hand touching the ground. The height of the kite at that instant is 12 feet less than twice the distance between Mark and Nina. The length of the kite string is 12 feet more than the distance between Mark and Nina. Find the length of the kite string.

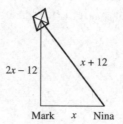

10. _____

11. Two ships left a dock at the same time. When they were 25 miles apart, the ship that sailed due south had gone 10 miles less than twice the distance traveled by the ship that sailed due west. Find the distance traveled by the ship that sailed due south.

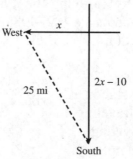

11. _____

Objective 4 Solve problems using given quadratic models.
For extra help, see Examples 4 and 5 on pages 404–405 of your text, the Section Lecture video for Section 6.6, and Exercise Solution Clip 35.

Solve each problem.

12. A ball is dropped from the roof of a 19.6 meter high building. Its height h (in meters) t seconds later is given by the equation $h = -4.9t^2 + 19.6$.

    (a) After how many seconds is the height 18.375 meters?

    (b) After how many seconds is the height 14.7 meters?

    (c) After how many seconds does the ball hit the ground?

12. **a.** _____

    b. _____

    c. _____

13. If an object is propelled upward from a height of 16 feet with an initial velocity of 48 feet per second, its height h (in feet) t seconds later is given by the equation $h = -16t^2 + 48t + 16$.

    (a) After how many seconds is the height 52 feet?

    (b) After how many seconds is the height 48 feet?

13. **a.** _____

    b. _____

14. A company determines that its daily revenue R (in dollars) for selling x items is modeled by the equation $R = x(150 - x)$. How many items must be sold for its revenue to be \$4400?

14.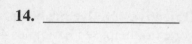

15. If a ball is batted at an angle of $35°$, the distance that the ball travels is given approximately by $D = 0.029v^2 + 0.021v - 1$, where v is the bat speed in miles per hour and D is the distance traveled in feet. Find the distance a batted ball will travel if the ball is batted with a velocity of 90 miles per hour. Round your answer to the nearest whole number.

15.

Chapter 7 RATIONAL EXPRESSIONS AND APPLICATIONS

7.1 The Fundamental Property of Rational Expressions

Learning Objectives
1 Find the numerical value of a rational expression.
2 Find the values of the variable for which a rational expression is undefined.
3 Write rational expressions in lowest terms.
4 Recognize equivalent forms of rational expressions.

Key Terms
Use the vocabulary terms listed below to complete each statement in exercises 1–3.

rational expression   **lowest terms**

fundamental property of rational expressions

1.   A _____ is the quotient of two polynomials with denominator not 0.

2.   The _____ states that $\frac{PK}{QK} = \frac{P}{Q}$ for $Q \neq 0$ and $K \neq 0$.

3.   A rational expression is in _____ if the greatest common factor of its numerator and denominator is 1.

Guided Examples

Review this example for Objective 1:
1. Find the value of the following rational expression for $x = -4$ and $x = 3$.

$$\frac{2x+5}{4x^2 - 25}$$

$x = -4$

$$\frac{2x+5}{4x^2-25} = \frac{2(-4)+5}{4(-4)^2 - 25} = \frac{-8+5}{4(16)-25}$$

$$= \frac{-3}{64-25} = \frac{-3}{39} = -\frac{1}{13}$$

$x = 3$

$$\frac{2x+5}{4x^2-25} = \frac{2(3)+5}{4(3)^2 - 25} = \frac{6+5}{4(9)-25}$$

$$= \frac{11}{36-25} = \frac{11}{11} = 1$$

Now Try:
1. Find the value of the following rational expression for
a. $x = 4$ and **b.** $x = -3$.

$$\frac{3x^2 - 2x}{2x}$$

a.   _____

b.   _____

Review this example for Objective 2:

2. Find any values of the variable for which each rational expression is undefined.

   a. $\dfrac{2z-3}{4z-7}$ **b.** $\dfrac{r}{r^2-3r+2}$

   c. $\dfrac{x-6}{x^2+1}$

   a. Set the denominator equal to 0, then solve.
   $$4z-7=0$$
   $$4z=7$$
   $$z=\frac{7}{4}$$
   The given expression is undefined for $\frac{7}{4}$, so $z \neq \frac{7}{4}$.

   b. $r^2-3r+2=0$ Set the denominator equal to 0.

   $(r-2)(r-1)=0$ Factor.

   $r-2=0$ or $r-1=0$ Zero-factor property

   $r=2$ $r=1$ Solve.

   The denominator equals 0 when $r=2$ or $r=1$, so $r \neq 2$, $r \neq 1$.

   c. $\dfrac{x-6}{x^2+1}$

   The denominator will not equal 0 for any value of x, so there are no values for which this expression is undefined.

Review these examples for Objective 3:

3. Write the rational expression in lowest terms.

   $$\frac{44x^4}{8x^6}$$

   Begin by factoring. Then group any factors common to the numerator and denominator, and use the fundamental property to divide out any common factors.

Now Try:

2. Find any values of the variable for which each rational expression is undefined.

   a. $\dfrac{9}{4x}$

   _____

   b. $\dfrac{z-3}{z^2+9}$

   _____

   c. $\dfrac{y+2}{y^4-16}$

   _____

Now Try:

3. Write the rational expression in lowest terms.

   $$\frac{15y^5}{35y^3}$$

   _____

$$\frac{44x^4}{8x^6} = \frac{2^2 \cdot 11 \cdot x \cdot x \cdot x \cdot x}{2^2 \cdot 2 \cdot x \cdot x \cdot x \cdot x \cdot x \cdot x}$$

$$= \frac{2^2 \cdot 11 \cdot (x \cdot x \cdot x \cdot x)}{2^2 \cdot 2 \cdot (x \cdot x \cdot x \cdot x) \cdot x \cdot x}$$

$$= \frac{11}{2x^2}$$

4. Write each rational expression in lowest terms.

   a. $\dfrac{2a+14}{7a+49}$

   b. $\dfrac{x^2-25}{x^2+10x+25}$

   a. $\dfrac{2a+14}{7a+49} = \dfrac{2(a+7)}{7(a+7)}$ Factor.

$$= \frac{2}{7}, \; a \ne -7$$

   b. $\dfrac{x^2-25}{x^2+10x+25} = \dfrac{(x-5)(x+5)}{(x+5)(x+5)}$ Factor.

$$\blacksquare \frac{x-5}{x+5} \qquad \begin{array}{l}\text{Fundamental}\\ \text{property}\end{array}$$

5. Write in lowest terms.

$$\frac{x^2-25}{25-x^2}$$

$$\frac{x^2-25}{25-x^2} = \frac{x^2-25}{(-1)\left(x^2-25\right)} = -1$$

6. Write each rational expression in lowest terms.

   a. $\dfrac{16-x^2}{2x-8}$

   b. $\dfrac{s-r}{s+r}$

4. Write each rational expression in lowest terms.

   a. $\dfrac{3r-12}{6r-24}$

   _____

   b. $\dfrac{3y^2-13y-10}{2y^2-9y-5}$

   _____

5. Write in lowest terms.

$$\frac{4r-s}{s-4r}$$

   _____

6. Write each rational expression in lowest terms.

   a. $\dfrac{4x^2-100}{20-4x}$

   _____

   b. $\dfrac{x-8}{x+8}$ _____

a. $\dfrac{16 - x^2}{2x - 8} = \dfrac{(4 - x)(4 + x)}{2(x - 4)}$

Factor the numerator
and denominator.

$= \dfrac{(4 - x)(4 + x)}{2(-1)(4 - x)}$

$x - 4 = (-1)(4 - x)$

$= \dfrac{4 + x}{-2} = -\dfrac{4 + x}{2}$

b. $\dfrac{s - r}{s + r}$

This expression is already in lowest
terms.

Review this example for Objective 4:

7. Write four equivalent forms of the rational
   expression.

   $-\dfrac{3x - 4}{x + 5}$

   Apply the negative sign to the numerator:

   $-\dfrac{3x - 4}{x + 5} = \dfrac{-(3x - 4)}{x + 5} = \dfrac{-3x + 4}{x + 5}$

   Apply the negative sign to the denominator:

   $-\dfrac{3x - 4}{x + 5} = \dfrac{3x - 4}{-(x + 5)} = \dfrac{3x - 4}{-x - 5}$

Now Try:

7. Write four equivalent forms of
   the rational expression.

   $-\dfrac{2x + 6}{x - 5}$

   _____

   _____

   _____

   _____

Objective 1 Find the numerical value of a rational expression.
For extra help, see Example 1 on page 420 of your text, the Section Lecture video for
Section 7.1, and Exercise Solution Clip 1.

Find the numerical value of each expression when (a) x = 4 and (b) x = −3.

1. $\dfrac{(-4 - x)^2}{x + 3}$

   1. a. _____

      b. _____

2. $\dfrac{2x^2 - 4}{x^2 - 2}$

   2. a. _____

      b. _____

3. $\dfrac{2x-5}{2+x-x^2}$

3. a. _____

b. _____

4. $\dfrac{-2x^2}{2x^2-x+2}$

4. a. _____

b. _____

Objective 2 Find the values of the variable for which a rational expression is undefined.
For extra help, see Example 2 on page 421 of your text, the Section Lecture video for Section 7.1, and Exercise Solution Clip 21.

Find all values for which the following expressions are undefined.

5. $\dfrac{x-4}{4x^2-16x}$

5. _____

6. $\dfrac{4x+3}{x^2+x-12}$

6. _____

7. $\dfrac{5x}{x^2-25}$

7. _____

8. $\dfrac{x+5}{x^3+9x^2+18x}$

8. _____

Objective 3 Write rational expressions in lowest terms.

For extra help, see Examples 3–6 on pages 422–424 of your text, the Section Lecture video for Section 7.1, and Exercise Solution Clips 31, 37, 85, and 87.

Write each rational expression in lowest terms.

9. $\dfrac{16r^2 - 4s^2}{8r + 4s}$

9. _____

10. $\dfrac{b^2 + 2b - 15}{b^2 + 9b + 20}$

10. _____

11. $\dfrac{2x^2 + 9x - 5}{2x^2 + 3x - 2}$

11. _____

12. $\dfrac{6r^2 - 7rs - 10s^2}{r^2 + 3rs - 10s^2}$

12. _____

Objective 4 Recognize equivalent forms of rational expressions.
For extra help, see Example 7 on page 425 of your text, the Section Lecture video for Section 7.1, and Exercise Solution Clip 95.

Write four equivalent forms of the following rational expressions.

13. $-\dfrac{2x-3}{x+2}$

13. _____

_____

_____

_____

14. $-\dfrac{4x+1}{5x-3}$

14. _____

_____

_____

_____

15. $-\dfrac{2x-1}{3x+5}$

15. _____

_____

_____

_____

Chapter 7 RATIONAL EXPRESSIONS AND APPLICATIONS

7.2 Multiplying and Dividing Rational Expressions

Learning Objectives
1   Multiply rational expressions.
2   Divide rational expressions.

Key Terms
Use the vocabulary terms listed below to complete each statement in exercises 1–2.

   rational expression   rational function

1.   A _____ is a function that is defined by

rational expression in the form $f(x) = \dfrac{P(x)}{Q(x)}$, where $Q(x) \neq 0$.

2.   The quotient of two polynomials with denominator not 0 is called a

_____.

Guided Examples

Review these examples for Objective 1:
1.   Multiply. Write the answer in lowest terms.

$$\frac{5r^3t^2}{6} \cdot \frac{3t^3}{25r^5}$$

Multiply the numerators and denominators, then use the fundamental property to write the rational expression in lowest terms. Alternatively, divide out common factors in the numerator and denominator before multiplying the rational expressions.

$$\frac{5r^3t^2}{6} \cdot \frac{3t^3}{25r^5} = \frac{5 \cdot 3 \cdot r^3 \cdot t^2 \cdot t^3}{6 \cdot 25 \cdot r^5} = \frac{t^5}{10r^2}$$

Alternate solution:

$$\frac{5r^3t^2}{6} \cdot \frac{3t^3}{25r^5} = \frac{5r^3t^2}{2 \cdot 3} \cdot \frac{3t^3}{5 \cdot 5r^3 \cdot r^2} = \frac{t^5}{10r^2}$$

Now Try:
1.   Multiply. Write the answer in lowest terms.

$$\frac{8m^4n^3}{3} \cdot \frac{5}{4mn^2}$$

_____

2. Multiply. Write the answer in lowest terms.

$$\frac{x+4}{4x^2} \cdot \frac{8x^5}{4(x+4)^2} = \frac{8(x+4)x^5}{4 \cdot 4x^2 (x+4)^2}$$

$$= \frac{x^3}{2(x+4)}$$

2. Multiply. Write the answer in lowest terms.

$$\frac{6y^3}{9(y+4)} \cdot \frac{4(y+4)^2}{9y}$$

_____

3. Multiply. Write the answer in lowest terms.

$$\frac{3x^2 - 12}{x^2 - x - 6} \cdot \frac{x^2 - 6x + 9}{2x - 4}$$

Begin by factoring. Then divide out common factors in the numerator and denominator before multiplying the rational expressions.

$$\frac{3x^2 - 12}{x^2 - x - 6} \cdot \frac{x^2 - 6x + 9}{2x - 4}$$

$$= \frac{3(x-2)(x+2)}{(x-3)(x+2)} \cdot \frac{(x-3)^2}{2(x-2)}$$

$$= \frac{3(x-3)}{2}$$

3. Multiply. Write the answer in lowest terms.

$$\frac{3x+12}{6x-30} \cdot \frac{x^2 - x - 20}{x^2 - 16}$$

_____

Review these examples for Objective 2:

4. Divide. Write the answer in lowest terms.

$$\frac{3x+7}{8x^6} \div \frac{3x+7}{2x^3}$$

Multiply the dividend by the reciprocal of the divisor.

$$\frac{3x+7}{8x^6} \div \frac{3x+7}{2x^3} = \frac{3x+7}{8x^6} \cdot \frac{2x^3}{3x+7}$$

$$= \frac{1}{4x^3}$$

Now Try:

4. Divide. Write the answer in lowest terms.

$$\frac{15y^5}{4(y-2)} \div \frac{3y^2}{6(y-2)}$$

_____

5. Divide. Write the answer in lowest terms.

$$\frac{3x}{x-4} \div \frac{x^2}{(x-4)^2}$$

$$\frac{3x}{x-4} \div \frac{x^2}{(x-4)^2} = \frac{3x}{x-4} \cdot \frac{(x-4)^2}{x^2}$$

$$= \frac{3(x-4)}{x}$$

6. Divide. Write the answer in lowest terms.

$$\frac{2x+2y}{8z} \div \frac{x^2\left(x^2-y^2\right)}{24}$$

$$\frac{2x+2y}{8z} \div \frac{x^2\left(x^2-y^2\right)}{24}$$

$$= \frac{2x+2y}{8z} \cdot \frac{24}{x^2\left(x^2-y^2\right)} \quad \text{Multiply by the reciprocal.}$$

$$= \frac{2(x+y)}{8z} \cdot \frac{3 \cdot 8}{x^2(x+y)(x-y)} \quad \text{Factor.}$$

$$= \frac{6}{z(x-y)} \quad \text{Multiply.}$$

7. Divide. Write the answer in lowest terms.

$$\frac{4m-12}{2m+10} \div \frac{9-m^2}{m^2-25}$$

$$\frac{4m-12}{2m+10} \div \frac{9-m^2}{m^2-25}$$

$$= \frac{4m-12}{2m+10} \cdot \frac{m^2-25}{9-m^2} \quad \text{Multiply by the reciprocal.}$$

$$= \frac{4(m-3)}{2(m+5)} \cdot \frac{(m+5)(m-5)}{(3+m)(3-m)} \quad \text{Factor.}$$

$$= \frac{-1 \cdot 2(m-5)}{3+m} \quad \begin{array}{l}\text{Divide out common}\\\text{factors, then multiply.}\end{array}$$

$$\text{Recall that } \frac{m-3}{3-m} = -1.$$

$$= -\frac{2(m-5)}{m+3}$$

5. Divide. Write the answer in lowest terms.

$$\frac{6z^3}{9zw} \div \frac{z^7}{21zw^2}$$

_____

6. Divide. Write the answer in lowest terms.

$$\frac{4m-12}{2m+10} \div \frac{m^2-9}{m^2-25}$$

_____

7. Divide. Write the answer in lowest terms.

$$\frac{6a(a+3)}{a-1} \div \frac{a^2(a+3)}{1-a^2}$$

_____

Name: _____ Date: _____

Instructor: _____ Section: _____

Objective 1 Multiply rational expressions.

For extra help, see Examples 1–3 on pages 429–430 of your text, the Section Lecture video for Section 7.2, and Exercise Solution Clips 3, 9, and 41.

Multiply. Write each answer in lowest terms.

1. $\dfrac{12-4z}{6} \cdot \dfrac{9}{4z-12}$

1. _____

2. $\dfrac{4r+4p}{8z^2} \cdot \dfrac{36z^6}{r^2+rp}$

2. _____

3. $\dfrac{x^2-4}{2x^2-2} \cdot \dfrac{x-x^2}{2x^2+4x}$

3. _____

4. $\dfrac{x^2+x-12}{x^2+7x+10} \cdot \dfrac{x^2+3x-10}{x^2+2x-8}$

4. _____

5. $\dfrac{a^2-3a+2}{a^2-1} \cdot \dfrac{a^2+2a-3}{a^2+a-6}$

5. _____

6. $\dfrac{3m^2-m-10}{2m^2-7m-4} \cdot \dfrac{4m^2-1}{6m^2+7m-5}$

6. _____

Name: Date:
Instructor: Section:

7. $\dfrac{2x^2+5x-12}{x^2-2x-24}\cdot\dfrac{x^2-9x+18}{9-4x^2}$

7. _____

Objective 2 Divide rational expressions.
For extra help, see Examples 4–7 on pages 431–432 of your text, the Section Lecture video for Section 7.2, and Exercise Solution Clips 17, 19, 33, and 35.

Divide. Write each answer in lowest terms.

8. $\dfrac{m^2+2mn+n^2}{m^2+m}\div\dfrac{m^2-n^2}{m^2-1}$

8. _____

9. $\dfrac{9a^2-1}{9a^2-6a+1}\div\dfrac{3a^2-11a-4}{12a^2+5a-3}$

9. _____

10. $\dfrac{12k^2-5k-3}{9k^2-1}\div\dfrac{16k^2-9}{12k^2+13k+3}$

10. _____

11. $\dfrac{2z^2-11z-21}{z^2-5z-14}\div\dfrac{4z^2-9}{z^2-6z-16}$

11. _____

12. $\dfrac{4(b-3)(b+2)}{b^2+3b+2} \div \dfrac{b^2-6b+9}{b^2+4b+4}$

12. _____

13. $\dfrac{27-3k^2}{3k^2+8k-3} \div \dfrac{k^2-6k+9}{6k^2-19k+3}$

13. _____

14. $\dfrac{2y^2-21y-11}{2-8y^2} \div \dfrac{y^2-12y+11}{4y^2+14y-8}$

14. _____

15. $\dfrac{2k^2+5k-12}{2k^2+k-3} \div \dfrac{k^2+8k+16}{2k^2+11k+12}$

15. _____

Chapter 7 RATIONAL EXPRESSIONS AND APPLICATIONS

7.3 Least Common Denominators

Learning Objectives
1   Find the least common denominator for a group of fractions.
2   Write equivalent rational expressions.

Key Terms

Use the vocabulary terms listed below to complete each statement in exercises 1–2.

least common denominator   equivalent expressions

1. $\dfrac{24x-8}{9x^2-1}$ and $\dfrac{8}{3x+1}$ are _____.

2. The _____ is the simplest expression that is divisible by all of the denominators in all of the expressions.

Guided Examples

Review these examples for Objective 1:

1. Find the least common denominator (LCD) for each pair of fractions.

    a. $\dfrac{5}{12}, \dfrac{7}{40}$ **b.** $\dfrac{3}{16x}, \dfrac{5}{36x}$

    Write each denominator in factored form with numerical coefficients in prime factored form.

a. $12 = 2^2 \cdot 3$	**b.** $16x = 2^4 \cdot x$
$40 = 2^3 \cdot 5$	$36x = 2^2 \cdot 3^2 \cdot x$

    Find the LCD by taking each different factor the greatest number of time it appears as a factor in any of the denominators.

$LCD = 2^3 \cdot 3 \cdot 5$	$LCD = 2^4 \cdot 3^2 \cdot x$
$= 120$	$= 144x$

Now Try:

1. Find the LCD for each pair of fractions.

    a. $\dfrac{1}{24}, \dfrac{5}{27}$ _____

    b. $\dfrac{1}{48y}, \dfrac{5}{32y}$ _____

2. Find the LCD for $\dfrac{5}{9x^3}$ and $\dfrac{7}{15x^2y}$.

Factor each denominator.

$9x^3 = 3^2 \cdot x^3$

$15x^2y = 3 \cdot 5 \cdot x^2 \cdot y$

Find the LCD by taking each different factor the greatest number of time it appears as a factor in any of the denominators.

$\text{LCD} = 3^2 \cdot 5 \cdot x^3 \cdot y = 45x^3y$

3. Find the LCD for the fractions in each list.

a. $\dfrac{3}{5a+10}, \dfrac{4}{a^2+2a}$

b. $\dfrac{2x}{x^2+5x+6}, \dfrac{3x}{x^2+4x+4}, \dfrac{5x}{x^2-9}$

c. $\dfrac{7}{x-y}, \dfrac{9}{y^2-x^2}$

Write each denominator in factored form with numerical coefficients in prime factored form. Then use each different factor the greatest number of times it appears.

a. $5a+10 = 5(a+2)$

$a^2+2a = a(a+2)$

$\text{LCD} = 5a(a+2)$

b. $x^2+5x+6 = (x+2)(x+3)$

$x^2+4x+4 = (x+2)^2$

$x^2-9 = (x+3)(x-3)$

$\text{LCD} = (x+2)^2(x+3)(x-3)$

c. $y^2-x^2 = (y-x)(y+x)$

Since $y-x$ and $x-y$ are opposites of each other, we can multiply one of the factors by -1 to obtain the other.

$\text{LCD} = -(x-y)(y+x)$ or

$\quad\quad\quad -(y-x)(y+x)$

2. Find the LCD for $\dfrac{5}{8x^2y^4}$ and

$\dfrac{5}{12x^3y^2}$.

_____

3. Find the LCD for the fractions in each list.

a. $\dfrac{4}{5r-25}, \dfrac{7}{15r^3}$

_____

b. $\dfrac{3p+2}{p^2-9}, \dfrac{2p+7}{p^2-p-12}$,

$\dfrac{4p-3}{p^2-7p+12}$

_____

c. $\dfrac{4}{a^2-b^2}, \dfrac{8}{b^2-a^2}$

_____

Review these examples for Objective 2:

4. Rewrite each rational expression with the indicated denominator.

   a. $\dfrac{3}{16} = \dfrac{?}{48}$ **b.** $\dfrac{7x}{32} = \dfrac{?}{96x}$

   For each example, first factor the denominator on the right. Then compare the denominator on the left with the one on the right to decide what factors are missing.

   a. $\dfrac{3}{16} = \dfrac{?}{3 \cdot 16}$

   A factor of 3 is missing, so multiply $\frac{3}{16}$ by $\frac{3}{3}$.

   $\dfrac{3}{16} = \dfrac{3}{16} \cdot \dfrac{3}{3} = \dfrac{9}{48}$

   b. $\dfrac{7x}{32} = \dfrac{?}{3 \cdot 32x}$

   Factors of 3 and x are missing, so multiply $\frac{7x}{32}$ by $\frac{3x}{3x}$.

   $\dfrac{7x}{32} = \dfrac{7x}{32} \cdot \dfrac{3x}{3x}$

   $= \dfrac{21x^2}{96x}$

5. Rewrite each rational expression with the indicated denominator.

   a. $\dfrac{6}{9m-2} = \dfrac{?}{45m-10}$

   b. $\dfrac{8z}{3z+3} = \dfrac{?}{12z^2+15z+3}$

   a. Factor the denominator on the right: $45m - 10 = 5(9m - 2)$. The missing factor is 5, so multiply the fraction on the left by $\frac{5}{5}$.

   $\dfrac{6}{9m-2} = \dfrac{6}{9m-2} \cdot \dfrac{5}{5} = \dfrac{30}{45m-10}$

Now Try:

4. Rewrite each rational expression with the indicated denominator.

   a. $\dfrac{5}{6} = \dfrac{?}{18}$ _____

   b. $\dfrac{4r}{7} = \dfrac{?}{28r}$ _____

5. Rewrite each rational expression with the indicated denominator.

   a. $\dfrac{11a+1}{2a-6} = \dfrac{?}{8a-24}$

   _____

   b. $\dfrac{2}{7p-35} = \dfrac{?}{14p^3-70p^2}$

   _____

b. Factor the denominator in each rational expression:

$$3z + 3 = 3(z+1)$$

$$12z^2 + 15z + 3 = 3\left(4z^2 + 5z + 1\right)$$
$$= 3(4z+1)(z+1)$$

The factor $4z + 1$ are missing, so multiply $\frac{8z}{3z+3}$ by $\frac{4z+1}{4z+1}$.

$$\frac{8z}{3z+3} \cdot \frac{4z+1}{4z+1} = \frac{8z(4z+1)}{(3z+3)(4z+1)}$$

$$= \frac{32z^2 + 8z}{12z^2 + 15z + 3}$$

Objective 1 Find the least common denominator for a group of fractions.

For extra help, see Examples 1–3 on pages 435–436 of your text, the Section Lecture video for Section 7.3, and Exercise Solution Clips 5, 13, and 19.

Find the least common denominator for each list of rational expressions.

1. $\dfrac{9}{14}, \dfrac{17}{20}, \dfrac{7}{15}$

1. _____

2. $\dfrac{5}{36ab^2}, \dfrac{7}{27a^2b}$

2. _____

3. $\dfrac{15}{7t - 28}, \dfrac{21}{24 - 6t}$

3. _____

4. $\dfrac{3m}{2m^2 + 9m - 5}, \dfrac{4}{m^2 + 5m}$

4. _____

5. $\dfrac{3z+1}{z^4+2z^3-8z^2}, \dfrac{5z+2}{z^3+8z^2+16z}$

5. _____

6. $\dfrac{11q-3}{2q^2-q-10}, \dfrac{21-q}{2q^2-9q+10}$

6. _____

7. $\dfrac{m+2}{m^3-2m^2}, \dfrac{3-m}{m^2+5m-14}$

7. _____

Objective 2 Write equivalent rational expressions.
For extra help, see Examples 4 and 5 on pages 437–438 of your text, the Section Lecture video for Section 7.3, and Exercise Solution Clips 51 and 57.

Rewrite each rational expression with the indicated denominator. Give the numerator of the new fraction.

8. $\dfrac{7m}{8n} = \dfrac{?}{24n^6}$

8. _____

9. $\dfrac{-3y}{4y+12} = \dfrac{?}{4(y+3)^2}$

9. _____

10. $\dfrac{5}{2r+8} = \dfrac{?}{2(r+4)(r^2+2r-8)}$

10. _____

11. $\dfrac{9}{y^2-4} = \dfrac{?}{(y+2)^2(y-2)}$

11. _____

12. $\dfrac{3}{5r-10} = \dfrac{?}{50r^2 - 100r}$

12. _____

13. $\dfrac{3}{k^2+3k} = \dfrac{?}{k^3+10k^2+21k}$

13. _____

14. $\dfrac{3x+1}{x^2-4} = \dfrac{?}{2x^2-8}$

14. _____

15. $\dfrac{2p^2}{p-9} = \dfrac{?}{p^2-81}$

15. _____

Chapter 7 RATIONAL EXPRESSIONS AND APPLICATIONS

7.4 Adding and Subtracting Rational Expressions

Learning Objectives
1. Add rational expressions having the same denominator.
2. Add rational expressions having different denominators.
3. Subtract rational expressions.

Key Terms

Use the vocabulary terms listed below to complete each statement in exercises 1–3.

same denominator   **different denominators**   **parentheses**

1. To add rational expressions with _____, rewrite each expression with the LCD of the rational expressions and add.

2. When subtracting rational expressions, be sure to use _____ after the subtraction sign.

3. To add rational expressions with the _____, add the numerators and keep the same denominator.

Guided Examples

Review this example for Objective 1:
1. Add. Write each answer in lowest terms.

   a. $\dfrac{2}{3x^2} + \dfrac{4}{3x^2}$ **b.** $\dfrac{-4x+3}{x-7} + \dfrac{2x+11}{x-7}$

   For each problem, the denominators are the same, so the sum is found by adding the two numerators and keeping the same denominator.

   a. $\dfrac{2}{3x^2} + \dfrac{4}{3x^2} = \dfrac{2+4}{3x^2} = \dfrac{6}{3x^2} = \dfrac{2}{x^2}$

   b. $\dfrac{-4x+3}{x-7} + \dfrac{2x+11}{x-7} = \dfrac{-2x+14}{x-7}$

   $\qquad\qquad = \dfrac{-2(x-7)}{x-7}$ Factor.

   $\qquad\qquad = -2$

Now Try:
1. Add. Write each answer in lowest terms.

   a. $\dfrac{4}{3w^2} + \dfrac{7}{3w^2}$ _____

   b. $\dfrac{b}{b^2-4} + \dfrac{2}{b^2-4}$

   _____

Review these examples for Objective 2:

2. Add. Write each answer in lowest terms.

    a. $\dfrac{5}{18} + \dfrac{7}{24}$ **b.** $\dfrac{6}{5r} + \dfrac{3}{4r}$

    For each problem, first find the LCD. Next, rewrite each rational expression as a fraction with the LCD as the denominator. Finally, add the numerators. The LCD is the denominator. Write the answer in lowest terms.

    a. $18 = 2 \cdot 3^2$

    $24 = 2^3 \cdot 3$

    $LCD = 2^3 \cdot 3^2 = 72$

    $\dfrac{5}{18} = \dfrac{5 \cdot 4}{18 \cdot 4} = \dfrac{20}{72}$

    $\dfrac{7}{24} = \dfrac{7 \cdot 3}{24 \cdot 3} = \dfrac{21}{72}$

    $\dfrac{5}{18} + \dfrac{7}{24} = \dfrac{20}{72} + \dfrac{21}{72} = \dfrac{41}{72}$

    b. $5r = 5 \cdot r$

    $4r = 2^2 \cdot r$

    $LCD = 2^2 \cdot 5 \cdot r = 20r$

    $\dfrac{6}{5r} = \dfrac{6 \cdot 4}{5r \cdot 4} = \dfrac{24}{20r}$

    $\dfrac{3}{4r} = \dfrac{3 \cdot 5}{4r \cdot 5} = \dfrac{15}{20r}$

    $\dfrac{6}{5r} + \dfrac{3}{4r} = \dfrac{24}{20r} + \dfrac{15}{20r} = \dfrac{39}{20r}$

3. Add. Write the answer in lowest terms.

    $\dfrac{4}{x^2 - 4} + \dfrac{3}{x - 2}$

    Since the denominators are different, find the LCD, and then write each rational expression with the LCD as the denominator.

Now Try:

2. Add. Write each answer in lowest terms.

    a. $\dfrac{8}{15} + \dfrac{3}{20}$ _____

    b. $\dfrac{2}{9z} + \dfrac{4}{5z}$ _____

3. Add. Write the answer in lowest terms.

    $\dfrac{1}{m - 3} + \dfrac{m - 2}{m^2 - 9}$

    _____

309

$x^2 - 4 = (x-2)(x+2)$

$x - 2$ is prime.

$LCD = (x-2)(x+2)$

$$\frac{4}{x^2-4} + \frac{3}{x-2}$$

$$= \frac{4}{(x-2)(x+2)} + \frac{3(x+2)}{(x-2)(x+2)}$$

$$= \frac{4}{(x-2)(x+2)} + \frac{3x+6}{(x-2)(x+2)}$$

$$= \frac{3x+10}{(x-2)(x+2)}$$

4. Add. Write the answer in lowest terms.

$$\frac{4z}{z^2+6z+8} + \frac{2z-1}{z^2+5z+6}$$

$$\frac{4z}{z^2+6z+8} + \frac{2z-1}{z^2+5z+6}$$

$$= \frac{4z}{(z+4)(z+2)} + \frac{2z-1}{(z+3)(z+2)}$$

Factor the denominators.

$$= \frac{4z(z+3)}{(z+4)(z+2)(z+3)} + \frac{(2z-1)(z+4)}{(z+4)(z+3)(z+2)}$$

The LCD is $(z+4)(z+2)(z+3)$.

$$= \frac{4z^2+12z}{(z+4)(z+2)(z+3)} + \frac{2z^2+7z-4}{(z+4)(z+3)(z+2)}$$

Multiply in the numerators.

$$= \frac{6z^2+19z-4}{(z+4)(z+2)(z+3)} \qquad \text{Add the numerators.}$$

The numerator cannot be factored, so the expression is in lowest terms.

4. Add. Write the answer in lowest terms.

$$\frac{2y+9}{y^2+6y+8} + \frac{y+3}{y^2+2y-8}$$

5. Add. Write the answer in lowest terms.

    $$\frac{4x}{x-4}+\frac{2}{4-x}$$

    The denominators are opposites, so multiply one of the fractions by $\frac{-1}{-1}$.

    $$\frac{4x}{x-4}+\frac{2}{4-x}=\frac{4x}{x-4}+\frac{2(-1)}{(4-x)(-1)}$$

    $$=\frac{4x}{x-4}+\frac{-2}{x-4}$$

    $$=\frac{4x-2}{x-4}$$

5. Add. Write the answer in lowest terms.

    $$\frac{7r-s}{s-4r}+\frac{2s}{4r-s}$$

    _____

Review these examples for Objective 3:

6. Subtract. Write the answer in lowest terms.

    $$\frac{16}{x-8}-\frac{x+8}{x-8}$$

    $$\frac{16}{x-8}-\frac{x+8}{x-8}=\frac{16-(x+8)}{x-8}$$

    Subtract numerators.
    Keep the same denominator.

    $$=\frac{16-x-8}{x-8}$$

    Distributive property

    $$=\frac{8-x}{x-8}=-1$$

    Combine like terms. $\frac{x-y}{y-x}=-1$

Now Try:

6. Subtract. Write the answer in lowest terms.

    $$\frac{6p}{p-4}-\frac{p+20}{p-4}$$

    _____

7. Subtract. Write the answer in lowest terms.

$$\frac{8}{x+8} - \frac{4}{x}$$

$$\frac{8}{x+8} - \frac{4}{x} = \frac{8x}{x(x+8)} - \frac{4(x+8)}{x(x+8)}$$

The LCD is $x(x+8)$.

$$= \frac{8x}{x(x+8)} - \frac{4x+32}{x(x+8)}$$

Distributive property

$$= \frac{8x-4x-32}{x(x-8)} \quad \text{Subtract.}$$

$$= \frac{4x-32}{x(x-8)} \quad \text{Combine like terms.}$$

$$= \frac{4(x-8)}{x(x-8)} = \frac{4}{x} \quad \begin{array}{l}\text{Factor the numerator.} \\ \text{Lowest terms}\end{array}$$

8. Subtract. Write the answer in lowest terms.

$$\frac{6p}{p-4} - \frac{p+20}{4-p}$$

The denominators are opposites, so choose one or the other as the common denominator. We choose $p - 4$.

$$\frac{6p}{p-4} - \frac{p+20}{4-p} = \frac{6p}{p-4} - \frac{(p+20)(-1)}{(4-p)(-1)}$$

Multiply $\frac{p+20}{4-p}$ by $\frac{-1}{-1}$.

$$= \frac{6p}{p-4} - \frac{-p-20}{p-4}$$

$$= \frac{6p-(-p-20)}{p-4}$$

Subtract numerators.

$$= \frac{6p+p+20}{p-4}$$

Distributive property

$$= \frac{7p+20}{p-4} \quad \text{Combine like terms.}$$

7. Subtract. Write the answer in lowest terms.

$$\frac{5}{s-6} - \frac{5}{s}$$

_____

8. Subtract. Write the answer in lowest terms.

$$\frac{16}{x-8} - \frac{x+8}{8-x}$$

_____

Name: 　　　　　　　　　　　　　　Date:

Instructor: 　　　　　　　　　　　Section:

9. Subtract. Write the answer in lowest terms.

$$\frac{m}{m^2-4}-\frac{1-m}{m^2+4m+4}$$

$$\frac{m}{m^2-4}-\frac{1-m}{m^2+4m+4}$$

$$=\frac{m}{(m-2)(m+2)}-\frac{(1-m)}{(m+2)^2} \quad \text{Factor the denominators.}$$

$$=\frac{m(m+2)}{(m-2)(m+2)^2}-\frac{(1-m)(m-2)}{(m-2)(m+2)^2}$$

The LCD is $(m-2)(m+2)^2$.

$$=\frac{m(m+2)-(1-m)(m-2)}{(m-2)(m+2)^2}$$

Subtract numerators.

$$=\frac{m^2+2m-\left(m-2-m^2+2m\right)}{(m-2)(m+2)^2} \quad \text{Multiply.}$$

$$=\frac{m^2+2m-m+2+m^2-2m}{(m-2)(m+2)^2}$$

Distributive property

$$=\frac{2m^2-m+2}{(m-2)(m+2)^2} \quad \text{Combine like terms.}$$

9. Subtract. Write the answer in lowest terms.

$$\frac{4z}{z^2+6z+8}-\frac{2z-1}{z^2+5z+6}$$

_____

Objective 1 Add rational expressions having the same denominator.

For extra help, see Example 1 on page 441 of your text, the Section Lecture video for Section 7.4, and Exercise Solution Clip 9.

Find each sum. Write each answer in lowest terms.

1. $\dfrac{3m+4}{2m^2-7m-15}+\dfrac{m+2}{2m^2-7m-15}$

1. _____

2. $\dfrac{4m}{m^2+3m+2}+\dfrac{8}{m^2+3m+2}$

2. _____

3. $\dfrac{2y-5}{2y^2-5y-3}+\dfrac{2-y}{2y^2-5y-3}$

3. _____

Objective 2 Add rational expressions having different denominators.
For extra help, see Examples 2–5 on pages 441–443 of your text, the Section Lecture video for Section 7.4, and Exercise Solution Clips 25, 41, 43, and 51.

Find each sum. Write each answer in lowest terms.

4. $\dfrac{3z-2}{5z+20}+\dfrac{2z+1}{3z+12}$

4. _____

5. $\dfrac{-4}{h+1}+\dfrac{2h}{1-h^2}$

5. _____

6. $\dfrac{7}{x-5}+\dfrac{4}{x+5}$

6. _____

7. $\dfrac{2s+3}{3s^2-14s+8}+\dfrac{4s+5}{2s^2-5s-12}$

7. _____

8. $\dfrac{5p-2}{2p^2+9p+9}+\dfrac{p+7}{6p^2+13p+6}$

8. _____

9. $\dfrac{1-3x}{4x^2-1}+\dfrac{3x-5}{2x^2+5x+2}$

9. _____

Objective 3 Subtract rational expressions.

For extra help, see Examples 6–9 on pages 444–445 of your text, the Section Lecture video for Section 7.4, and Exercise Solution Clip 15, 39, 53, and 63.

Find each difference. Write each answer in lowest terms.

10. $\dfrac{2x}{x^2+3x-10}-\dfrac{x+2}{x^2+3x-10}$

10. _____

11. $\dfrac{7}{x+4}-\dfrac{5}{3x+12}$

11. _____

12. $\dfrac{6}{x-y}-\dfrac{4+y}{y-x}$

12. _____

13. $\dfrac{6}{2q^2+5q+2}-\dfrac{5}{2q^2-3q-2}$

13. _____

14. $\dfrac{4y}{y^2+4y+3}-\dfrac{3y+1}{y^2-y-2}$

14. _____

15. $\dfrac{4x-1}{2x^2+5x-3}-\dfrac{x+3}{6x^2+x-2}$

15. _____

Chapter 7 RATIONAL EXPRESSIONS AND APPLICATIONS

7.5 Complex Fractions

Learning Objectives
1   Simplify a complex fraction by writing it as a division problem (Method 1).
2   Simplify a complex fraction by multiplying numerator and denominator by the least common denominator (Method 2).

Key Terms
Use the vocabulary terms listed below to complete each statement in exercises 1–2.

complex fraction   LCD

1.   A _____ is a rational expression with one or more fractions in the numerator, or denominator, or both.

2.   To simplify a complex fraction, multiply the numerator and denominator by the _____ of all the fractions within the complex fraction.

Guided Examples

Review these examples for Objective 1:
1.   Simplify each complex fraction.

a. $\dfrac{\dfrac{5}{6}+\dfrac{1}{2}}{\dfrac{3}{4}+\dfrac{2}{9}}$

b. $\dfrac{3+\dfrac{4}{a}}{\dfrac{1}{a}+\dfrac{2}{5}}$

For each problem, write each numerator as a single fraction and write each denominator as a single fraction. Then write the equivalent complex fraction as a division. Use the rule for division and the fundamental property.

a. $\dfrac{5}{6}+\dfrac{1}{2}=\dfrac{5}{6}+\dfrac{1(3)}{2(3)}=\dfrac{5}{6}+\dfrac{3}{6}=\dfrac{8}{6}=\dfrac{4}{3}$

$\dfrac{3}{4}+\dfrac{2}{9}=\dfrac{3(9)}{4(9)}+\dfrac{2(4)}{9(4)}=\dfrac{27}{36}+\dfrac{8}{36}=\dfrac{35}{36}$

$\dfrac{\dfrac{5}{6}+\dfrac{1}{2}}{\dfrac{3}{4}+\dfrac{2}{9}}=\dfrac{\dfrac{4}{3}}{\dfrac{35}{36}}=\dfrac{4}{3}\div\dfrac{35}{36}=\dfrac{4}{3}\cdot\dfrac{36}{35}=\dfrac{48}{35}$

Now Try:
1.   Simplify each complex fraction.

a. $\dfrac{\dfrac{1}{2}-\dfrac{1}{3}}{\dfrac{1}{3}+\dfrac{1}{6}}$ _____

b. $\dfrac{3x+\dfrac{1}{3}}{\dfrac{2}{3}+\dfrac{3}{x}}$ _____

b. $3 + \dfrac{4}{a} = \dfrac{3a}{a} + \dfrac{4}{a} = \dfrac{3a+4}{a}$

$\dfrac{1}{a} + \dfrac{2}{5} = \dfrac{1(5)}{5a} + \dfrac{2a}{5a} = \dfrac{2a+5}{5a}$

$\dfrac{3 + \dfrac{4}{a}}{\dfrac{1}{a} + \dfrac{2}{5}} = \dfrac{\dfrac{3a+4}{a}}{\dfrac{2a+5}{5a}} = \dfrac{3a+4}{a} \div \dfrac{2a+5}{5a}$

$ = \dfrac{3a+4}{a} \cdot \dfrac{5a}{2a+5} = \dfrac{5(3a+4)}{2a+5}$

2. Simplify the complex fraction.

$$\dfrac{\dfrac{rs}{3r^2}}{\dfrac{s^2}{3}}$$

The numerator and denominator are single fractions, so use the definition of division and then the fundamental property.

$\dfrac{\dfrac{rs}{3r^2}}{\dfrac{s^2}{3}} = \dfrac{rs}{3r^2} \div \dfrac{s^2}{3} = \dfrac{rs}{3r^2} \cdot \dfrac{3}{s^2} = \dfrac{1}{rs}$

3. Simplify the complex fraction.

$$\dfrac{\dfrac{4}{x-4} + 3}{5 - \dfrac{3}{x-4}}$$

Start by writing the numerator and the denominator as single fractions.

$\dfrac{\dfrac{4}{x-4} + 3}{5 - \dfrac{3}{x-4}} = \dfrac{\dfrac{4}{x-4} + \dfrac{3(x-4)}{x-4}}{\dfrac{5(x-4)}{x-4} - \dfrac{3}{x-4}} = \dfrac{\dfrac{4 + 3(x-4)}{x-4}}{\dfrac{5(x-4)-3}{x-4}}$

$ = \dfrac{\dfrac{4 + 3x - 12}{x-4}}{\dfrac{5x - 20 - 3}{x-4}} = \dfrac{\dfrac{3x-8}{x-4}}{\dfrac{5x-23}{x-4}}$

2. Simplify the complex fraction.

$$\dfrac{\dfrac{49m^3}{18n^5}}{\dfrac{21m}{27n^2}}$$

_____

3. Simplify the complex fraction.

$$\dfrac{\dfrac{2}{a+2} - 4}{\dfrac{1}{a+2} - 3}$$

_____

Now multiply the numerator by the reciprocal of the denominator.

$$\frac{\dfrac{3x-8}{x-4}}{\dfrac{5x-23}{x-4}} = \frac{3x-8}{x-4} \cdot \frac{x-4}{5x-23} = \frac{3x-8}{5x-23}$$

Review these examples for Objective 2:

4. Simplify each complex fraction.

a. $\dfrac{\dfrac{7}{6}-\dfrac{1}{2}}{\dfrac{1}{4}+\dfrac{5}{9}}$ b. $\dfrac{3-\dfrac{6}{a}}{\dfrac{1}{a}+\dfrac{3a}{5}}$

For each problem, find the LCD for all denominators in the complex fraction. Then multiply the numerator and denominator of the complex fraction by the LCD. Use the distributive property, if necessary.

a. The LCD for 6, 2, 4, and 9 is 36.

$$\frac{\dfrac{7}{6}-\dfrac{1}{2}}{\dfrac{1}{4}+\dfrac{5}{9}} = \frac{36\left(\dfrac{7}{6}-\dfrac{1}{2}\right)}{36\left(\dfrac{1}{4}+\dfrac{5}{9}\right)} = \frac{42-18}{9+20} = \frac{24}{29}$$

b. The LCD for a and 5 is $5a$.

$$\frac{3-\dfrac{5}{a}}{\dfrac{1}{a}+\dfrac{3a}{5}} = \frac{5a\left(3-\dfrac{5}{a}\right)}{5a\left(\dfrac{1}{a}+\dfrac{3a}{5}\right)} = \frac{15a-25}{3a^2+5}$$

Now Try:

4. Simplify each complex fraction.

a. $\dfrac{\dfrac{3}{4}+\dfrac{2}{5}}{\dfrac{7}{10}-\dfrac{2}{5}}$ _____

b. $\dfrac{\dfrac{2}{y}+6}{\dfrac{1}{12}+\dfrac{y}{4}}$ _____

5. Simplify the complex fraction.

$$\dfrac{\dfrac{1}{3x^3}+\dfrac{3}{4x^2}}{\dfrac{1}{2x^2}-\dfrac{1}{x}}$$

The LCD for all the denominators is $12x^3$.

$$\dfrac{12x^3\left(\dfrac{1}{3x^3}+\dfrac{3}{4x^2}\right)}{12x^3\left(\dfrac{1}{2x^2}-\dfrac{1}{x}\right)}=\dfrac{4+9x}{6x-12x^2}$$

$$=\dfrac{9x+4}{-6x(x+2)}$$

6. Simplify each complex fraction.

a. $\dfrac{\dfrac{6}{k+1}-\dfrac{5}{k-3}}{\dfrac{3}{k-3}+\dfrac{2}{k+1}}$ **b.** $\dfrac{\dfrac{15}{10k+10}}{\dfrac{5}{3k+3}}$

a. $\dfrac{\dfrac{6}{k+1}-\dfrac{5}{k-3}}{\dfrac{3}{k-3}+\dfrac{2}{k+1}}$

There are sums and differences in the numerator and denominator. Use Method 2. The LCD is $(k+1)(k-3)$.

$$\dfrac{\dfrac{6}{k+1}-\dfrac{5}{k-3}}{\dfrac{3}{k-3}+\dfrac{2}{k+1}}$$

$$=\dfrac{(k+1)(k-3)\left(\dfrac{6}{k+1}-\dfrac{5}{k-3}\right)}{(k+1)(k-3)\left(\dfrac{3}{k-3}+\dfrac{2}{k+1}\right)}$$

$$=\dfrac{6(k-3)-5(k+1)}{3(k+1)+2(k-3)}$$

$$=\dfrac{6k-18-5k-5}{3k+3+2k-6}=\dfrac{k-23}{5k-3}$$

5. Simplify the complex fraction.

$$\dfrac{\dfrac{2}{3x^2}-\dfrac{3}{2x^3}}{\dfrac{5}{2x}+\dfrac{1}{4x^2}}\quad\underline{\hspace{3cm}}$$

6. Simplify each complex fraction.

a. $\dfrac{\dfrac{4}{t+2}+\dfrac{5}{t-1}}{\dfrac{3}{t+2}-\dfrac{1}{t-1}}$

$$\underline{\hspace{4cm}}$$

b. $\dfrac{\dfrac{4}{x+4}}{\dfrac{3}{x^2-16}}\quad\underline{\hspace{2cm}}$

b. $\dfrac{\dfrac{15}{10k+10}}{\dfrac{5}{3k+3}}$

This is a quotient of two rational expressions. Use Method 1.

$$\dfrac{\dfrac{15}{10k+10}}{\dfrac{5}{3k+3}} = \dfrac{15}{10k+10} \cdot \dfrac{3k+3}{5}$$

$$= \dfrac{15}{10(k+1)} \cdot \dfrac{3(k+1)}{5}$$

$$= \dfrac{9}{10}$$

Objective 1 Simplify a complex fraction by writing it as a division problem (Method 1).
For extra help, see Examples 1–3 on pages 449–450 of your text, the Section Lecture video for Section 7.5, and Exercise Solution Clip 7.

Simplify each complex fraction by writing it as a division problem.

1. $\dfrac{\dfrac{3}{4}-\dfrac{1}{2}}{\dfrac{1}{4}+\dfrac{5}{8}}$

1. _____

2. $\dfrac{\dfrac{r-s}{12}}{\dfrac{r^2-s^2}{6}}$

2. _____

3. $\dfrac{2-\dfrac{1}{y-2}}{3-\dfrac{2}{y-2}}$

3. _____

4. $\dfrac{\dfrac{a+2}{a-2}}{\dfrac{1}{a^2-4}}$

4. _____

5. $\dfrac{\dfrac{3}{w-4}-\dfrac{3}{w+4}}{\dfrac{1}{w+4}+\dfrac{1}{w^2-16}}$

5. _____

6. $\dfrac{\dfrac{5}{rs^2}-\dfrac{2}{rs}}{\dfrac{3}{rs}-\dfrac{4}{r^2s}}$

6. _____

7. $\dfrac{\dfrac{3a+4}{a}}{\dfrac{1}{a}+\dfrac{2}{5}}$

7. _____

Objective 2 Simplify a complex fraction by multiplying by the least common denominator (Method 2).

For extra help, see Examples 4–6 on pages 451–452 of your text, the Section Lecture video for Section 7.5, and Exercise Solution Clips 23, 27, and 31.

Simplify each complex fraction by multiplying by the least common denominator.

8. $\dfrac{\dfrac{1}{2}+\dfrac{3}{8}}{\dfrac{3}{4}-\dfrac{9}{8}}$

8. _____

9. $\dfrac{2x - y^2}{x + \dfrac{y^2}{x}}$

9. _____

10. $\dfrac{r + \dfrac{3}{r}}{\dfrac{5}{r} + rt}$

10. _____

11. $\dfrac{\dfrac{x-2}{x+2}}{\dfrac{x}{x-2}}$

11. _____

12. $\dfrac{2s + \dfrac{3}{s}}{\dfrac{1}{s} - 3s}$

12. _____

13. $\dfrac{\dfrac{1}{h} - 4}{\dfrac{1}{2} + 2h}$

13. _____

14. $\dfrac{\dfrac{4}{x} - \dfrac{1}{2}}{\dfrac{5}{x} + \dfrac{1}{3}}$

14. _____

15. $\dfrac{\dfrac{1}{m-1} + 4}{\dfrac{2}{m-1} - 4}$

15. _____

Chapter 7 RATIONAL EXPRESSIONS AND APPLICATIONS

7.6 Solving Equations with Rational Expressions

Learning Objectives
1 Distinguish between operations with rational expressions and equations with terms that are rational expressions.
2 Solve equations with rational expressions.
3 Solve a formula for a specified variable.

Key Terms
Use the vocabulary terms listed below to complete each statement in exercises 1–2.

proposed solution **extraneous solution (value)**

1. A proposed solution that is not an actual solution of the original equation is called an
_____.

2. A value of the variable that appears to be a solution after both sides of a rational equation are multiplied by a variable expression is called a _____.

Guided Examples

Review this example for Objective 1:
1. Identify each of the following as an *expression* or an *equation*. Then simplify the expression or solve the equation.

a. $\dfrac{9}{8}p - \dfrac{1}{8}p = \dfrac{1}{2}$ b. $\dfrac{2}{3}t - \dfrac{5}{3}t$

a. This is an *equation* because there is an equals symbol.

$$\frac{9}{8}p - \frac{1}{8}p = \frac{1}{2}$$

$$8\left(\frac{9}{8}p - \frac{1}{8}p\right) = 8\left(\frac{1}{2}\right) \quad \begin{array}{l}\text{Multiply by the}\\ \text{LCD, 8, to clear}\\ \text{the fractions.}\end{array}$$

$$9p - p = 4 \quad \text{Distributive property}$$

$$8p = 4$$

$$p = \frac{4}{8} = \frac{1}{2}$$

Now Try:
1. Identify each of the following as an *expression* or an *equation*. Then simplify the expression or solve the equation.

a. $\dfrac{3x}{5} - \dfrac{4x}{3} = \dfrac{22}{15}$

_____

b. $\dfrac{4x}{5} - \dfrac{5x}{10}$ _____

b. $\dfrac{2}{3}t - \dfrac{5}{4}t$

This is an *expression* because there is no equals sign. The LCD is $3 \cdot 4 = 12$.

$$\frac{2}{3}t - \frac{5}{4}t = \frac{2 \cdot 4}{3 \cdot 4}t - \frac{5 \cdot 3}{4 \cdot 3}t$$

$$= \frac{8}{12}t - \frac{15}{12}t$$

$$= -\frac{7}{12}t$$

Review these examples for Objective 2:

2. Solve, and check the solution.

$$\frac{3x - 2}{8} - \frac{x}{3} = \frac{1}{12}$$

$$\frac{3x - 2}{8} - \frac{x}{3} = \frac{1}{12}$$

$$24\left(\frac{3x - 2}{8} - \frac{x}{3}\right) = 24\left(\frac{1}{12}\right) \quad \text{Multiply by the LCD, 24.}$$

$$24\left(\frac{3x - 2}{8}\right) - 24\left(\frac{x}{3}\right) = 2 \quad \text{Distributive property}$$

$$3(3x - 2) - 8x = 2 \quad \text{Multiply.}$$

$$9x - 6 - 8x = 2 \quad \text{Distributive property}$$

$$x - 6 = 2 \quad \text{Combine like terms.}$$

$$x = 8 \quad \text{Add 6.}$$

Check:

$$\frac{3x - 2}{8} - \frac{x}{3} = \frac{1}{12}$$

$$\frac{3(8) - 2}{8} - \frac{8}{3} \overset{?}{=} \frac{1}{12} \qquad x = 8$$

$$\frac{22}{8} - \frac{8}{3} \overset{?}{=} \frac{1}{12}$$

$$\frac{22 \cdot 3}{8 \cdot 3} - \frac{8 \cdot 8}{3 \cdot 8} \overset{?}{=} \frac{1 \cdot 2}{12 \cdot 2} \qquad \text{The LCD is 24.}$$

$$\frac{66}{24} - \frac{64}{24} \overset{?}{=} \frac{2}{24}$$

$$\frac{2}{24} = \frac{2}{24} \quad \checkmark$$

The solution set is $\{8\}$.

Now Try:

2. Solve, and check the solution.

$$\frac{2m - 1}{3} - \frac{3m}{4} = \frac{5}{6} \underline{\qquad\qquad}$$

3. Solve, and check the proposed solution.

$$\frac{x-2}{x-7} - 6 = \frac{5}{x-7}$$

Note that x cannot equal 7 because 7 causes both denominators to equal 0. Begin by multiplying both sides by the LCD, $x - 7$.

$$\frac{x-2}{x-7} - 6 = \frac{5}{x-7}$$

$$(x-7)\left(\frac{x-2}{x-7} - 6\right) = (x-7)\left(\frac{5}{x-7}\right)$$

$$(x-7)\left(\frac{x-2}{x-7}\right) + (x-7)(-6)$$

$$= (x-7)\left(\frac{5}{x-7}\right)$$

Distributive property

$$x - 2 - 6x + 42 = 5 \quad \text{Multiply.}$$
$$-5x + 40 = 5 \quad \text{Combine like terms.}$$
$$-5x = -35 \quad \text{Subtract 40.}$$
$$x = 7 \quad \text{Divide by } -5.$$

As noted, x cannot equal 7 since replacing x with 7 in the original equation causes the denominators to equal 0. Thus, 7 must be rejected as a solution, and the solution set is $\emptyset$.

4. Solve, and check the proposed solution.

$$\frac{1}{x^2 - 2x} = \frac{3}{x^2 - 4}$$

Factor the denominators to find the LCD.

$$x^2 - 2x = x(x-2)$$
$$x^2 - 4 = (x-2)(x+2)$$
$$\text{LCD} = x(x-2)(x+2)$$

Note that x cannot equal -2 or 2 because either of those solutions will cause a denominator to equal 0.

$$x(x-2)(x+2)\frac{1}{x(x-2)}$$

$$= x(x-2)(x+2)\frac{3}{(x-2)(x+2)}$$

Multiply by the LCD.

3. Solve, and check the proposed solution.

$$\frac{x}{x-3} = \frac{3}{x-3} - 1$$

_____

4. Solve, and check the proposed solution.

$$\frac{6}{x^2 + 2x} = \frac{3}{3x + 6}$$

_____

$x + 2 = 3x$ Divide out the common factors.
$\quad 2 = 2x$ Subtract x.
$\quad 1 = x$ Divide by 2.

Check:

$$\frac{1}{x^2 - 2x} = \frac{3}{x^2 - 4}$$

$$\frac{1}{1^2 - 2(1)} \overset{?}{=} \frac{3}{1^2 - 4} \qquad x = 1$$

$$\frac{1}{-1} \overset{?}{=} \frac{3}{-3}$$

$$-1 = -1 \quad \checkmark$$

The solution set is $\{1\}$.

5. Solve, and check the proposed solution.

$$\frac{r+5}{r^2 - 16} = \frac{3}{r-4} + \frac{1}{r+4}$$

The LCD for all the denominators is
$(r-4)(r+4)$. Note that 4 and -4 cannot be
solutions of the equation.

$$(r-4)(r+4)\frac{r+5}{(r-4)(r+4)}$$

$$= (r-4)(r+4)\left(\frac{3}{r-4} + \frac{1}{r+4}\right)$$

Multiply by the LCD.

$$(r-4)(r+4)\frac{r+5}{(r-4)(r+4)}$$

$$= (r-4)(r+4)\left(\frac{3}{r-4}\right)$$

$$+ (r-4)(r+4)\left(\frac{1}{r+4}\right)$$

Distributive property

$$r + 5 = 3(r+4) + 1(r-4)$$

Divide out the common factors.

$$r + 5 = 3r + 12 + r - 4$$

Distributive property

$r + 5 = 4r + 8$ Combine like terms.

$\quad -3r = 3$ Subtract $4r$; subtract 5.

$\quad\quad r = -1$ Divide by -3.

Check that $\{-1\}$ is the solution set.

5. Solve, and check the proposed
   solution.

$$\frac{2}{x+1} - \frac{4}{x-1} = \frac{8}{x^2 - 1}$$

_____

Name:

Instructor:

Date:

Section:

6. Solve, and check the proposed solution.

$$\frac{3}{x^2-1}-\frac{3x}{2x+2}=-\frac{2}{3}$$

Factor the denominators on the left and then determine the LCD.

$x^2-1=(x+1)(x-1)$

$2x+2=2(x+1)$

$\text{LCD}=3\cdot 2(x+1)(x-1)=6(x+1)(x-1)$

Multiply both sides by the LCD to clear the fractions.

$$6(x+1)(x-1)\left(\frac{3}{(x+1)(x-1)}-\frac{3x}{2(x+1)}\right)$$
$$=6(x+1)(x-1)\left(-\frac{2}{3}\right)$$

$$6(x+1)(x-1)\frac{3}{(x+1)(x-1)}$$
$$+6(x+1)(x-1)\left(-\frac{3x}{2(x+1)}\right)$$
$$=6(x+1)(x-1)\left(-\frac{2}{3}\right)$$

Distributive property

$6\cdot 3+3(x-1)(-3x)=-4(x+1)(x-1)$

Divide out common factors.

$18-9x^2+9x=-4x^2+4$ Multiply.

$-5x^2+9x+14=0$

Subtract; combine terms.

$5x^2-9x-14=0$ Multiply by -1.

$(5x-14)(x+1)=0$ Factor.

$5x-14=0$ or $x+1=0$ Zero-factor

$x=\dfrac{14}{5}$ $\qquad x=-1$ property

Since -1 make a denominator equal 0, -1 is *not* a solution. Check that $\left\{\frac{14}{5}\right\}$ is a solution.

6. Solve, and check the proposed solution.

$$\frac{9}{n}-\frac{8}{n+1}=10$$

_____

329

7. Solve, and check the proposed solution.

$$\frac{9}{x^2 - x - 12} = -\frac{3}{4 - x} - \frac{x}{x + 3}$$

Determine the LCD.

$$x^2 - x - 12 = (x - 4)(x + 3)$$
$$4 - x = -(x - 4)$$
$$\text{LCD} = -(x - 4)(x + 3)$$

Multiply both sides by the LCD to clear the fractions.

$$-(x - 4)(x + 3)\frac{9}{(x - 4)(x + 3)}$$

$$= -(x - 4)(x + 3)\left(-\frac{3}{-(x - 4)} - \frac{x}{x + 3}\right)$$

$$-(x - 4)(x + 3)\frac{9}{(x - 4)(x + 3)}$$

$$= -(x - 4)(x + 3)\left(-\frac{3}{-(x - 4)}\right)$$

$$-(x - 4)(x + 3)\left(-\frac{x}{x + 3}\right)$$

Distributive property

$$-9 = -3(x + 3) - (x - 4)(-x)$$

Divide out common factors.

$$-9 = -3x - 9 + x^2 - 4x$$

Distributive property

$$x^2 - 7x = 0 \qquad \text{Combine terms.}$$
$$x(x - 7) = 0 \qquad \text{Factor.}$$
$$x = 0 \quad \text{or} \quad x - 7 = 0 \quad \text{Zero-factor property}$$
$$x = 7$$

Neither solution makes a denominator equal 0. Check that $\{0, 7\}$ is the solution set.

7. Solve, and check the proposed solution.

$$\frac{1}{z^2 + 5z + 6} + \frac{1}{12(z + 2)} = \frac{-1}{z^2 - 2z - 8}$$

_____

Review these examples for Objective 3:

8. Solve each formula for the specified variable.

   a. $P = \dfrac{r+c}{n}$ for c.

   b. $h = \dfrac{2A}{B+b}$ for B

   a. We must isolate c.

   $$P = \dfrac{r+c}{n}$$
   $$Pn = r + c \qquad \text{Multiply by } n.$$
   $$Pn - r = c \qquad \text{Subtract } r.$$

   b. $\qquad h = \dfrac{2A}{B+b}$

   $$(B+b)h = 2A \qquad \text{Clear the fraction.}$$
   $$Bh + bh = 2A \qquad \text{Distributive property}$$
   $$Bh = 2A - bh \qquad \text{Subtract } bh.$$
   $$B = \dfrac{2A - bh}{h} \qquad \text{Divide by } h.$$

9. Solve the following formula for R_1.

   $$\dfrac{1}{R} = \dfrac{1}{R_1} + \dfrac{1}{R_2}$$

   Start by multiplying by the LCD, RR_1R_2.

   $$RR_1R_2\left(\dfrac{1}{R}\right) = RR_1R_2\left(\dfrac{1}{R_1} + \dfrac{1}{R_2}\right)$$

   $$RR_1R_2\left(\dfrac{1}{R}\right) = RR_1R_2\left(\dfrac{1}{R_1}\right) + RR_1R_2\left(\dfrac{1}{R_2}\right)$$

   $$\qquad\qquad\qquad\qquad \text{Distributive property}$$
   $$R_1R_2 = RR_2 + RR_1 \quad \text{Simplify.}$$

   $$R_1R_2 - RR_1 = RR_2 \quad \begin{array}{l}\text{Subtract } RR_1 \text{ to get} \\ \text{both terms with } R_1 \\ \text{on same side.}\end{array}$$

   $$R_1(R_2 - R) = RR_2 \quad \text{Factor.}$$

   $$R_1 = \dfrac{RR_2}{R_2 - R} \quad \text{Divide by } R_2 - R.$$

Now Try:

8. Solve each formula for the specified variable.

   a. $F = \dfrac{Gm_1m_2}{d^2}$ for G

   _____

   b. $A = \dfrac{R_1R_2}{R_1 + R_r}$ for R_r

   _____

9. Solve $\dfrac{1}{R} = \dfrac{1}{R_1} + \dfrac{1}{R_2}$ for R.

   _____

Name: _____   Date: _____

Instructor: _____   Section: _____

Objective 1 Distinguish between operations with rational expressions and equations with terms that are rational expressions.

For extra help, see Example 1 on page 456 of your text, the Section Lecture video for Section 7.6, and Exercise Solution Clip 1.

Identify each of the following as an expression *or an* equation. *Then simplify the expression or solve the equation.*

1. $\dfrac{2x}{3} + \dfrac{2x}{5} = \dfrac{64}{15}$

1. _____

2. $\dfrac{2x}{5} + \dfrac{7x}{3}$

2. _____

Objective 2 Solve equations with rational expressions.

For extra help, see Examples 2–7 on pages 457–461 of your text, the Section Lecture video for Section 7.6, and Exercise Solution Clips 33, 43, 49, 59, 69, and 71.

Solve each equation and check your answers.

3. $\dfrac{p}{p-2} = \dfrac{2}{p-2} + 1$

3. _____

4. $\dfrac{4}{5x} + \dfrac{3}{2x} = \dfrac{23}{50}$

4. _____

5. $\dfrac{4}{n+2} - \dfrac{2}{n} = \dfrac{1}{6}$

5. _____

6. $\dfrac{x}{3x+16} = \dfrac{4}{x}$

6. _____

7. $\dfrac{8}{2m+4} + \dfrac{2}{3m+6} = \dfrac{7}{9}$

7. _____

8. $\dfrac{2}{z-1} + \dfrac{3}{z+1} - \dfrac{17}{24} = 0$

8. _____

9. $\dfrac{2}{m-3} + \dfrac{12}{9-m^2} = \dfrac{3}{m+3}$

9. _____

10. $\dfrac{-16}{n^2-8n+12} = \dfrac{3}{n-2} + \dfrac{n}{n-6}$

10. _____

11. $\dfrac{3x}{x+1} - \dfrac{6}{x^2-1} = 2$

11. _____

Objective 3 Solve a formula for a specified variable.
For extra help, see Examples 8 and 9 on pages 461–462 of your text, the Section Lecture video for Section 7.6, and Exercise Solution Clip 83.

Solve each formula for the specified variable.

12. $S = \dfrac{a_1}{1-r}$ for r **12.** _____

13. $\dfrac{V_1 P_1}{T_1} = \dfrac{V_2 P_2}{T_2}$ for T_1 **13.** _____

14. $\dfrac{1}{f} = \dfrac{1}{d_0} + \dfrac{1}{d_1}$ for f **14.** _____

15. $A = \dfrac{1}{2} h\left(b_1 + b_2\right)$ for b_2 **15.** _____

Chapter 7 RATIONAL EXPRESSIONS AND APPLICATIONS

7.7 Applications of Rational Expressions

Learning Objectives
1   Solve problems about numbers.
2   Solve problems about distance, rate, and time.
3   Solve problems about work.

Key Terms
Use the vocabulary terms listed below to complete each statement in exercises 1–4.

   read   **check**   **rate of work**   **smaller**

1.   If t is the amount of time needed to complete a job, then $\frac{1}{t}$ is the

     _____, or the amount of the job completed for one unit of time.

2.   The last step when solving a problem is to _____ the solution.

3.   The first step when solving a problem is to _____ the problem.

4.   The rate traveling upstream is always _____ than the rate traveling
     downstream.

Guided Examples

Review this example for Objective 1:
1.   If the same number is added to the
     numerator and denominator of the fraction
     $\frac{5}{9}$, the resulting fraction is equivalent to $\frac{2}{3}$.
     Find the number.

     Step 1 **Read** the problem carefully. We are
     asked to find a number.

     Step 2 **Assign a variable.**
     Let x = the number added to the numerator
     and the denominator.

     Step 3 **Write an equation.**
     $$\frac{5+x}{9+x} = \frac{2}{3}$$

Now Try:
1.   If a certain number is added to
     the numerator and twice that
     number is subtracted from the
     denominator of the fraction $\frac{3}{5}$,
     the result is equal to 5. Find the
     number.

     _____

Step 4 **Solve.**

$$\frac{5+x}{9+x} = \frac{2}{3}$$

$$3(9+x)\left(\frac{5+x}{9+x}\right) = 3(9+x)\left(\frac{2}{3}\right) \quad \text{Multiply by the LCD, } 3(9+x).$$

$$3(5+x) = (9+x)2 \quad \text{Divide out common factors.}$$

$$15+3x = 18+2x \quad \text{Distributive property}$$

$$x = 3 \quad \text{Subtract } 2x, 15.$$

Step 5 **State the answer.**
The number is 3.

Step 6 **Check** the solution in the words of the original problem. If 3 is added to both the numerator and the denominator of $\frac{5}{9}$, the result is $\frac{8}{12} = \frac{2}{3}$, as required.

Review this example for Objective 2:

2. Mark can row 5 miles per hour in still water. It takes him as long to row 4 miles upstream as 16 miles downstream. How fast is the current?

Step 1 **Read** the problem carefully. We must find the speed of the current.

Step 2 **Assign a variable.**
Let x = the speed of the current.
We can summarize the given information in a table. Use the formula $d = rt$.

	r	d	t
upstream	$5-x$	4	$\frac{4}{5-x}$
downstream	$5+x$	16	$\frac{16}{5+x}$

Step 3 **Write an equation.**
$$\frac{4}{5-x} = \frac{16}{5+x}$$

Now Try:

2. A boat goes 6 miles per hour in still water. It takes as long to go 40 miles upstream as 80 miles downstream. Find the speed of the current.

_____

Step 4 **Solve.**

$$\frac{4}{5-x} = \frac{16}{5+x}$$

$$(5-x)(5+x)\left(\frac{4}{5-x}\right)$$

$$= (5-x)(5+x)\left(\frac{16}{5+x}\right)$$

Multiply by the LCD,
$(5-x)(5+x)$.

$4(5+x) = (5-x)16$ Divide out
common factors.

$20 + 4x = 80 - 16x$ Distributive
property

$20x = 60$ Add $16x$;
subtract 20.

$x = 3$ Divide by 20.

Step 5 **State the answer.**

The speed of the current is 3 miles per hour.

Step 6 **Check** the solution in the words of the original problem. Mike can row upstream $5 - 3 = 2$ miles per hour. It will take him $\frac{4}{2} = 2$ hours to row four miles upstream. Mike can row downstream $5 + 3 = 8$ miles per hour. It will takes him $\frac{16}{8} = 2$ hours to row 16 miles downstream.

The time upstream equals the time downstream, as required.

Review this example for Objective 3:

3. Kelly can clean the house in 6 hours, but it takes Linda 4 hours. How long would it take them to clean the house if they worked together?

   Step 1 **Read the problem carefully.** We must find how long will it take them to clean the house if they work together.

   Step 2 **Assign a variable.**

   Let x = the time working together.

   We can summarize the given information in a table.

Now Try:

3. Nina can wash the walls in a certain room in 2 hours and Mark can wash these walls in 5 hours. How long would it take them to complete the task if they work together?

   _____

   337

	Rate	Time working together	Part of the job
Kelly	$\dfrac{1}{6}$	x	$\dfrac{x}{6}$
Linda	$\dfrac{1}{4}$	x	$\dfrac{x}{4}$

Step 3 **Write an equation.**

$$\frac{x}{6}+\frac{x}{4}=1$$

Step 4 **Solve.**

$$\frac{x}{6}+\frac{x}{4}=1$$

$$24\left(\frac{x}{6}+\frac{x}{4}\right)=24\cdot1$$

Multiply by the LCD, 24.

$$24\left(\frac{x}{6}\right)+24\left(\frac{x}{4}\right)=24\cdot1 \qquad \text{Distributive property}$$

$$4x+6x=24 \qquad \text{Multiply.}$$

$$10x=24 \qquad \text{Add.}$$

$$x=\frac{24}{10}=2.4 \qquad \text{Divide by 10.}$$

Step 5 **State the answer.**

It will take Kelly and Linda 2.4 hours to clean the house working together.

Step 6 **Check** to be sure that answer is correct.

Objective 1 Solve problems about numbers.

For extra help, see Example 1 on page 467 of your text, the Section Lecture video for Section 7.7, and Exercise Solution Clip 3.

Solve each problem.

1. If two times a number is added to one-half of its reciprocal, the result is $\frac{13}{6}$. Find the number.

1. _____

2. In a certain fraction, the numerator is 4 less than the denominator. If 5 is added to both the numerator and the denominator, the resulting fraction is equal to $\frac{7}{9}$. Find the original fraction.

2. _____

3. The sum of a number and its reciprocal is $\frac{13}{6}$. Find the number.

3. _____

4. If twice the reciprocal of a number is added to the number, the result is $\frac{9}{2}$. Find the number.

4. _____

5. If three times a number is subtracted from twice its reciprocal, the result is -1. Find the number.

5. _____

Objective 2 Solve problems about distance, rate, and time.
For extra help, see Example 2 on pages 468–469 of your text, the Section Lecture video for Section 7.7, and Exercise Solution Clip 21.

Solve each problem.

6. Dipti flew her plane 600 miles against the wind in the same time it took her to fly 900 miles with the wind. If the speed of the wind was 30 miles per hour, what was the speed of the plane?

6. _____

7. A ship goes 120 miles downriver in $2\frac{2}{3}$ hours less than it takes to go the same distance upriver. If the speed of the current is 6 miles per hour, find the speed of the ship.

7. _____

8. A plane traveling 450 miles per hour can go 1000 miles with the wind in $\frac{1}{2}$ hour less than when traveling against the wind. Find the speed of the wind.

8. _____

9. On Saturday, Pablo jogged 6 miles. On Monday, jogging at the same speed, it took him 30 minutes longer to cover 10 miles. How fast did Pablo jog?

9. _____

10. A plane made the trip from Redding to Los Angeles, a distance of 560 miles, in 1.5 hours less than it took to fly from Los Angeles to Portland, a distance of 1130 miles. Find the rate of the plane. (Assume there is no wind in either direction.)

10. _____

Objective 3 Solve problems about work.
For extra help, see Example 3 on page 470 of your text and the Section Lecture video for Section 7.7.

Solve each problem.

11. Phil can install the carpet in a room in 3 hours, but Lil needs 8 hours. How long will it take them to complete this task if they work together?

11. _____

12. Chuck can weed the garden in $\frac{1}{2}$ hour, but David takes 2 hours. How long does it take them to weed the garden if they work together?

12. _____

13. Jack can paint a certain room in $1\frac{1}{2}$ hours, but Joe needs 4 hours to paint the same room. How long does it take them to paint the room if they work together?

13. _____

14. Working together, Arlene and Al can install carpeting in   14. _____
    a room in three hours. Working alone, it would take
    Arlene $\frac{2}{3}$ as long as Al to install the carpeting. How
    long would it take Al to do the job alone?

15. John can seal an asphalt driveway in 3 times the amount   15. _____
    of the time it takes Fred. Working together, it takes them
    $1\frac{1}{2}$ hours to seal the driveway. How long would it have
    taken Fred working alone?

Chapter 7 RATIONAL EXPRESSIONS AND APPLICATIONS

7.8 Variation

Learning Objectives
1 Solve direct variation problems.
2 Solve inverse variation problems.

Key Terms
Use the vocabulary terms listed below to complete each statement in exercises 1–3.

    direct variation   **constant of variation**   **inverse variation**

1. The constant k in the direct variation equation $y = kx$ is called the

    _____.

2. If one quantity is a constant multiple of another, then the two quantities are a(n)

    _____.

3. Two variables are said to be a(n) _____ if one quantity increases while the other quantity decreases.

Guided Examples

Review these examples for Objective 1:	Now Try:
1. If y varies directly as x, and $y = 40$ when $x = 5$. Find y when x is 9.	1. If p varies directly as q, and $p = 84$ when $q = 12$, find p when q is 4.

    $y = kx$ Equation for direct variation
   $40 = k \cdot 5$ Substitute given values.
    $8 = k \leftarrow$ Constant of variation
Since $y = kx$ and $k = 8$, we have $y = 8x$.
Now find y when x is 9.
 $y = 8 \cdot 9 = 72$
Thus, $y = 72$ when $x = 9$.

---

2. The surface area of a sphere varies directly as the square of its radius. If the surface area of a sphere with a radius of 12 inches is 576π square inches, find the surface area of a sphere with a radius of 3 inches.

$$A = kr^2 \qquad \text{Equation for direct variation}$$

$$576\pi = k \cdot 12^2 \quad \text{Substitute given values.}$$

$$576\pi = 144k$$

$$\frac{576\pi}{144} = 4\pi = k \quad \text{Constant of variation}$$

Since $A = kr^2$ and $k = \dfrac{576\pi}{144} = 4\pi$ we have

$A = 4\pi r^2$.

Now find A when r is 3.

$$A = 4\pi \cdot 3^2 = 4\pi \cdot 9 = 36\pi$$

The surface area of a sphere with a radius of 3 inches is 36π square inches.

2. The area of a circle varies directly as the square of the radius. A circle with a radius of 5 centimeters has an area of 78.5 square centimeters. Find the area if the radius changes to 7 centimeters.

_____

Review these examples for Objective 2:

3. If y varies inversely as x, and $y = 10$ when $x = 3$, find y when $x = 12$.

$$y = \frac{k}{x} \qquad \text{Equation for indirect variation}$$

$$10 = \frac{k}{3} \qquad \text{Substitute given values.}$$

$$30 = k \leftarrow \text{Constant of variation}$$

Since $y = \frac{k}{x}$ and $k = 30$, we have $y = \frac{30}{x}$.

Now find y when x is 12.

$$y = \frac{30}{12} = \frac{5}{2}$$

Thus, $y = \frac{5}{2}$ when $x = 12$.

Now Try:

3. If y varies inversely as x, and $y = 20$ when $x = 4$, find y when $x = 12$.

_____

4. The current in a simple electrical circuit varies inversely as the resistance. If the current is 50 amps (an *ampere* is a unit for measuring current) when the resistance is 10 ohms (an *ohm* is a unit for measuring resistance), find the current if the resistance is 5 ohms.

Let c = the current in amps and let r = the resistance in ohms.

$$c = \frac{k}{r} \quad \text{Equation for indirect variation}$$

$$50 = \frac{k}{10} \quad \text{Substitute given values.}$$

$$500 = k \leftarrow \text{Constant of variation}$$

Now use $c = \frac{500}{r}$.

$$c = \frac{500}{5} = 100$$

The current is 100 amps when the resistance is 5 ohms.

4. The speed of a pulley varies inversely as its diameter. One kind of pulley, with a diameter of 3 inches, turns at 150 revolutions per minute. Find the speed of a similar pulley with a diameter of 5 inches

_____

Objective 1 Solve direct variation problems.
For extra help, see Examples 1 and 2 on pages 476–477 of your text, the Section Lecture video for Section 7.8, and Exercise Solution Clips 19 and 41.

Solve each problem.

1. If y varies directly as x, and $y = 20$ when $x = 2$, find y when x is 5.

1. _____

2. If w varies directly as the square of v, and $w = 24$ when $v = 6$, find w when v is 10.

2. _____

3. What you would weigh on the moon varies directly with your weight on earth. A 120 pound person would weigh about 20 pounds on the moon. How much would a 150 pound person weigh on the moon?

3. _____

4. For a given height, the area of a triangle varies directly as its base. Find the area of a triangle with a base of 4 centimeters, if the area is 9.6 square centimeters when the base is 3 centimeters.

4. _____

5. The area of a circle varies directly as the square of the radius. A circle with a radius of 7 centimeters has an area of 153.86 square centimeters. Find the area if the radius changes to 11 centimeters.

5. _____

6. The pressure exerted by water at a given point varies directly with the depth of the point beneath the surface of the water. At a depth of 10 feet, the pressure is 4.34 lb per sq in. Find the pressure on a scuba diver at 35 feet.

6. _____

7. With constant resistance, the power used in a simple electric circuit varies directly as the square of the current. If the power used is 37,500 watts when the current is 50 amps, find the power if the current is reduced to 40 amps.

7. _____

Objective 2 Solve inverse variation problems.

For extra help, see Examples 3 and 4 on pages 478–479 of your text, the Section Lecture video for Section 7.8, and Exercise Solution Clips 23 and 37.

Solve each problem.

8. If t varies inversely as r, and $t = 4$ when $r = 60$, find t when $r = 45$.

        8. _____

9. If n varies inversely as a^2, and $n = 12$ when $a = \frac{1}{2}$, find n when a is 3.

        9. _____

10. For a specified distance, time varies inversely with speed. If Ramona walks a certain distance on a treadmill in 40 minutes at 4.2 miles per hour, how long will it take her to walk the same distance at 3.5 miles per hour?

      10. _____

11. If the temperature is constant, the pressure of a gas in a container varies inversely as the volume of the container. If the pressure is 140 pounds per square foot when a gas is in a container of 1000 cubic feet, what is the volume of the container when the gas exerts a pressure of 700 pounds per square feet?

      11. _____

12. The weight of an object varies inversely as the square of its distance from the center of the earth. If an object 8000 miles from the center of the earth weighs 90 pounds, find its weight when it is 12,000 miles from the center of the earth.

12. _____

13. The illumination produced by a light source varies inversely as the square of the distance from the source. If the illumination produced 5 feet from a light source is 60 foot-candles, find the illumination produced 10 feet from the same source.

13. _____

14. The force with which Earth attracts an object above Earth's surface varies inversely as the square of the object's distance from the center of Earth. If an object 5000 miles from the center of Earth is attracted with a force of 102.4 lb, find the force of attraction of an object 5800 miles from the center of Earth.

14. _____

15. With constant power, the resistance used in a simple electric circuit varies inversely as the square of the current. If the resistance is 120 ohms when the current is 12 amps, find the resistance if the current is reduced to 9 amps.

15. _____

Chapter 8 ROOTS AND RADICALS

8.1 Evaluating Roots

Learning Objectives

1   Find square roots.
2   Decide whether a given root is rational, irrational, or not a real number.
3   Find decimal approximations for irrational square roots.
4   Use the Pythagorean theorem.
5   Use the distance formula.
6   Find the cube, fourth, and other roots.

Key Terms

Use the vocabulary terms listed below to complete each statement in exercises 1–12.

square root	principal root	negative square root
radical	radicand	radical expression
perfect square	cube root	fourth root
index (order)	perfect cube	perfect fourth power

1.   5 is the _____ of 625.

2.   A _____ is a radical sign and the number or expression that appears under it.

3.   The _____ of a positive number with even index n is the positive nth root of the number.

4.   A _____ is the number or expression that appears inside a radical sign.

5.   A number with a rational square root is called a _____.

6.   The symbol $-\sqrt{}$ is used for the _____ of a number.

7.   In a radical of the form $\sqrt[n]{a}$, the number n is the _____.

8.   The number b is a _____ of a if $b^3 = a$.

9.   Since $2^4 = 16$, 16 is a _____.

10.   The _____ of a number is a number that, when multiplied by itself, gives the original number.

11.   A number with a rational cube root is called a _____.

12.   A _____ is an algebraic expression containing a radical.

Guided Examples

Review these examples for Objective 1:

1.   Find all square roots of 256.

$$\sqrt{256} = 16, -16$$

2.   Find each square root.

   a. $\sqrt{900}$ **b.** $-\sqrt{144}$

   c. $\sqrt{\dfrac{49}{81}}$

   a. The radical $\sqrt{900}$ represents the positive or principal square root of 900. Since $30^2 = 900$, $\sqrt{900} = 30$.

   b. $-\sqrt{144}$ represents the negative square root of 144. Therefore, $-\sqrt{144} = -12$.

   c. $\sqrt{\dfrac{49}{81}} = \dfrac{7}{9}$

3.   Find the square of each radical expression.

   a. $\sqrt{900}$ **b.** $-\sqrt{144}$

   c. $\sqrt{x^2 - 4}$

   a. $\left(\sqrt{900}\right)^2 = 900$

   b. The square of a negative number is positive, so $\left(-\sqrt{144}\right)^2 = 144$.

   c. $\left(\sqrt{x^2 - 4}\right)^2 = x^2 - 4$

Now Try:

1.   Find all square roots of 400.

   _____

2.   Find each square root.

   a. $\sqrt{484}$ _____

   b. $-\sqrt{324}$ _____

   c. $-\sqrt{\dfrac{289}{10,000}}$ _____

3.   Find the square of each radical expression.

   a. $\sqrt{484}$ _____

   b. $-\sqrt{324}$ _____

   c. $\sqrt{3x^2 + 4}$ _____

Review this example for Objective 2:

4. Tell whether each square root is *rational*, *irrational*, or *not a real number*.

   a. $\sqrt{21}$ **b.** $\sqrt{25}$

   c. $\sqrt{-100}$

   a. Because 21 is not a perfect square, $\sqrt{21}$ is *irrational*.

   b. The number 25 is a perfect square, so $\sqrt{25}$ is a *rational* number.

   c. There is no real number whose square is -100, so $\sqrt{-100}$ is *not a real number*.

Now Try:

4. Tell whether each square root is *rational*, *irrational*, or *not a real number*.

   a. $\sqrt{35}$ _____

   b. $\sqrt{121}$ _____

   c. $\sqrt{-49}$ _____

Review this example for Objective 3:

5. Find a decimal approximation for each square root. Round answers to the nearest thousandth.

   a. $\sqrt{21}$ **b.** $-\sqrt{420}$

   a. $\sqrt{21} \approx 4.583$

   b. $-\sqrt{420} \approx -20.494$

Now Try:

5. Find a decimal approximation for each square root. Round answers to the nearest thousandth.

   a. $\sqrt{35}$ _____

   b. $-\sqrt{138}$ _____

Review these examples for Objective 4:

6. Find the length of the unknown side in each right triangle with sides a, b, and c. Give any decimal approximations to the nearest thousandth.

   a. $a = 40, b = 9$ **b.** $a = 12, c = 25$

   a. $\quad a^2 + b^2 = c^2 \quad$ Pythagorean theorem

   $\quad\quad 40^2 + 9^2 = c^2 \quad a = 40, b = 9$

   $\quad\quad 1600 + 81 = c^2 \quad$ Square.

   $\quad\quad\quad\quad 1681 = c^2 \quad$ Add.

   $\quad\quad\quad\quad\quad 41 = c \quad \sqrt{1681} = 41$

   b. $\quad a^2 + b^2 = c^2 \quad\quad$ Pythagorean theorem

   $\quad\quad 12^2 + b^2 = 25^2 \quad\quad a = 12, c = 25$

   $\quad\quad 144 + b^2 = 625 \quad\quad$ Square.

   $\quad\quad\quad\quad b^2 = 451 \quad\quad$ Subtract 144.

   $\quad\quad\quad\quad b = \sqrt{451} \approx 21.237$

Now Try:

6. Find the length of the unknown side in each right triangle with sides a, b, and c. Give any decimal approximations to the nearest thousandth.

   a. $a = 11, b = 60$ _____

   b. $b = 6, c = 18$ _____

7. Susan started to drive due south at the same time John started to drive due west. John drove 21 miles in the same time that Susan drove 28 miles. How far apart were they at that time?

Draw a diagram to help interpret the problem.

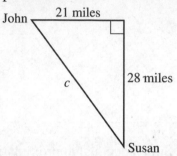

We want to know how far apart John and Susan are after he has traveled 21 miles and she has traveled 28 miles. Since he drove due west and she drove due south, a right triangle is formed, with legs 21 and 28. We are seeking the length of the hypotenuse. Write an equation using the Pythagorean theorem.

$c^2 = a^2 + b^2$

$c^2 = 21^2 + 28^2$ $a = 21,\ c = 28$

$c^2 = 1225$ Add.

$c = \sqrt{1225} = 35$

Choose the positive square root because c represents a distance.

Susan and John are 35 miles apart.

7. Stan is flying a kite on 60 feet of string. How high is the kite above his head (vertically) if the horizontal distance between Stan and the kite is 36 feet?

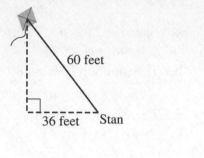

_____

Review this example for Objective 5:

8. Find the distance between $(2, -2)$ and $(-6, 1)$.

Use the distance formula.
Let $(x_1, y_1) = (2, -2)$ and
$(x_2, y_2) = (-6, 1)$.

$d = \sqrt{(x_2 - x_1)^2 + (y_2 - y_1)^2}$

$ = \sqrt{(-6 - 2)^2 + (1 - (-2))^2}$

$ = \sqrt{(-8)^2 + 3^2}$

$ = \sqrt{64 + 9} = \sqrt{73}$

Now Try:

8. Find the distance between $(-7, -6)$ and $(-7, 6)$.

_____

Review this example for Objective 6:

9. Find each root.

    a. $\sqrt[3]{-64}$ **b.** $\sqrt[4]{-625}$

    c. $-\sqrt[5]{243}$

    a. $\sqrt[3]{-64} = -4$ because $(-4)^3 = -64$.

    b. For a fourth root to be a real number, the radicand must be nonnegative. There is no real number that equals $\sqrt[4]{-625}$.

    c. First find $\sqrt[5]{243}$. Because $3^5 = 243$, $\sqrt[5]{243} = 3$. Then, $-\sqrt[5]{243} = -3$.

Now Try:

9. Find each root.

    a. $-\sqrt[3]{-216}$ _____

    b. $\sqrt[4]{256}$ _____

    c. $\sqrt[5]{1024}$ _____

Objective 1 Find square roots.
For extra help, see Examples 1–3 on pages 494–495 of your text, the Section Lecture video for Section 8.1, and Exercise Solution Clips 7, 19, and 31.

Find all square roots of each number.

1. Find the square roots of $\frac{49}{144}$.

    1. _____

2. Find $-\sqrt{49}$.

    2. _____

3. Find the square of $\sqrt{3x^2 + 2x}$.

    3. _____

Objective 2 Decide whether a given root is rational, irrational, or not a real number.
For extra help, see Example 4 on page 496 of your text, the Section Lecture video for Section 8.1, and Exercise Solution Clips 43 and 49.

Tell whether the square root is rational, irrational, or not a real number.

4. $\sqrt{8.1}$

    4. _____

5. $-\sqrt{-144}$

    5. _____

Objective 3 Find decimal approximations for irrational square roots.
For extra help, see Example 5 on page 497 of your text, the Section Lecture video for Section 8.1, and Exercise Solution Clip 49.

Use a calculator to find a decimal approximation for the root. Round answers to the nearest thousandth.

6. $\sqrt{640}$ 6.

7. $-\sqrt{26,358}$ 7.

Objective 4 Use the Pythagorean theorem.
For extra help, see Examples 6 and 7 on pages 497–498 of your text, the Section Lecture video for Section 8.1, and Exercise Solution Clips 65 and 73.

8. Find the length of the unknown side of the right triangle   8. _____
   with leg 16 and hypotenuse 17. Round your answer to
   the nearest thousandth.

Use the Pythagorean theorem to solve each problem. If necessary, round your answer to the nearest thousandth.

9. A ladder 25 feet long leans against a wall. The foot of   9. _____
   the ladder is 7 feet from the base of the wall. How high
   up the wall does the top of the ladder rest?

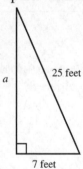

25 feet

a

7 feet

10. A plane flies due east for 35 miles and then due south   **10.** _____
until it is 37 miles from its starting point. How far south
did the plane fly?

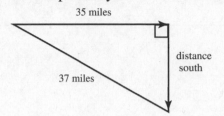

Objective 5 Use the distance formula.

For extra help, see Example 8 on page 499 of your text, the Section Lecture video for
Section 8.1, and Exercise Solution Clip 89.

Find the distance between each pair of points.

11. $(6,6),(3,8)$ **11.** _____

12. $(7,-11),(9,2)$ **12.** _____

Objective 6 Find the cube, fourth, and other roots.

For extra help, see Examples 9 and 10 on page 500 of your text, the Section Lecture video for
Section 8.1, and Exercise Solution Clips 97 and 105.

Find each root that is a real number.

13. $\sqrt[3]{-64}$ **13.** _____

14. $\sqrt[4]{-10,000}$ **14.** _____

15. $-\sqrt[5]{-32}$ **15.** _____

Chapter 8 ROOTS AND RADICALS

8.2 Multiplying, Dividing, and Simplifying Radicals

Learning Objectives
1   Multiply square root radicals.
2   Simplify radicals by using the product rule.
3   Simplify radicals by using the quotient rule.
4   Simplify radicals involving variables.
5   Simplify other roots.

Key Terms

Use the vocabulary terms listed below to complete each statement in exercises 1–3.

    simplified form   **product rule for radicals**   **quotient rule for radicals**

1.   The _____ allows you to rewrite the square root of a quotient as the quotient of two square roots and visa versa.

2.   A radical with index 2 is in _____ if the radicand does not contain perfect square factors.

3.   The _____ allows you to rewrite the square root of a product as the product of two square roots and visa versa.

Guided Examples

Review this example for Objective 1:

1.   Use the product rule for radicals to find each product.

    a. $\sqrt{13} \cdot \sqrt{5}$ **b.** $\sqrt{7} \cdot \sqrt{7}$

    c. $\sqrt{5} \cdot \sqrt{2r}$

    a. $\sqrt{13} \cdot \sqrt{5} = \sqrt{13 \cdot 5} = \sqrt{65}$

    b. $\sqrt{7} \cdot \sqrt{7} = 7$

    c. $\sqrt{5} \cdot \sqrt{2r} = \sqrt{5 \cdot 2r} = \sqrt{10r}$

Now Try:

1.   Use the product rule for radicals to find each product.

    a. $\sqrt{2} \cdot \sqrt{7}$ _____

    b. $\sqrt{13} \cdot \sqrt{13}$ _____

    c. $\sqrt{3x} \cdot \sqrt{7},\ x > 0$ _____

Review these examples for Objective 2:

2.   Simplify each radical.

    a. $\sqrt{24}$ **b.** $\sqrt{288}$

    c. $\sqrt{35}$

Now Try:

2.   Simplify each radical.

    a. $\sqrt{84}$ _____

    b. $\sqrt{162}$ _____

a. $\sqrt{24} = \sqrt{4 \cdot 6}$ Factor; 4 is a perfect square.

      $= \sqrt{4} \cdot \sqrt{6}$ Product rule

      $= 2\sqrt{6}$ $\sqrt{4} = 2$

b. $\sqrt{288} = \sqrt{144 \cdot 2}$ Factor; 144 is a perfect square.

      $= \sqrt{144} \cdot \sqrt{2}$ Product rule

      $= 12\sqrt{2}$ $\sqrt{144} = 12$

c. The number 35 has no perfect square factors, so $\sqrt{35}$ cannot be simplified further.

c. $\sqrt{95}$ _____

3. Find each product and simplify.

   a. $\sqrt{3} \cdot \sqrt{15}$ **b.** $3\sqrt{2} \cdot 2\sqrt{6}$

   a. $\sqrt{3} \cdot \sqrt{15} = \sqrt{3 \cdot 15}$ Product rule

              $= \sqrt{3 \cdot 3 \cdot 5}$ Factor 15.

              $= \sqrt{3} \cdot \sqrt{3} \cdot \sqrt{5}$ Product rule

              $= 3\sqrt{5}$ Multiply.

   b. $3\sqrt{2} \cdot 2\sqrt{6} = 3 \cdot 2 \cdot \sqrt{2 \cdot 6}$ Commutative property; product rule

              $= 6\sqrt{12}$ Multiply.

              $= 6\sqrt{4 \cdot 3}$ Factor.

              $= 6\sqrt{4} \cdot \sqrt{3}$ Product rule

              $= 6 \cdot 2 \cdot \sqrt{3}$ $\sqrt{4} = 2$

              $= 12\sqrt{3}$ Multiply.

3. Find each product and simplify.

   a. $\sqrt{6} \cdot \sqrt{8}$ _____

   b. $5\sqrt{3} \cdot 4\sqrt{6}$ _____

Review these examples for Objective 3:

4. Simplify each radical.

   a. $\sqrt{\dfrac{16}{9}}$ **b.** $\dfrac{\sqrt{75}}{\sqrt{3}}$

   c. $\sqrt{\dfrac{7}{81}}$

   a. $\sqrt{\dfrac{16}{9}} = \dfrac{\sqrt{16}}{\sqrt{9}} = \dfrac{4}{3}$

   b. $\dfrac{\sqrt{75}}{\sqrt{3}} = \sqrt{\dfrac{75}{3}} = \sqrt{25} = 5$

Now Try:

4. Simplify each radical.

   a. $\sqrt{\dfrac{36}{49}}$ _____

   b. $\dfrac{\sqrt{162}}{\sqrt{2}}$ _____

   c. $\sqrt{\dfrac{125}{64}}$ _____

c. $\sqrt{\dfrac{7}{81}} = \dfrac{\sqrt{7}}{\sqrt{81}} = \dfrac{\sqrt{7}}{9}$

5. Simplify $\dfrac{5\sqrt{48}}{10\sqrt{8}}$.

$\dfrac{5\sqrt{48}}{10\sqrt{8}} = \dfrac{5}{10} \cdot \dfrac{\sqrt{48}}{\sqrt{8}}$ Multiplication of fractions

$= \dfrac{5}{10} \cdot \sqrt{\dfrac{48}{8}}$ Quotient rule

$= \dfrac{1}{2}\sqrt{6}$ Divide.

5. Simplify $\dfrac{6\sqrt{30}}{18\sqrt{15}}$.

_____

6. Simplify $\sqrt{\dfrac{15}{8}} \cdot \sqrt{\dfrac{1}{2}}$

$\sqrt{\dfrac{15}{8}} \cdot \sqrt{\dfrac{1}{2}} = \sqrt{\dfrac{15 \cdot 1}{8 \cdot 2}}$ Product rule

$= \sqrt{\dfrac{15}{16}}$ Multiply fractions.

$= \dfrac{\sqrt{15}}{\sqrt{16}}$ Quotient rule

$= \dfrac{\sqrt{15}}{4}$ Simplify.

6. Simplify $\sqrt{\dfrac{3}{36}} \cdot \sqrt{\dfrac{16}{27}}$

_____

Review this example for Objective 4:

7. Simplify each radical. Assume that all variables represent positive real numbers.

a. $\sqrt{81x^6}$ **b.** $\sqrt{56x^7}$

c. $\sqrt{\dfrac{7}{x^4}},\ x \neq 0$

a. $\sqrt{81x^6} = \sqrt{81}\sqrt{x^6} = 9x^3$

b. $\sqrt{56x^7} = \sqrt{56} \cdot \sqrt{x^7}$

$= \sqrt{4 \cdot 14} \cdot \sqrt{x^6 \cdot x}$

$= \sqrt{4} \cdot \sqrt{14} \cdot \sqrt{x^6} \cdot \sqrt{x}$

$= 2x^3\sqrt{14x}$

c. $\sqrt{\dfrac{7}{x^4}} = \dfrac{\sqrt{7}}{\sqrt{x^4}} = \dfrac{\sqrt{7}}{x^2},\ x \neq 0$

Now Try:

7. Simplify each radical. Assume that all variables represent positive real numbers.

a. $\sqrt{100y^{12}}$ _____

b. $\sqrt{48x^7}$ _____

c. $\sqrt{\dfrac{10}{49x^8}}$ _____

Name: Date:
Instructor: Section:

Review these examples for Objective 5:	Now Try:

Review these examples for Objective 5:

8. Simplify each radical.

   a. $\sqrt[3]{128}$ **b.** $\sqrt[4]{243}$

   c. $\sqrt[4]{\dfrac{16}{81}}$

   a. $\sqrt[3]{128} = \sqrt[3]{64 \cdot 2} = \sqrt[3]{64} \cdot \sqrt[3]{2} = 4\sqrt[3]{2}$

   b. $\sqrt[4]{243} = \sqrt[4]{81 \cdot 3} = \sqrt[4]{81} \cdot \sqrt[4]{3} = 3\sqrt[4]{3}$

   c. $\sqrt[4]{\dfrac{16}{81}} = \dfrac{2}{3}$

9. Simplify each radical.

   a. $\sqrt[3]{y^{27}}$ **b.** $\sqrt[3]{243x^5}$

   c. $\sqrt[3]{\dfrac{z^8}{81}}$

   a. $\sqrt[3]{y^{27}} = y^9$

   b. $\sqrt[3]{243x^5} = \sqrt[3]{243 \cdot x^5} = \sqrt[3]{243} \cdot \sqrt[3]{x^5}$
   $$= \sqrt[3]{27 \cdot 9} \cdot \sqrt[3]{x^3 \cdot x^2}$$
   $$= \sqrt[3]{27} \cdot \sqrt[3]{9} \cdot \sqrt[3]{x^3} \cdot \sqrt[3]{x^2}$$
   $$= 3x\sqrt[3]{9x^2}$$

   c. $\sqrt[3]{\dfrac{z^8}{27}} = \dfrac{\sqrt[3]{z^8}}{\sqrt[3]{27}} = \dfrac{\sqrt[3]{z^6 \cdot z^2}}{3} = \dfrac{z^2\sqrt[3]{z^2}}{3}$

Now Try:

8. Simplify each radical.

   a. $\sqrt[3]{162}$ _____

   b. $\sqrt[4]{1875}$ _____

   c. $\sqrt[3]{\dfrac{512}{27}}$ _____

9. Simplify each radical.

   a. $\sqrt[3]{x^{18}}$ _____

   b. $\sqrt[3]{32n^7}$ _____

   c. $\sqrt[3]{\dfrac{y^4}{125}}$ _____

Objective 1 Multiply square root radicals.
For extra help, see Example 1 on page 505 of your text, the Section Lecture video for Section 8.2, and Exercise Solution Clip 1.

Use the product rule for radicals to find each product.

1. $\sqrt{7a} \cdot \sqrt{5b}$, $a \geq 0$, $b \geq 0$

   1. _____

2. $\sqrt{3} \cdot \sqrt{3}$

   2. _____

    359

Objective 2 Simplify radicals by using the product rule.
For extra help, see Examples 2 and 3 on pages 505–506 of your text, the Section Lecture video for Section 8.2, and Exercise Solution Clips 25 and 43.

Simplify each radical.

3. $\sqrt{250}$ 3. _____

4. $-2\sqrt{54}$ 4. _____

Find each product and simplify.

5. $\sqrt{6} \cdot \sqrt{15}$ 5. _____

6. $\sqrt{11} \cdot \sqrt{33}$ 6. _____

7. $\sqrt{18} \cdot \sqrt{24}$ 7. _____

Objective 3 Simplify radicals by using the quotient rule.
For extra help, see Examples 4–6 on page 507 of your text, the Section Lecture video for Section 8.2, and Exercise Solution Clips 57, 65 and 67.

Use the quotient rule and product rule, as necessary, to simplify each expression.

8. $\sqrt{\dfrac{13}{121}}$ 8. _____

9. $\dfrac{9\sqrt{200}}{3\sqrt{2}}$ 9. _____

10. $\sqrt{\dfrac{2}{125}} \cdot \sqrt{\dfrac{2}{5}}$ 10. _____

11. $\sqrt{\dfrac{5}{2}} \cdot \sqrt{\dfrac{16}{32}}$ 11. _____

Name: _____   Date: _____

Instructor: _____   Section: _____

Objective 4 Simplify radicals involving variables.

For extra help, see Example 7 on page 508 of your text, the Section Lecture video for Section 8.2, and Exercise Solution Clip 69.

Simplify each radical. Assume that all variables represent positive real numbers.

12. $\sqrt{18p^2q^6}$ 12. _____

13. $\sqrt{\dfrac{6}{x^2}}$ 13. _____

Objective 5 Simplify other roots.

For extra help, see Examples 8 and 9 on pages 509–510 of your text, the Section Lecture video for Section 8.2, and Exercise Solution Clips 97 and 111.

Simplify each expression.

14. $\sqrt[5]{-64}$ 14. _____

15. $\sqrt[3]{\dfrac{64}{27}}$ 15. _____

Chapter 8 ROOTS AND RADICALS

8.3 Adding and Subtracting Radicals

Learning Objectives
1 Add and subtract radicals.
2 Simplify radical sums and differences.
3 Simplify more complicated radical expressions.

Key Terms
Use the vocabulary terms listed below to complete each statement in exercises 1–2.

> **like radicals**   **unlike radicals**

1. Two or more radicals that are multiples of the same root of the same number are
    _____.

2. Radicals that cannot be added or subtracted are _____.

Guided Examples

Review this example for Objective 1:
1. Add or subtract, as indicated.

   a. $2\sqrt{5} + 8\sqrt{5}$ b. $8\sqrt{7} - 13\sqrt{7}$

   c. $\sqrt{5} + \sqrt{2}$

   a. $2\sqrt{5} + 8\sqrt{5} = (2 + 8)\sqrt{5} = 10\sqrt{5}$

   b. $8\sqrt{7} - 13\sqrt{7} = (8 - 13)\sqrt{7} = -5\sqrt{7}$

   c. $\sqrt{5} + \sqrt{2}$
       The radicands are different, so we cannot use the distributive property to add the radicals.

Now Try:
1. Add or subtract, as indicated.

   a. $3\sqrt{2} + 5\sqrt{2}$ _____

   b. $6\sqrt{13} - 10\sqrt{13}$ _____

   c. $\sqrt{3} + \sqrt{7}$ _____

Review this example for Objective 2:
2. Add or subtract, as indicated.

   a. $\sqrt{24} + \sqrt{6}$ b. $3\sqrt{40} - \sqrt{200}$

   c. $3\sqrt{54} - 4\sqrt{96}$ d. $7\sqrt[3]{54} - 6\sqrt[3]{128}$

Now Try:
2. Add or subtract, as indicated.

   a. $\sqrt{18} + \sqrt{32}$ _____

   b. $2\sqrt{27} - 8\sqrt{12}$ _____

a. $\sqrt{24} + \sqrt{6}$

$\quad = \sqrt{4 \cdot 6} + \sqrt{6}$ Factor; 4 is a perfect square.

$\quad = \sqrt{4} \cdot \sqrt{6} + \sqrt{6}$ Product rule

$\quad = 2\sqrt{6} + 1\sqrt{6}$ $\quad \sqrt{4} = 2$

$\quad = (2 + 1)\sqrt{6}$ Distributive property

$\quad = 3\sqrt{6}$ Add.

b. $3\sqrt{40} - \sqrt{200}$

$\quad = \sqrt{4 \cdot 5} - \sqrt{100 \cdot 2}$ Factor; 4 and 100 are perfect squares.

$\quad = \sqrt{4} \cdot \sqrt{5} - \sqrt{100} \cdot \sqrt{2}$ Product rule

$\quad = 2\sqrt{5} - 10\sqrt{2}$ $\quad \sqrt{4} = 2; \sqrt{100} = 10$

The last expression contains unlike radicals. They cannot be combined.

c. $3\sqrt{54} - 4\sqrt{96}$

$\quad = 3\sqrt{9 \cdot 6} - 4\sqrt{16 \cdot 6}$ Factor; 9 and 16 are perfect squares.

$\quad = 3\left(\sqrt{9} \cdot \sqrt{6}\right) - 4\left(\sqrt{16} \cdot \sqrt{6}\right)$ Product rule

$\quad = 3\left(3\sqrt{6}\right) - 4\left(4\sqrt{6}\right)$ $\sqrt{9} = 3; \sqrt{16} = 4$

$\quad = 9\sqrt{6} - 16\sqrt{6}$ Multiply.

$\quad = -7\sqrt{6}$ Subtract like radicals.

d. $7\sqrt[3]{54} - 6\sqrt[3]{128}$

$\quad = 7\sqrt[3]{27 \cdot 2} - 6\sqrt[3]{64 \cdot 2}$ Factor; 27 and 64 are perfect cubes.

$\quad = 7\left(\sqrt[3]{27} \cdot \sqrt[3]{2}\right) - 6\left(\sqrt[3]{64} \cdot \sqrt[3]{2}\right)$

Product rule

$\quad = 7\left(3\sqrt[3]{2}\right) - 6\left(4\sqrt[3]{2}\right)$ $\sqrt[3]{27} = 3; \sqrt[3]{64} = 4$

$\quad = 21\sqrt[3]{2} - 24\sqrt[3]{2}$ Multiply.

$\quad = -3\sqrt[3]{2}$ Subtract like radicals.

c. $5\sqrt{128} + 3\sqrt{27}$

_____

d. $2\sqrt[4]{243} - 3\sqrt[4]{48}$

_____

Review this example for Objective 3:

3. Simplify. Assume that all variables represent nonnegative real numbers.

a. $3\sqrt{6} \cdot \sqrt{10} - 7\sqrt{15}$

b. $6\sqrt{54w} - 4\sqrt{6w}$

c. $4\sqrt{5r^2} \cdot \sqrt{40} - 7\sqrt{8r^2}$

d. $2\sqrt[3]{54x^4} + 3x\sqrt[3]{2x}$

Now Try:

3. Simplify. Assume that all variables represent nonnegative real numbers.

a. $4\sqrt{6} \cdot \sqrt{3} - 2\sqrt{2}$

_____

a. $3\sqrt{6} \cdot \sqrt{10} - 7\sqrt{15}$

$\quad = 3\sqrt{6 \cdot 10} - 7\sqrt{15}$ Product rule

$\quad = 3\sqrt{60} - 7\sqrt{15}$ Multiply.

$\quad = 3\sqrt{4 \cdot 15} - 7\sqrt{15}$ Factor.

$\quad = 3\sqrt{4} \cdot \sqrt{15} - 7\sqrt{15}$ Product rule

$\quad = 6\sqrt{15} - 7\sqrt{15}$ $\sqrt{4} = 2$; multiply.

$\quad = -\sqrt{15}$ Subtract like radicals.

b. $6\sqrt{54w} - 4\sqrt{6w}$

$\quad = 6\sqrt{9 \cdot 6w} - 4\sqrt{6w}$ Factor.

$\quad = 6\sqrt{9} \cdot \sqrt{6w} - 4\sqrt{6w}$ Product rule

$\quad = 18\sqrt{6w} - 4\sqrt{6w}$ $\sqrt{9} = 3$; multiply.

$\quad = 14\sqrt{6w}$ Subtract like radicals.

c. $4\sqrt{5r^2} \cdot \sqrt{40} - 7\sqrt{8r^2}$

$\quad = 4\sqrt{5r^2 \cdot 40} - 7\sqrt{8r^2}$ Product rule

$\quad = 4\sqrt{200r^2} - 7\sqrt{8r^2}$ Multiply.

$\quad = 4\sqrt{100r^2 \cdot 2} - 7\sqrt{4r^2 \cdot 2}$ Factor.

$\quad = 4\sqrt{100r^2} \cdot \sqrt{2} - 7\sqrt{4r^2} \cdot \sqrt{2}$

         Product rule

$\quad = 40r\sqrt{2} - 14r\sqrt{2}$ $\sqrt{100r^2} = 10r$;

                       $\sqrt{4r^2} = 2r$; multiply.

$\quad = 26r\sqrt{2}$ Subtract like radicals.

d. $2\sqrt[3]{54x^4} + 3x\sqrt[3]{2x}$

$\quad = 2\sqrt[3]{27x^3 \cdot 2x} + 3x\sqrt[3]{2x}$ Factor.

$\quad = 2\sqrt[3]{27x^3} \cdot \sqrt[3]{2x} + 3x\sqrt[3]{2x}$ Product rule

$\quad = 6x\sqrt[3]{2x} + 3x\sqrt[3]{2x}$ $\sqrt[3]{27x^3} = 3x$;

                        multiply.

$\quad = 9x\sqrt[3]{2x}$ Add like radicals.

b. $4x\sqrt{5} - 3\sqrt{25x} \cdot \sqrt{3x}$

_____

c. $11\sqrt{5w} \cdot \sqrt{30w} - 8w\sqrt{24}$

_____

d. $4\sqrt[3]{r^5} + 3\sqrt[3]{27r^5}$

_____

Objective 1 Add and subtract radicals.

For extra help, see Example 1 on page 513 of your text, the Section Lecture video for Section 8.3, and Exercise Solution Clip 5.

Add or subtract as indicated.

1. $6\sqrt{2} - 4\sqrt{2}$

                                                         1. _____

2. $5\sqrt[3]{10} + \sqrt[3]{10}$ **2.** _____

3. $5\sqrt{2} + \sqrt{3}$ **3.** _____

4. $6\sqrt{3} - 2\sqrt{3} + 4\sqrt{3}$ **4.** _____

5. $3\sqrt{5} - 9\sqrt{5} + \sqrt{5}$ **5.** _____

Objective 2 Simplify radical sums and differences.

For extra help, see Example 2 on pages 513–514 of your text, the Section Lecture video for Section 8.3, and Exercise Solution Clip 13.

Simplify and add or subtract terms wherever possible. Assume that all variables represent nonnegative real numbers.

6. $\sqrt{80} + \sqrt{45}$ **6.** _____

7. $4\sqrt{128} - 2\sqrt{32}$ **7.** _____

8. $-7\sqrt{63} - 9\sqrt{28} + 6\sqrt{28}$ **8.** _____

9. $\dfrac{1}{3}\sqrt{27} - \dfrac{3}{4}\sqrt{48}$ **9.** _____

10. $\dfrac{1}{2}\sqrt{8} - \dfrac{7}{2}\sqrt{32}$ **10.** _____

Objective 3 Simplify more complicated radical expressions.
For extra help, see Example 3 on pages 514–515 of your text, the Section Lecture video for Section 8.3, and Exercise Solution Clip 31.

Perform the indicated operations. Assume that all variables represent nonnegative real numbers.

11. $\sqrt{5} \cdot \sqrt{7} + 3\sqrt{35}$

11. _____

12. $7\sqrt{x} \cdot \sqrt{9} + \sqrt{9x}$

12. _____

13. $3\sqrt{2y} \cdot \sqrt{6} - 5\sqrt{6y}$

13. _____

14. $\sqrt{6k^2} \cdot \sqrt{3k} + k\sqrt{2k}$

14. _____

15. $3\sqrt{12y} - 2\sqrt{5y} \cdot \sqrt{15}$

15. _____

Chapter 8   ROOTS AND RADICALS

8.4   Rationalizing the Denominator

Learning Objectives
1   Rationalize denominators with square roots.
2   Write radicals in simplified form.
3   Rationalize denominators with cube roots.

Key Terms

Use the vocabulary terms listed below to complete each statement in exercises 1–2.

    simplified form   **rationalizing the denominator**

1.   The process of changing the denominator of a fraction from a radical to a rational number is called _____.

2.   A radical is considered to be in _____ if (i) the radicand contains no factor (except 1) that is a perfect square, a perfect cube, and so on, (ii) the radicand has no fractions, and (iii) no denominator contains a radical.

Guided Examples

Review this example for Objective 1:

1.   Rationalize each denominator.

     a. $\dfrac{12}{\sqrt{3}}$ **b.** $\dfrac{6}{\sqrt{30}}$

     a. $\dfrac{12}{\sqrt{3}} = \dfrac{12}{\sqrt{3}} \cdot \dfrac{\sqrt{3}}{\sqrt{3}}$ Multiply by $\dfrac{\sqrt{3}}{\sqrt{3}} = 1$.

         $= \dfrac{12\sqrt{3}}{3}$ $\sqrt{3} \cdot \sqrt{3} = 3$

         $= 4\sqrt{3}$ Lowest terms

     b. $\dfrac{6}{\sqrt{30}} = \dfrac{6}{\sqrt{30}} \cdot \dfrac{\sqrt{30}}{\sqrt{30}}$ Multiply by $\dfrac{\sqrt{30}}{\sqrt{30}} = 1$.

         $= \dfrac{6\sqrt{30}}{30}$ $\sqrt{30} \cdot \sqrt{30} = 30$

         $= \dfrac{\sqrt{30}}{5}$ Lowest terms

Now Try:

1.   Rationalize each denominator.

     a. $\dfrac{10}{\sqrt{15}}$ _____

     b. $\dfrac{6}{\sqrt{28}}$ _____

Review these examples for Objective 2:

2. Simplify $\sqrt{\dfrac{50}{3}}$.

$$\sqrt{\dfrac{50}{3}} = \dfrac{\sqrt{50}}{\sqrt{3}}$$

$$= \dfrac{\sqrt{50}}{\sqrt{3}} \cdot \dfrac{\sqrt{3}}{\sqrt{3}} \qquad \text{Rationalize the denominator.}$$

$$= \dfrac{\sqrt{25 \cdot 2} \cdot \sqrt{3}}{3} \qquad \sqrt{3} \cdot \sqrt{3} = 3$$

$$= \dfrac{\sqrt{25} \cdot \sqrt{2} \cdot \sqrt{3}}{3} \qquad \text{Factor.}$$

$$= \dfrac{5\sqrt{6}}{3} \qquad \begin{array}{l} \sqrt{25} = 5; \\ \text{product rule} \end{array}$$

3. Simplify $\sqrt{\dfrac{5}{21}} \cdot \sqrt{\dfrac{2}{7}}$.

$$\sqrt{\dfrac{5}{21}} \cdot \sqrt{\dfrac{2}{7}}$$

$$= \sqrt{\dfrac{5}{21} \cdot \dfrac{2}{7}} \qquad \text{Product rule}$$

$$= \sqrt{\dfrac{10}{147}} \qquad \text{Multiply fractions.}$$

$$= \dfrac{\sqrt{10}}{\sqrt{147}} \qquad \text{Quotient rule}$$

$$= \dfrac{\sqrt{10}}{\sqrt{49 \cdot 3}} \qquad \text{Factor.}$$

$$= \dfrac{\sqrt{10}}{\sqrt{49} \cdot \sqrt{3}} \qquad \text{Product rule}$$

$$= \dfrac{\sqrt{10}}{7\sqrt{3}} \qquad \sqrt{49} = 7$$

$$= \dfrac{\sqrt{10}}{7\sqrt{3}} \cdot \dfrac{\sqrt{3}}{\sqrt{3}} \qquad \text{Rationalize the denominator.}$$

$$= \dfrac{\sqrt{10} \cdot \sqrt{3}}{7 \cdot 3} \qquad \sqrt{3} \cdot \sqrt{3} = 3$$

$$= \dfrac{\sqrt{30}}{21} \qquad \begin{array}{l} \text{Product rule} \\ \text{Multiply.} \end{array}$$

Now Try:

2. Simplify $\sqrt{\dfrac{32}{27}}$.

_____

3. Simplify $\sqrt{\dfrac{3}{2}} \cdot \sqrt{\dfrac{5}{6}}$.

_____

4. Simplify. Assume that all variables represent positive real numbers.

a. $\dfrac{\sqrt{8a}}{\sqrt{b}}$ b. $\sqrt{\dfrac{2x^4y^3}{15}}$

a. $\dfrac{\sqrt{8a}}{\sqrt{b}} = \dfrac{\sqrt{8a}}{\sqrt{b}} \cdot \dfrac{\sqrt{b}}{\sqrt{b}}$ Rationalize the denominator.

$= \dfrac{\sqrt{8ab}}{b}$ Product rule

$= \dfrac{\sqrt{4 \cdot 2ab}}{b}$ Factor.

$= \dfrac{2\sqrt{2ab}}{b}$ $\sqrt{4} = 2$

b. $\sqrt{\dfrac{2x^4y^3}{15}}$

$= \dfrac{\sqrt{2x^4y^3}}{\sqrt{15}}$ Quotient rule

$= \dfrac{\sqrt{2x^4y^3}}{\sqrt{15}} \cdot \dfrac{\sqrt{15}}{\sqrt{15}}$ Rationalize the denominator.

$= \dfrac{\sqrt{30x^4y^3}}{15}$ Product rule $\sqrt{15} \cdot \sqrt{15} = 15$

$= \dfrac{\sqrt{30\left(x^2\right)^2 y^2 y}}{15}$ Factor.

$= \dfrac{x^2y\sqrt{30y}}{15}$ Product rule $\sqrt{\left(x^2\right)^2} = x^2,$ $\sqrt{y^2} = y$

---

4. Simplify. Assume that all variables represent positive real numbers.

a. $\dfrac{\sqrt{12x}}{\sqrt{z}}$ _____

b. $\sqrt{\dfrac{72qt^5}{20t^6}}$ _____

Review this example for Objective 3:

5. Rationalize each denominator.

a. $\dfrac{\sqrt[3]{3}}{\sqrt[3]{8}}$ b. $\sqrt[3]{\dfrac{11}{6}}$

c. $\sqrt[3]{\dfrac{2}{3ab}}$, $a \neq 0$, $b \neq 0$

Now Try:

5. Rationalize each denominator.

a. $\dfrac{\sqrt[3]{9}}{\sqrt[3]{4}}$ _____

b. $\sqrt[3]{\dfrac{1}{25}}$ _____

a. We multiply the numerator and denominator by the number of factors of 8 to make the radicand in the denominator a perfect cube in order to eliminate the radical in the denominator.

$$\frac{\sqrt[3]{3}}{\sqrt[3]{8}} = \frac{\sqrt[3]{3}}{\sqrt[3]{8}} \cdot \frac{\sqrt[3]{8 \cdot 8}}{\sqrt[3]{8 \cdot 8}} = \frac{\sqrt[3]{3 \cdot 8 \cdot 8}}{\sqrt[3]{8 \cdot 8 \cdot 8}} = \frac{\sqrt[3]{192}}{8}$$

$$= \frac{\sqrt[3]{64 \cdot 3}}{8} = \frac{\sqrt[3]{64} \cdot \sqrt[3]{3}}{8} = \frac{4\sqrt[3]{3}}{8} = \frac{\sqrt[3]{3}}{2}$$

b. $\sqrt[3]{\dfrac{11}{6}} = \dfrac{\sqrt[3]{11}}{\sqrt[3]{6}} = \dfrac{\sqrt[3]{11}}{\sqrt[3]{6}} \cdot \dfrac{\sqrt[3]{6 \cdot 6}}{\sqrt[3]{6 \cdot 6}}$

$$= \frac{\sqrt[3]{11 \cdot 6 \cdot 6}}{\sqrt[3]{6 \cdot 6 \cdot 6}} = \frac{\sqrt[3]{396}}{6}$$

c. $\sqrt[3]{\dfrac{2}{3ab}}, \, a \neq 0, \, b \neq 0$

$$= \frac{\sqrt[3]{2}}{\sqrt[3]{3ab}}$$

$$= \frac{\sqrt[3]{2}}{\sqrt[3]{3ab}} \cdot \frac{\sqrt[3]{3 \cdot 3 \cdot a \cdot a \cdot b \cdot b}}{\sqrt[3]{3 \cdot 3 \cdot a \cdot a \cdot b \cdot b}}$$

$$= \frac{\sqrt[3]{2 \cdot 3 \cdot 3 \cdot a \cdot a \cdot b \cdot b}}{\sqrt[3]{3 \cdot 3 \cdot 3 \cdot a \cdot a \cdot a \cdot b \cdot b \cdot b}}$$

$$= \frac{\sqrt[3]{18a^2 b^2}}{\sqrt[3]{(3ab)^3}}$$

$$= \frac{\sqrt[3]{2a^2 b^2}}{3ab}$$

c. $\sqrt[3]{\dfrac{3a^4}{7b^2}}, \, a \neq 0, \, b \neq 0$

_____

Objective 1 Rationalize denominators with square roots.
For extra help, see Example 1 on page 517 of your text, the Section Lecture video for Section 8.4, and Exercise Solution Clip 5.

Rationalize each denominator.

1. $\dfrac{5}{\sqrt{5}}$

1. _____

2. $\dfrac{-2}{\sqrt{7}}$

2. _____

3. $\dfrac{-4}{\sqrt{12}}$

3. _____

4. $\dfrac{\sqrt{3}}{\sqrt{8}}$

4. _____

5. $\dfrac{\sqrt{4}}{\sqrt{24}}$

5. _____

Objective 2 Write radicals in simplified form.

For extra help, see Examples 2–4 on pages 518–519 of your text, the Section Lecture video for Section 8.4, and Exercise Solution Clips 19, 45, 59 and 67.

Perform the indicated operations and write all answers in simplest form. Rationalize all denominators. Assume that all variables represent positive real numbers.

6. $\sqrt{\dfrac{5}{7}}$

6. _____

7. $\sqrt{\dfrac{1}{5}} \cdot \sqrt{\dfrac{9}{35}}$

7. _____

8. $\dfrac{\sqrt{5y}}{\sqrt{x}}$

8. _____

9. $\sqrt{\dfrac{5a^2b^3}{6}}$

9. _____

10. $\dfrac{\sqrt{k^2m^4}}{\sqrt{k^5}}$

10. _____

Name: Date:
Instructor: Section:

Objective 3 Rationalize denominators with cube roots.
For extra help, see Example 5 on page 520 of your text, the Section Lecture video for Section 8.4, and Exercise Solution Clip 77.

Rationalize each denominator. Assume that all variables in the denominator represent nonzero real numbers.

11. $\sqrt[3]{\dfrac{2}{5}}$ 11. _____

12. $\dfrac{\sqrt[3]{6}}{\sqrt[3]{9}}$ 12. _____

13. $\dfrac{3}{\sqrt[3]{49}}$ 13. _____

14. $\sqrt[3]{\dfrac{2}{5r^2}}$ 14. _____

15. $\sqrt[3]{\dfrac{x^2}{25y}}$ 15. _____

Chapter 8 ROOTS AND RADICALS

8.5 More Simplifying and Operations with Radicals

Learning Objectives
1   Simplify products of radical expressions.
2   Use conjugates to rationalize denominators of radical expressions.
3   Write radical expressions with quotients in lowest terms.

Key Terms

Use the vocabulary terms listed below to complete each statement in exercises 1–4.

conjugates   **special products**

FOIL   **rationalize the denominator**

1. The _____ method is used to find the products of sums of radicals.

2. Two expressions are called _____ if they are of the form $a + b$ and $a - b$.

3. The expressions $(x + y)^2$, $(x - y)^2$ and $(x + y)(x - y)$ are _____ that can be applied to radicals.

4. To _____ of $\dfrac{3}{\sqrt{5}}$, multiply both the numerator and the denominator by $\sqrt{5}$.

Guided Examples

Review these examples for Objective 1:
1. Find each product and simplify.

   a. $\sqrt{3}\left(\sqrt{75} + \sqrt{6}\right)$

   b. $\left(-2\sqrt{3} - 3\sqrt{5}\right)\left(5\sqrt{5} - \sqrt{3}\right)$

   c. $\left(3\sqrt{6} + \sqrt{5}\right)\left(\sqrt{6} - \sqrt{5}\right)$

Now Try:
1. Find each product and simplify.

   a. $\sqrt{5}\left(\sqrt{12} + 4\sqrt{7}\right)$

   _____

   b. $\left(4\sqrt{2} + 3\sqrt{3}\right)\left(\sqrt{2} - 7\sqrt{3}\right)$

   _____

   c. $\left(\sqrt{5} - \sqrt{8}\right)\left(\sqrt{3} + \sqrt{2}\right)$

   _____

a. $\sqrt{3}\left(\sqrt{75}+\sqrt{6}\right)$

$=\sqrt{3}\left(5\sqrt{3}+\sqrt{6}\right)$ Simplify inside the parenthesis.

$=5\cdot\sqrt{3}\cdot\sqrt{3}+\sqrt{3}\cdot\sqrt{6}$

Distributive property

$=5\cdot3+\sqrt{18}$ $\sqrt{3}\cdot\sqrt{3}=3$; product rule

$=15+\sqrt{18}$ Multiply.

$=15+3\sqrt{2}$ $\sqrt{18}=\sqrt{9\cdot2}=3\sqrt{2}.$

b. $\left(-2\sqrt{3}-3\sqrt{5}\right)\left(5\sqrt{5}-\sqrt{3}\right)$

$=-2\sqrt{2}\left(5\sqrt{5}\right)-2\sqrt{3}\left(-\sqrt{3}\right)$
$\quad-3\sqrt{5}\left(5\sqrt{5}\right)-3\sqrt{5}\left(-\sqrt{3}\right)$

Use the FOIL method to multiply.

$=-10\sqrt{10}+2\cdot3-15\cdot5+3\sqrt{15}$

Product rule

$=-10\sqrt{10}+6-75+3\sqrt{15}$

Multiply.

$=-10\sqrt{10}-69+3\sqrt{15}$

Combine like terms.

c. $\left(3\sqrt{6}+\sqrt{5}\right)\left(\sqrt{6}-\sqrt{5}\right)$

$=3\sqrt{6}\left(\sqrt{6}\right)+3\sqrt{6}\left(-\sqrt{5}\right)$
$\quad+\sqrt{5}\left(\sqrt{6}\right)+\sqrt{5}\left(-\sqrt{5}\right)$ FOIL

$=3\cdot6-3\sqrt{30}+\sqrt{30}-5$ Product rule

$=18-3\sqrt{30}+\sqrt{30}-5$ Multiply.

$=13-2\sqrt{30}$ Combine like terms.

2. Find each product and simplify.

a. $\left(\sqrt{7}+6\right)^{2}$

b. $\left(\sqrt{x}-\sqrt{2}\right)^{2},\ x\geq0$

c. $\left(\sqrt{6}+\sqrt{5}\right)\left(\sqrt{6}-\sqrt{5}\right)$

a. $\left(\sqrt{7}+6\right)^{2}=\left(\sqrt{7}\right)^{2}+2\cdot\sqrt{7}\cdot6+6^{2}$

$(x+y)^{2}=x^{2}+2xy+y^{2}$

$=7+12\sqrt{7}+36$ Multiply.

$=43+12\sqrt{7}$ Combine like terms.

2. Find each product and simplify.

a. $\left(\sqrt{3}+5\right)^{2}$

_____

b. $\left(3-2\sqrt{x}\right)^{2},\ x\geq0$

_____

c. $\left(2\sqrt{3}-5\sqrt{2}\right)\left(2\sqrt{3}+5\sqrt{2}\right)$

_____

b. $\left(\sqrt{x}-\sqrt{2}\right)^2$, $x \geq 0$

$$=\left(\sqrt{x}\right)^2 - 2 \cdot \sqrt{x} \cdot \sqrt{2} + \left(\sqrt{2}\right)^2$$

$$(x-y)^2 = x^2 - 2xy + y^2$$

$$= x - 2\sqrt{2x} + 4 \quad \text{Multiply.}$$

c. $\left(\sqrt{6}+\sqrt{5}\right)\left(\sqrt{6}-\sqrt{5}\right)$

$$=\left(\sqrt{6}\right)^2 - \left(\sqrt{5}\right)^2$$

$$(x+y)(x-y) = x^2 - y^2$$

$$= 6 - 5 = 1$$

Review this example for Objective 2:

3. Simplify by rationalizing each denominator.

a. $\dfrac{5}{5+\sqrt{2}}$ **b.** $\dfrac{2-\sqrt{3}}{2-\sqrt{5}}$

c. $\dfrac{2}{3-\sqrt{x}}$, $x \neq 0$

a. $\dfrac{5}{5+\sqrt{2}} = \dfrac{5\left(5-\sqrt{2}\right)}{\left(5+\sqrt{2}\right)\left(5-\sqrt{2}\right)}$

Multiply the numerator and denominator by the conjugate of the denominator.

$$= \dfrac{5\left(5-\sqrt{2}\right)}{5^2 - \left(\sqrt{2}\right)^2}$$

$$(x+y)(x-y) = x^2 - y^2$$

$$= \dfrac{25-5\sqrt{2}}{25-2} = \dfrac{25-5\sqrt{2}}{23}$$

b. $\dfrac{2-\sqrt{3}}{2-\sqrt{5}} = \dfrac{\left(2-\sqrt{3}\right)\left(2+\sqrt{5}\right)}{\left(2-\sqrt{5}\right)\left(2+\sqrt{5}\right)}$

Multiply the numerator and denominator by the conjugate of the denominator.

$$= \dfrac{4+2\sqrt{5}-2\sqrt{3}-\sqrt{15}}{2^2 - \left(\sqrt{5}\right)^2}$$

FOIL;

$$(x+y)(x-y) = x^2 - y^2$$

Now Try:

3. Simplify by rationalizing each denominator.

a. $\dfrac{4}{\sqrt{3}+2}$

_____

b. $\dfrac{\sqrt{3}+\sqrt{2}}{\sqrt{3}-\sqrt{2}}$

_____

c. $\dfrac{3}{\sqrt{x}-3}$

_____

$$= \frac{4 + 2\sqrt{5} - 2\sqrt{3} - \sqrt{15}}{4 - 5}$$

$$= \frac{4 + 2\sqrt{5} - 2\sqrt{3} - \sqrt{15}}{-1}$$

$$= -4 - 2\sqrt{5} + 2\sqrt{3} + \sqrt{15}$$

c. $\dfrac{2}{3 - \sqrt{x}}, \ x \neq 0$

$$= \frac{2(3 + \sqrt{x})}{(3 - \sqrt{x})(3 + \sqrt{x})}$$

Multiply the numerator and
denominator by the
conjugate of the denominator.

$$= \frac{6 + 2\sqrt{x}}{3^2 - (\sqrt{x})^2} = \frac{6 + 2\sqrt{x}}{9 - x}$$

Review this example for Objective 3:

4. Write the quotient in lowest terms.

$$\frac{72\sqrt{2} - 16\sqrt{7}}{24}$$

$$\frac{72\sqrt{2} - 16\sqrt{7}}{24}$$

$$= \frac{8(9\sqrt{2} - 2\sqrt{7})}{24} \quad \text{Factor.}$$

$$= \frac{9\sqrt{2} - 2\sqrt{7}}{3}$$

Divide numerator and
denominator by 8.

Now Try:

4. Write the quotient in lowest terms.

$$\frac{12 + 6\sqrt{6}}{8}$$ _____

Objective 1 Simplify products of radical expressions.

For extra help, see Examples 1–3 on pages 524–525 of your text, the Section Lecture video for Section 8.5, and Exercise Solution Clips 5, 15, and 21.

Simplify each expression.

1. $\sqrt{2}(\sqrt{2} - \sqrt{5})$

1. _____

2. $\sqrt{7}\left(2\sqrt{8}-9\sqrt{7}\right)$

2. _____

3. $\left(4\sqrt{5}+\sqrt{3}\right)\left(\sqrt{2}-\sqrt{7}\right)$

3. _____

4. $\left(3+\sqrt{5}\right)\left(3-\sqrt{5}\right)$

4. _____

5. $\left(7-4\sqrt{3}\right)^2$

5. _____

Objective 2 Use conjugates to rationalize denominators of radical expressions.
For extra help, see Example 4 on page 526 of your text, the Section Lecture video for Section 8.5, and Exercise Solution Clip 47.

Rationalize each denominator.

6. $\dfrac{2}{7+\sqrt{3}}$

6. _____

7. $\dfrac{4+\sqrt{5}}{\sqrt{5}}$

7. _____

8. $\dfrac{3}{\sqrt{2}-\sqrt{5}}$

8. _____

9. $\dfrac{\sqrt{2}+\sqrt{5}}{\sqrt{3}-\sqrt{2}}$

9. _____

10. $\dfrac{\sqrt{5}+2}{\sqrt{x}-3}$, $x \neq 0$ 10. _____

Objective 3 Write radical expressions with quotients in lowest terms.

For extra help, see Example 5 on page 527 of your text, the Section Lecture video for Section 8.5, and Exercise Solution Clip 69.

Write each quotient in lowest terms.

11. $\dfrac{\sqrt{28}-\sqrt{8}}{6}$ 11. _____

12. $\dfrac{2\sqrt{28}+12}{14}$ 12. _____

13. $\dfrac{6+\sqrt{8}}{2}$ 13. _____

14. $\dfrac{3+\sqrt{27}}{9}$ 14. _____

15. $\dfrac{135\sqrt{3}+25}{5}$ 15. _____

Chapter 8   ROOTS AND RADICALS

8.6   Solving Equations with Radicals

Learning Objectives

1   Solve radical equations having square root radicals.
2   Identify equations with no solutions.
3   Solve equations by squaring a binomial.
4   Solve radical equations having cube root radicals.

Key Terms

Use the vocabulary terms listed below to complete each statement in exercises 1–3.

> **radical equation**   **squaring property**   **extraneous solution**

1. The _____ guarantees that, if each side of a given equation is squared, then all the solutions of the original equation are among the solutions of the squared equation.

2. An equation with a variable in the radicand is a _____.

3. A number is called an _____ if, when checked in the original equation, a false statement results.

Guided Examples

Review these examples for Objective 1:

1. Solve $\sqrt{12-x}=8$.

$$\sqrt{12-x}=8$$

$$\left(\sqrt{12-x}\right)^2=8^2 \quad \text{Squaring property}$$

$$12-x=64 \quad \left(\sqrt{a}\right)^2=a$$

$$-x=52 \quad \text{Subtract 12.}$$

$$x=-52 \quad \text{Multiply by } -1.$$

Check:

$$\sqrt{12-x}=8$$

$$\sqrt{12-(-52)}\overset{?}{=}8$$

$$\sqrt{64}\overset{?}{=}8$$

$$8=8 \;\checkmark$$

Now Try:

1. Solve $\sqrt{7x-6}=8$.

         _____

2. Solve $\sqrt{6r-6}=2\sqrt{r}$.

$$\sqrt{6r-6}=2\sqrt{r}$$

$$\left(\sqrt{6r-6}\right)^2=\left(2\sqrt{r}\right)^2 \quad \text{Squaring property}$$

$$\left(\sqrt{6r-6}\right)^2=2^2\left(\sqrt{r}\right)^2 \quad (ab)^2=a^2b^2$$

$$6r-6=4r$$

$$-6=-2r \qquad \text{Subtract } 6r.$$

$$3=r \qquad \text{Divide by } -2.$$

Check:

$$\sqrt{6r-6}=2\sqrt{r}$$

$$\sqrt{6(3)-6}\overset{?}{=}2\sqrt{3}$$

$$\sqrt{12}\overset{?}{=}2\sqrt{3}$$

$$2\sqrt{3}=2\sqrt{3} \checkmark$$

2. Solve $\sqrt{2+4k}=3\sqrt{k}$.

_____

Review these examples for Objective 2:

3. Solve $\sqrt{x}=-5$.

$$\sqrt{x}=-5$$

$$\left(\sqrt{x}\right)^2=(-5)^2 \quad \text{Squaring property}$$

$$x=5 \qquad (ab)^2=a^2b^2$$

Check:

$$\sqrt{x}=-5$$

$$\sqrt{5}\overset{?}{=}-5$$

$$5=-5 \quad \text{False}$$

A proposed solution that is not an actual solution of the original equation is called an *extraneous solution* and must be rejected. Therefore, $\sqrt{x}=-5$ has no solution. The solution set is $\varnothing$.

4. Solve $r=\sqrt{r^2+6r+12}$.

$$r=\sqrt{r^2+6r+12}$$

$$r^2=\left(\sqrt{r^2+6r+12}\right)^2 \quad \text{Squaring property}$$

$$r^2=r^2+6r+12 \qquad \left(\sqrt{a}\right)^2=a$$

Now Try:

3. Solve $\sqrt{x}+2=0$.

_____

4. Solve $s=\sqrt{s^2+4s+4}$

_____

$$0 = 6r + 12 \qquad \text{Subtract } r^2.$$
$$-6r = 12 \qquad \text{Subtract } 6r.$$
$$r = -2 \qquad \text{Divide by } -6.$$

Check:

$$r = \sqrt{r^2 + 6r + 12}$$

$$-2 \overset{?}{=} \sqrt{(-2)^2 + 6(-2) + 12}$$

$$-2 \overset{?}{=} \sqrt{4 + (-12) + 12}$$

$$-2 \overset{?}{=} \sqrt{4}$$

$$-2 = 2 \quad \text{False}$$

A proposed solution that is not an actual solution of the original equation is called an *extraneous solution* and must be rejected.

Therefore, $\sqrt{x} = -5$ has no solution. The solution set is $\varnothing$.

Review these examples for Objective 3:

5. Solve $\sqrt{x+3} = x - 3$.

$$\sqrt{x+3} = x - 3$$

$$\left(\sqrt{x+3}\right)^2 = (x-3)^2 \qquad \text{Square each side.}$$

$$x + 3 = x^2 - 6x + 9$$

$$0 = x^2 - 7x + 6 \quad \text{Standard form}$$

$$0 = (x-1)(x-6) \qquad \text{Factor.}$$

$$x - 1 = 0 \quad \text{or} \quad x - 6 = 0 \quad \text{Zero-factor property}$$

$$x = 1 \qquad\qquad x = 6$$

Check:

$$\sqrt{x+3} = x - 3$$

$$\sqrt{1+3} \overset{?}{=} 1 - 3 \qquad \bigg| \qquad \sqrt{6+3} \overset{?}{=} 6 - 3$$

$$\sqrt{4} \overset{?}{=} -2 \qquad \bigg| \qquad \sqrt{9} \overset{?}{=} 3$$

$$2 = -2 \quad \text{False} \qquad \bigg| \qquad 3 = 3 \quad \checkmark$$

Only 6 is a valid solution. (1 is an extraneous solution.) The solution set is $\{6\}$.

Now Try:

5. Solve $\sqrt{x+11} = x - 1$.

_____

6. Solve $\sqrt{2x-1}+2=x$.

$$\sqrt{2x-1}+2=x$$

$$\sqrt{2x-1}=x-2 \qquad \text{Isolate the radical.}$$

$$\left(\sqrt{2x-1}\right)^2=(x-2)^2 \qquad \text{Square both sides.}$$

$$2x-1=x^2-4x+4$$

$$0=x^2-6x+5 \qquad \text{Standard form}$$

$$0=(x-1)(x-5) \qquad \text{Factor.}$$

$$x-1=0 \quad \text{or} \quad x-5=0 \qquad \text{Zero-factor property}$$

$$x=1 \qquad\qquad x=5$$

Check:

$$\sqrt{2x-1}+2=x$$

$$\sqrt{2(1)-1}+2\overset{?}{=}1 \qquad \Big| \qquad \sqrt{2(5)-1}+2\overset{?}{=}5$$

$$\sqrt{1}+2\overset{?}{=}1 \qquad \Big| \qquad \sqrt{9}+2\overset{?}{=}5$$

$$1+2\overset{?}{=}1 \qquad \Big| \qquad 3+2\overset{?}{=}5$$

$$3=1 \text{ False} \qquad \Big| \qquad 5=5 \;\checkmark$$

Only 5 is a valid solution. (1 is an extraneous solution.) The solution set is $\{5\}$.

7. Solve $\sqrt{3x}-4=\sqrt{x-2}$.

$$\sqrt{3x}-4=\sqrt{x-2}$$

$$\left(\sqrt{3x}-4\right)^2=\left(\sqrt{x-2}\right)^2$$

$$\text{Square both sides.}$$

$$3x-8\sqrt{3x}+16=x-2$$

$$-8\sqrt{3x}=-2x-18$$

$$\text{Isolate the radical.}$$

$$\left(-8\sqrt{3x}\right)^2=(-2x-18)^2$$

$$\text{Square both sides.}$$

$$192x=4x^2+72x+324$$

$$0=4x^2-120x+324$$

$$\text{Standard form}$$

$$0=4(x-27)(x-3)$$

$$\text{Factor.}$$

$$0=(x-27)(x-3)$$

$$\text{Divide by 4.}$$

6. Solve $5\sqrt{p}-4=p+2$.

_____

7. Solve $\sqrt{3x+4}=\sqrt{9x}-2$.

_____

$x - 27 = 0$ or $x - 3 = 0$ Zero-factor property

$x = 27$ $x = 3$

Check:

$$\sqrt{3x} - 4 = \sqrt{x-2}$$

$$\sqrt{3(27)} - 4 \overset{?}{=} \sqrt{27-2} \quad \bigg| \quad \sqrt{3(3)} - 4 \overset{?}{=} \sqrt{3-2}$$

$$\sqrt{81} - 4 \overset{?}{=} \sqrt{25} \quad\quad\quad \sqrt{9} - 4 \overset{?}{=} \sqrt{1}$$

$$9 - 4 \overset{?}{=} 5 \quad\quad\quad\quad\quad 3 - 4 \overset{?}{=} 1$$

$$5 = 5 \ \checkmark \quad\quad\quad\quad\quad -1 = 1 \ \text{False}$$

Only 27 is a valid solution. (3 is an extraneous solution.) The solution set is $\{27\}$.

Review this example for Objective 4:

8. Solve each equation.

   a. $\sqrt[3]{2x+5} = \sqrt[3]{7x}$

   b. $\sqrt[3]{x^2-6} = \sqrt[3]{4x-1}$

   a. $\sqrt[3]{2x+5} = \sqrt[3]{7x}$

       $\left(\sqrt[3]{2x+5}\right)^3 = \left(\sqrt[3]{7x}\right)^3$ Cube both sides.

         $2x + 5 = 7x$

            $5 = 5x$ Subtract $2x$.

            $1 = x$ Divide by 5.

   Check:

$$\sqrt[3]{2x+5} = \sqrt[3]{7x}$$

$$\sqrt[3]{2(1)+5} \overset{?}{=} \sqrt[3]{7(1)}$$

$$\sqrt[3]{7} = \sqrt[3]{7} \ \checkmark$$

   The solution set is $\{1\}$.

   b. $\sqrt[3]{x^2-6} = \sqrt[3]{4x-1}$

       $\left(\sqrt[3]{x^2-6}\right)^3 = \left(\sqrt[3]{4x-1}\right)^3$

                         Cube both sides.

         $x^2 - 6 = 4x - 1$

       $x^2 - 4x - 5 = 0$ Standard form

      $(x-5)(x+1) = 0$ Factor.

   $x - 5 = 0$ or $x + 1 = 0$ Zero-factor

      $x = 5$ $x = -1$ property

Now Try:

8. Solve each equation.

   a. $\sqrt[3]{5x+3} = 2$ _____

   b. $\sqrt[3]{5x-4} = \sqrt[3]{x^2}$

             _____

Check:

$$\sqrt[3]{x^2 - 6} = \sqrt[3]{4x - 1}$$

$$\sqrt[3]{5^2 - 6} \overset{?}{=} \sqrt[3]{4(5) - 1}$$

$$\sqrt[3]{19} = \sqrt[3]{19} \checkmark$$

$$\sqrt[3]{x^2 - 6} = \sqrt[3]{4x - 1}$$

$$\sqrt[3]{(-1)^2 - 6} \overset{?}{=} \sqrt[3]{4(-1) - 1}$$

$$\sqrt[3]{-5} = \sqrt[3]{-5} \checkmark$$

The solution set is $\{-1, 5\}$.

Objective 1 Solve radical equations having square root radicals.
For extra help, see Examples 1 and 2 on pages 531–532 of your text, the Section Lecture video for Section 8.6, and Exercise Solution Clips 3 and 15.

Solve each equation.

1. $\sqrt{x} - 9 = 0$

1. _____

2. $\sqrt{3x + 1} = 3$

2. _____

3. $\sqrt{3x + 3} = 2\sqrt{3x}$

3. _____

4. $\sqrt{5t + 2} = \sqrt{6t - 1}$

4. _____

Objective 2 Identify equations with no solutions.

For extra help, see Examples 3 and 4 on pages 532–533 of your text, the Section Lecture video for Section 8.6, and Exercise Solution Clips 11 and 21.

Solve each equation that has a solution.

5. $\sqrt{y+3} = -4$ 5. _____

6. $\sqrt{2m+3} = 3\sqrt{m+5}$ 6. _____

7. $\sqrt{d^2+4d+12} + d = 0$ 7. _____

Objective 3 Solve equations by squaring a binomial.

For extra help, see Examples 5–7 on pages 533–535 of your text, the Section Lecture video for Section 8.6, and Exercise Solution Clips 35, 49, and 55.

Find all solutions for each equation.

8. $\sqrt{5x+1} = x+1$ 8. _____

9. $t+1 = \sqrt{t^2+4} - 1$ 9. _____

10. $\sqrt{b-4} = b-6$

10. _____

11. $3\sqrt{p+6} = p+6$

11. _____

12. $\sqrt{x-1} + \sqrt{2x} = 3$

12. _____

Objective 4 Solve radical equations having cube root radicals.
For extra help, see Example 8 on page 535 of your text, the Section Lecture video for Section 8.6, and Exercise Solution Clip 61.

Solve each equation.

13. $\sqrt[3]{2x-1} = \sqrt[3]{3x}$

13. _____

14. $\sqrt[3]{5x+2} = \sqrt[3]{2x+7}$

14. _____

15. $\sqrt[3]{-2x^2 + 5x - 1} = \sqrt[3]{-8x^2 - 2x + 2}$

15. _____

Chapter 8 ROOTS AND RADICALS

8.7 Using Rational Numbers as Exponents

Learning Objectives

1   Define and use expressions of the form $a^{1/n}$.

2   Define and use expressions of the form $a^{m/n}$.

3   Apply the rules for exponents using rational exponents.

4   Use rational exponents to simplify radicals.

Key Terms

Use the vocabulary terms listed below to complete each statement in exercises 1–4.

     rational exponent $a^{1/n}$ $a^{m/n}$ $a^{-m/n}$

1.   The equivalent form of $\left(\sqrt[n]{a}\right)^m$ is _____.

2.   The equivalent form of $\dfrac{1}{a^{m/n}}$ is _____.

3.   A _____ is an exponent that is a rational number.

4.   The equivalent form of $\sqrt[n]{a}$ is _____.

Guided Examples

Review this example for Objective 1:

1.   Simplify.

    a. $81^{1/2}$ **b.** $125^{1/3}$

    c. $81^{1/4}$

    a. $81^{1/2} = \sqrt{81} = 9$

    b. $125^{1/3} = \sqrt[3]{125} = 5$

    c. $81^{1/4} = \sqrt[4]{81} = 3$

Now Try:

1.   Simplify.

    a. $625^{1/2}$ _____

    b. $625^{1/4}$ _____

    c. $243^{1/5}$ _____

     387

Review these examples for Objective 2:

2. Evaluate.

    a. $100^{3/2}$ **b.** $125^{2/3}$

    c. $-729^{5/6}$

    a. $100^{3/2} = \left(100^{1/2}\right)^3 = 10^3 = 1000$

    b. $125^{2/3} = \left(125^{1/3}\right)^2 = 5^2 = 25$

    c. $-729^{5/6} = -\left(729^{1/6}\right)^5 = -3^5 = -243$

3. Evaluate.

    a. $100^{-3/2}$ **b.** $216^{-2/3}$

    a. $100^{-3/2} = \dfrac{1}{100^{3/2}} = \dfrac{1}{\left(100^{1/2}\right)^3}$

                  $= \dfrac{1}{10^3} = \dfrac{1}{1000}$

    b. $216^{-2/3} = \dfrac{1}{216^{2/3}} = \dfrac{1}{\left(216^{1/3}\right)^2}$

                  $= \dfrac{1}{6^2} = \dfrac{1}{36}$

Now Try:

2. Evaluate.

    a. $27^{2/3}$ _____

    b. $16^{3/4}$ _____

    c. $36^{3/2}$ _____

3. Evaluate.

    a. $343^{-2/3}$ _____

    b. $-125^{-4/3}$ _____

Review these examples for Objective 3:

4. Simplify. Write each answer in exponential form with only positive exponents.

    a. $13^{4/5} \cdot 13^{6/5}$ **b.** $\dfrac{8^{3/4}}{8^{-1/4}}$

    c. $\left(\dfrac{81}{256}\right)^{3/4}$ **d.** $\dfrac{3^{2/3} \cdot 3^{-1/3}}{3^{-1/2}}$

    a. $13^{4/5} \cdot 13^{6/5} = 13^{4/5+6/5} = 13^{10/5}$

                        $= 13^2 = 169$

    b. $\dfrac{8^{3/4}}{8^{-1/4}} = 8^{3/4-(-1/4)} = 8^{4/4} = 8^1 = 8$

Now Try:

4. Simplify. Write each answer in exponential form with only positive exponents.

    a. $5^{3/4} \cdot 5^{9/4}$ _____

    b. $\dfrac{5^{3/7}}{5^{4/7}}$ _____

    c. $\left(-\dfrac{32}{3125}\right)^{3/5}$ _____

    d. $\dfrac{6^{3/5} \cdot 6^{-8/5}}{6^{-2}}$ _____

c. $\left(\dfrac{81}{256}\right)^{3/4} = \dfrac{81^{3/4}}{256^{3/4}} = \dfrac{\left(81^{1/4}\right)^3}{\left(256^{1/4}\right)^3}$

$\qquad\qquad = \dfrac{3^3}{4^3} = \dfrac{27}{64}$

d. $\dfrac{3^{2/3} \cdot 3^{-1/3}}{3^{-1/2}} = \dfrac{3^{2/3+(-1/3)}}{3^{-1/2}} = \dfrac{3^{1/3}}{3^{-1/2}}$

$\qquad\qquad = 3^{1/3-(-1/2)} = 3^{5/6}$

5. Simplify. Write each answer in exponential form with only positive exponents. Assume that all variables represent positive numbers.

a. $\left(x^{7/3}y^{-4/3}\right)^6$ **b.** $\left(\dfrac{c^6}{x^3}\right)^{2/3}$

c. $\dfrac{w^{7/4}w^{-1/2}}{w^{5/4}}$

a. $\left(x^{7/3}y^{-4/3}\right)^6 = \left(x^{7/3}\right)^6\left(y^{-4/3}\right)^6$

$\qquad\qquad = x^{(7/3)(6)}y^{(-4/3)(6)}$

$\qquad\qquad = x^{14}y^{-8} = \dfrac{x^{14}}{y^8}$

b. $\left(\dfrac{c^6}{x^3}\right)^{2/3} = \dfrac{\left(c^6\right)^{2/3}}{\left(x^3\right)^{2/3}} = \dfrac{c^{6(2/3)}}{x^{3(2/3)}} = \dfrac{c^4}{x^2}$

c. $\dfrac{w^{7/4}w^{-1/2}}{w^{5/4}} = \dfrac{w^{(7/4-1/2)}}{w^{5/4}} = \dfrac{w^{5/4}}{w^{5/4}} = 1$

5. Simplify. Write each answer in exponential form with only positive exponents. Assume that all variables represent positive numbers.

a. $\left(x^2y^{-5}\right)^{2/5}$ _____

b. $\left(\dfrac{x^{3/8}}{x^{3/2}}\right)^8$ _____

c. $\dfrac{\left(x^{-1}y^{2/3}\right)^3}{\left(x^{1/3}y^{1/2}\right)^2}$ _____

Review this example for Objective 4:
6. Simplify each radical by first writing it in exponential form.

a. $\sqrt[4]{36^2}$ **b.** $\sqrt[3]{x^{12}}$

a. $\sqrt[4]{36^2} = \left(36^2\right)^{1/4} = 36^{1/2} = 6$

b. $\sqrt[3]{x^{12}} = \left(x^{12}\right)^{1/3} = x^4, \; x \ge 0$

Now Try:
6. Simplify each radical by first writing it in exponential form.

a. $\sqrt[4]{100^2}$ _____

b. $\sqrt[20]{x^5}$ _____

Objective 1 Define and use expressions of the form $a^{1/n}$.

For extra help, see Example 1 on page 540 of your text, the Section Lecture video for Section 8.7, and Exercise Solution Clip 5.

Evaluate the expression.

1. $169^{1/2}$

1. _____

2. $8^{1/3}$

2. _____

3. $512^{1/3}$

3. _____

Objective 2 Define and use expressions of the form $a^{m/n}$.

For extra help, see Examples 2 and 3 on page 541 of your text, the Section Lecture video for Section 8.7, and Exercise Solution Clips 13 and 25.

Evaluate the expression.

4. $64^{3/2}$

4. _____

5. $125^{2/3}$

5. _____

6. $9^{-3/2}$

6. _____

7. $1000^{-2/3}$

7. _____

Objective 3 Apply the rules for exponents using rational exponents.

For extra help, see Examples 4 and 5 on pages 541–542 of your text, the Section Lecture video for Section 8.7, and Exercise Solution Clips 31 and 49.

Simplify the expression. Write the answer in exponential form with only positive exponents. Assume that all variables represent positive numbers.

8. $8^{-5/8} \cdot 8^{-3/8}$ 8. _____

9. $\dfrac{6^{4/5}}{6^{-3/5}}$ 9. _____

10. $\left(r^4 s^{2/3}\right)^{3/2}$ 10. _____

11. $\left(\dfrac{r^{1/2}}{s^{-1/2}}\right)^{2/3}$ 11. _____

12. $\dfrac{z^{-2/3} \cdot z^{7/3}}{z^{2/3}}$ 12. _____

13. $\left(a^{-1}\right)^{1/2}\left(a^{-3}\right)^{-1/2}$ 13. _____

Name: Date:
Instructor: Section:

Objective 4 Use rational exponents to simplify radicals.
For extra help, see Example 6 on page 542 of your text, the Section Lecture video for
Section 8.7, and Exercise Solution Clip 59.

*Simplify the radical by first writing it in exponential form. Give the answer as an
integer or a radical in simplest form. Assume that all variables represent nonnegative
numbers.*

14. $\sqrt[9]{125^3}$ 14. _____

15. $\sqrt[8]{a^2}$ 15. _____

Chapter 9 QUADRATIC EQUATIONS

9.1 Solving Quadratic Equations by the Square Root Property

Learning Objectives
1 Review the zero-factor property.
2 Solve equations of the form $x^2 = k$, where $k > 0$.
3 Solve equations of the form $(ax + b)^2 = k$, where $k > 0$.
4 Use formulas involving squared variables.

Key Terms

Use the vocabulary terms listed below to complete each statement in exercises 1–2.

quadratic equation   **square root property**

1. A _____ is an equation that can be written in the form
    $ax^2 + bx + c = 0$.

2. An equation of the form $x^2 = k$ can be solved using the _____.

Guided Examples

Review this example for Objective 1:
1. Solve each equation by the zero-factor
    property.

    a. $x^2 + 3x - 4 = 0$ **b.** $x^2 = 49$

    a. $x^2 + 3x - 4 = 0$
            $(x + 4)(x - 1) = 0$ Factor.

            $x + 4 = 0$ or $x - 1 = 0$ Zero-factor
            $x = -4$ $x = 1$ property; solve.
    The solution set is $\{-4, 1\}$.

    b. $x^2 = 49$

              $x^2 - 49 = 0$ Subtract 49.
        $(x + 7)(x - 7) = 0$ Factor.

        $x + 7 = 0$ or $x - 7 = 0$ Zero-factor
          $x = -7$ $x = 7$ property; solve.
    The solution set is $\{-7, 7\}$.

Now Try:
1. Solve each equation by the
    zero-factor property.

    a. $x^2 + 2x - 63 = 0$

    _____

    b. $x^2 = 81$ _____

Review this example for Objective 2:

2. Solve each equation. Write radicals in simplified form.

    a. $x^2 = 121$ **b.** $x^2 = 23$

    c. $x^2 = -25$ **d.** $3x^2 + 11 = 35$

    a. By the square root property, if
        $x^2 = 121$, then $x = \sqrt{121} = 11$ or
        $x = -\sqrt{121} = -11$. The solution set is
        $\{\pm 11\}$.

    b. The solutions are $x = \sqrt{23}$ or
        $x = -\sqrt{23}$, so the solution set is
        $\left\{\pm\sqrt{23}\right\}$.

    c. Since -25 is a negative number, there
        is no real number solution of this
        equation.

    d. $3x^2 + 11 = 35$

    $\qquad 3x^2 = 24$ Subtract 11.

    $\qquad\quad x^2 = 8$ Divide by 3.

    $\qquad\quad x = \pm\sqrt{8}$ Square root property

    $\qquad\quad x = \pm 2\sqrt{2}$ Simplify

Now Try:

2. Solve each equation. Write radicals in simplified form.

    a. $x^2 = 225$ _____

    b. $x^2 = 47$ _____

    c. $x^2 = -36$ _____

    d. $12 - 4y^2 = 9$ _____

Review these examples for Objective 3:

3. Solve $(x - 2)^2 = 25$.

    $(x - 2)^2 = 25$
    $x - 2 = \sqrt{25}$ or $x - 2 = -\sqrt{25}$
    $\qquad\qquad\qquad$ Square root property
    $x - 2 = 5 \qquad\quad x - 2 = -5 \quad \sqrt{25} = \pm 5$
    $\quad\ x = 7 \qquad\qquad\ x = -3 \quad$ Add 2.

    The solution set is $\{-3, 7\}$.

4. Solve $(3x + 4)^2 = 40$.

    $(3x + 4)^2 = 40$
    $3x + 4 = \sqrt{40}$ or $3x + 4 = -\sqrt{40}$
    $\qquad\qquad\qquad$ Square root property
    $3x + 4 = 2\sqrt{10} \qquad 3x + 4 = -2\sqrt{10}$
    $\qquad\qquad\quad \sqrt{40} = \sqrt{4 \cdot 10} = 2\sqrt{10}$

Now Try:

3. Solve $(2x - 1)^2 = 16$.

    _____

4. Solve $(2x + 5)^2 = 32$.

    _____

$3x = 2\sqrt{10} - 4$ or $3x = -2\sqrt{10} - 4$

Subtract 4.

$$x = \frac{2\sqrt{10} - 4}{3} \qquad x = \frac{-2\sqrt{10} - 4}{3}$$

Divide by 3.

5. Solve $(m+1)^2 = -36$

Because the square root of -36 is not a real number, the solution set is $\varnothing$.

5. Solve $(4a+5)^2 = -12$

_____

Review this example for Objective 4:

6. The formula $s = \dfrac{v^2}{2a}$ is used to find the stopping distance (s), in feet, of a car traveling at v ft/sec which starts decelerating at a constant rate of a ft/sec^2 Use this information to find the velocity of a car, in ft/sec, with stopping distance 80 feet and deceleration rate 40 ft/sec^2 before the brakes are applied.

$$s = \frac{v^2}{2a}$$

$$80 = \frac{v^2}{2(40)} \qquad s = 80,\ a = 40$$

$$80 = \frac{v^2}{80} \qquad \text{Multiply.}$$

$$6400 = v^2 \qquad \text{Multiply by 80.}$$

$$\pm 80 = v \qquad \text{Square root property}$$

We reject the negative solution since v represents velocity. The velocity of the car was 80 ft/sec.

Now Try:

6. Use the formula at the left to determine the velocity of a car with stopping distance 120 feet and deceleration rate 20 ft/sec^2. Round your answer to two decimal places.

_____

Objective 1 Review the zero-factor property.
For extra help, see Example 1 on page 554 of your text and the Section Lecture video for Section 9.1.

Solve each equation using the zero-factor property.

1. $x^2 - 16x + 63 = 0$

1. _____

2. $2x^2 + 5x - 3 = 0$

2. _____

3. $3x^2 = 5x + 28$

3. _____

Objective 2 Solve equations of the form $x^2 = k$, where $k > 0$.
For extra help, see Example 2 on page 555 of your text, the Section Lecture video for Section 9.1, and Exercise Solution Clip 11.

Solve each equation by using the square root property. Express all radicals in simplest form.

4. $r^2 = 900$

4. _____

5. $a^2 = 24$

5. _____

6. $c^2 + 36 = 0$

6. _____

7. $h^2 = \dfrac{90}{289}$

7. _____

Objective 3 Solve equations of the form $(ax + b)^2 = k$, where $k > 0$.
For extra help, see Examples 3–5 on pages 555–556 of your text, the Section Lecture video for Section 9.1, and Exercise Solution Clips 37, 39, and 49.

Solve each equation by using the square root property. Express all radicals in simplest form

8. $(y + 2)^2 = 16$

8. _____

9. $(n+3)^2 = 18$ 9. _____

10. $(3p-2)^2 - 28 = 0$ 10. _____

11. $(7p-4)^2 = -81$ 11. _____

Objective 4 Use formulas involving squared variables.
For extra help, see Example 6 on page 557 of your text, the Section Lecture video for Section 9.1, and Exercise Solution Clip 63.

The formula $V = \pi r^2 h$ is used to find the volume (V) of a right circular cylinder, given its radius r and its height h. Use this information to find the radius for each circular cone with the given dimensions. Round answers to two decimal places, if necessary.

12. $V = 200\pi$ feet^3, $h = 5$ feet   12. _____

13. $V = 3500\pi$ feet^3, $h = 70$ feet   13. _____

The formula $V = \dfrac{1}{3}\pi r^2 h$ is used to find the volume (V) of a right circular cone, given its radius r and its height h. Use this information to find the radius for each cylinder with the given dimensions. Use 3.14 as the value for π. Round answers to two decimal places, if necessary.

14. $V = 782.5$ feet^3, $h = 31.3$ feet   14. _____

15. $V = 90$ in.^{3}, $h = 2.5$ in.   15. _____

Chapter 9 QUADRATIC EQUATIONS

9.2 Solving Quadratic Equations by Completing the Square

Learning Objectives

1   Solve quadratic equations by completing the square when the coefficient of the second-degree term is 1.
2   Solve quadratic equations by completing the square when the coefficient of the second-degree term is not 1.
3   Simplify the terms of an equation before solving.
4   Solve applied problems that require quadratic equations.

Key Terms

Use the vocabulary terms listed below to complete each statement in exercises 1–3.

completing the square   square root property   perfect square trinomial

1.   A _____ can be written in the form $x^2 + 2kx + k^2$ or $x^2 - 2kx + k^2$.

2.   The process of rewriting a quadratic equation written in standard form to allow for the solution to be found by the square root property is called

   _____.

3.   The _____ says that, if k is positive and $a^2 = k$, then $a = \pm\sqrt{k}$.

Guided Examples

Review these examples for Objective 1:

1.   Complete each trinomial so that it is a perfect square. Then factor the trinomial.

   a. $x^2 + 10x +$ _____   **b.** $x^2 - 4x +$ _____

   a. The perfect square trinomial will have the form $x^2 + 2kx + k^2$. Thus, the middle term, $10x$, must equal $2kx$.
   $$10x = 2kx$$
   $$5 = k \quad \text{Divide by } 2x.$$
   Therefore, $k = 5$ and $k^2 = 25$. The required trinomial is $x^2 + 10x + 25$, which factors as $(x+5)^2$.

Now Try:

1.   Complete each trinomial so that it is a perfect square. Then factor the trinomial.

   a. $x^2 + 16x +$ _____

   _____

   b. $x^2 - 24x +$ _____

   _____

b. The perfect square trinomial will have the form $x^2 - 2kx + k^2$. Thus, the middle term, $-4x$, must equal $2kx$.

$-4x = -2kx$

$2 = k$ Divide by $-2x$.

Therefore, $k = 2$ and $k^2 = 4$. The required trinomial is $x^2 - 4x + 4$, which factors as $(x-2)^2$.

2. Solve $x^2 - 10x = -11$.

To solve this equation with the square root property, the quantity on the left must be written as a perfect square trinomial.

$2kx = -10x$, so $k = -5$ and $k^2 = 25$. Thus, the required perfect square trinomial is $x^2 - 10x + 25$, which factors as $(x-5)^2$. Therefore, if we add 25 to each side of $x^2 - 10x = -11$, the equation will have a perfect square trinomial on the left side.

$x^2 - 10x = -11$

$x^2 - 10x + 25 = -11 + 25$ Add 25.

$(x-5)^2 = 14$ Factor; add.

Now use the square root property to solve for x.

$x - 5 = -\sqrt{14}$ or $x - 5 = \sqrt{14}$

$x = 5 - \sqrt{14}$ $x = 5 + \sqrt{14}$

Check by substituting $5 - \sqrt{14}$ and $5 + \sqrt{14}$ for x in the original equation. The solution set is $\{5 \pm \sqrt{14}\}$.

3. Solve $x^2 - 12x + 24 = 0$.

$x^2 - 12x + 24 = 0$

$x^2 - 12x = -24$ Subtract 24.

$x^2 - 12x + 36 = -24 + 36$

             $2kx = -12$, so $k = -6$

             and $k^2 = 36$.

$(x-6)^2 = 12$ Factor.

2. Solve $r^2 + 8r = -4$.

_____

3. Solve $x^2 - 6x + 1 = 0$.

_____

Now use the square root property to solve for x.

$$x - 6 = -\sqrt{12} \quad \text{or} \quad x - 6 = \sqrt{12}$$
$$x - 6 = -2\sqrt{3} \qquad\qquad x - 6 = 2\sqrt{3}$$
$$x = 6 - 2\sqrt{3} \qquad\qquad x = 6 + 2\sqrt{3}$$

The solution set is $\left\{6 \pm 2\sqrt{3}\right\}$.

Review these examples for Objective 2:	**Now Try:**

4. Solve $4n^2 + 4n - 15 = 0$

4. Solve $9x^2 + 6x - 8 = 0$

In order to complete the square, the coefficient of n^2 must be 1, so start by dividing each side of the equation by 4.

$$4n^2 + 4n - 15 = 0$$

$$n^2 + n - \frac{15}{4} = 0$$

$$n^2 + n = \frac{15}{4} \quad \text{Add } \tfrac{15}{4}.$$

Now complete the square on the left by taking half the coefficient of n and squaring it, then adding the result to both sides of the equation. The coefficient of n is 1, so add $\left(\frac{1}{2}\right)^2 = \frac{1}{4}$ to each side.

$$n^2 + n + \frac{1}{4} = \frac{15}{4} + \frac{1}{4}$$

$$\left(n + \frac{1}{2}\right)^2 = \frac{16}{4} = 4$$

Solve the equation by using the square root property.

$$n + \frac{1}{2} = -\sqrt{4} \quad \text{or} \quad n + \frac{1}{2} = \sqrt{4}$$

$$n + \frac{1}{2} = -2 \qquad\qquad n + \frac{1}{2} = 2$$

$$n = -\frac{5}{2} \qquad\qquad n = \frac{3}{2}$$

Check by substituting each value of n into the original equation. The solution set is $\left\{-\frac{5}{2}, \frac{3}{2}\right\}$.

5. Solve $3n^2 + 7n + 2 = 0$

$3n^2 + 7n + 2 = 0$

$n^2 + \dfrac{7}{3}n + \dfrac{2}{3} = 0$ Divide by 3.

$n^2 + \dfrac{7}{3}n = -\dfrac{2}{3}$ Subtract $\dfrac{2}{3}$.

Now complete the square on the left by taking half the coefficient of n and squaring it, then adding the result to both sides of the equation. The coefficient of n is $\dfrac{7}{3}$, so add

$\left(\dfrac{7}{6}\right)^2 = \dfrac{49}{36}$ to each side.

$n^2 + \dfrac{7}{3}n + \dfrac{49}{36} = -\dfrac{2}{3} + \dfrac{49}{36}$

$\left(n + \dfrac{7}{6}\right)^2 = \dfrac{25}{36}$

Solve the equation by using the square root property.

$n + \dfrac{7}{6} = -\sqrt{\dfrac{25}{36}}$ or $n + \dfrac{7}{6} = \sqrt{\dfrac{25}{36}}$

$n + \dfrac{7}{6} = -\dfrac{5}{6}$ $n + \dfrac{7}{6} = \dfrac{5}{6}$

$n = -\dfrac{12}{6} = -2$ $n = -\dfrac{2}{6} = -\dfrac{1}{3}$

Check by substituting each value of n into the original equation. The solution set is $\left\{-2, -\dfrac{1}{3}\right\}$.

6. Solve $4x^2 - 12x = -25$

$4x^2 - 12x = -25$

$x^2 - 3x = -\dfrac{25}{4}$ Divide by 4.

Now complete the square on the left by taking half the coefficient of x and squaring it, then adding the result to both sides of the equation. The coefficient of x is -3, so add

$\left(-\dfrac{3}{2}\right)^2 = \dfrac{9}{4}$ to each side.

5. Solve $16x^2 - 56x = 207$.

_____

6. Solve $4x^2 - 8x + 5 = 0$.

_____

$$x^2 - 3x + \frac{9}{4} = -\frac{25}{4} + \frac{9}{4}$$

$$\left(x - \frac{3}{2}\right)^2 = -\frac{16}{4} = -4$$

We cannot use the square root property to solve this equation since the right side is −4, and the square root is not a real number. Therefore, the solution set is ∅.

Review this example for Objective 3:	Now Try:
7. Solve $(b-1)(b+7) = 12$.	7. Solve $(j+3)(j-2) = 5$.

$$(b-1)(b+7) = 12$$

$$b^2 + 6b - 7 = 12 \qquad \text{Use FOIL.}$$

$$b^2 + 6b = 19 \qquad \text{Add 7.}$$

$$b^2 + 6b + 9 = 19 + 9 \quad \text{Complete the square.}$$

$$(b+3)^2 = 28 \qquad \text{Factor.}$$

Solve the equation by using the square root property.

$$b + 3 = -\sqrt{28} \qquad \text{or} \quad b + 3 = \sqrt{28}$$

$$b + 3 = -2\sqrt{7} \qquad\qquad b + 3 = 2\sqrt{7}$$

$$b = -3 - 2\sqrt{7} \qquad\qquad b = -3 + 2\sqrt{7}$$

The solution set is $\left\{-3 \pm 2\sqrt{7}\right\}$.

Review this example for Objective 4:	Now Try:
8. A certain projectile is located at a distance of $d = 3t^2 - 6t + 1$ feet from its starting point after t seconds. How many seconds will it take the projectile to travel 10 feet?	8. A ball is thrown downward from a tower 280 feet high. The distance the object has fallen at time t in seconds is given by $s = 16t^2 + 64t$. How long will it take for the object to fall 100 feet? (Round your answer to the nearest tenth.)

Let $d = 10$ in the formula and solve for t.

$$d = 3t^2 - 6t + 1$$

$$10 = 3t^2 - 6t + 1$$

$$9 = 3t^2 - 6t \qquad \text{Subtract 1.}$$

$$3 = t^2 - 2t \qquad \text{Divide by 3.}$$

$$3 + 1 = t^2 - 2t + 1 \quad \text{Complete the square.}$$

$$4 = (t-1)^2 \qquad \text{Factor.}$$

Solve the equation by using the square root property.

$$t - 1 = -\sqrt{4} \quad \text{or} \quad t - 1 = \sqrt{4}$$
$$t - 1 = -2 \qquad\qquad t - 1 = 2$$
$$t = -1 \qquad\qquad\quad t = 3$$

Since t represents time, we reject the negative solution. It will take 3 seconds for the projectile to travel 10 feet.

Objective 1   Solve quadratic equations by completing the square when the coefficient of the second-degree term is 1.

For extra help, see Examples 1–3 on pages 560–561 of your text, the Section Lecture video for Section 9.2, and Exercise Solution Clips 1, 13, and 15.

Solve each equation by completing the square.

1. $x^2 + 2x = 63$

1. _____

2. $z^2 + 3z - \frac{7}{4} = 0$

2. _____

3. $b^2 + 5b - 5 = 0$

3. _____

4. $m^2 - 6m = -12$

4. _____

**Objective 2   Solve quadratic equations by completing the square when the coefficient
of the second-degree term is not 1.**

For extra help, see Examples 4–6 on pages 562–564 of your text, the Section Lecture
video for Section 9.2, and Exercise Solution Clips 19 and 21.

Solve each equation by completing the square.

5. $2x^2 - 13x + 20 = 0$

5. _____

6. $3r^2 = 6r + 2$

6. _____

7. $6x^2 - x = 15$

7. _____

8. $3x^2 - 2x + 4 = 0$

8. _____

Objective 3 Simplify the terms of an equation before solving.

For extra help, see Example 7 on page 564 of your text, the Section Lecture video for
Section 9.2, and Exercise Solution Clip 29.

Simplify each of the following equations and then solve by completing the square.

9. $3p^2 = 3p + 5$

9. _____

10. $4y^2 + 6y = 2y + 3$

10. _____

21. $6y^2 + 3y = 4y^2 + y - 5$ **11.** _____

12. $(s+3)(s+1) = 1$ **12.** _____

Objective 4 Solve applied problems that require quadratic equations.

For extra help, see Example 8 on page 565 of your text, the Section Lecture video for Section 9.2, and Exercise Solution Clip 37.

Solve each problem. Round answers to the nearest tenth if necessary.

13. The time t in seconds for a car to skid 48 feet is given   **13.** _____
(approximately) by $48 = 64t - 16t^2$. Solve this equation
for t. Are both answers reasonable?

14. If Pablo throws an object upward from a height of 32 feet   **14.** _____
with an initial velocity of 48 feet per second, then its
height h (in feet) after t seconds is given by the formula
$h = -16t^2 + 48t + 32$. At what times will it be 50 feet
above the ground?

15. George and Albert have found that the profit (in dollars)   **15.** _____
from their cigar shop is given by the formula
$P = -10x^2 + 100x + 300$, where x is the number of units
of cigars sold daily. How many units should be sold for a
profit of $460?

Chapter 9 QUADRATIC EQUATIONS

9.3   Solving Quadratic Equations by the Quadratic Formula

Learning Objectives
1   Identify the values of a, b, and c in a quadratic equation.
2   Use the quadratic formula to solve quadratic equations.
3   Solve quadratic equations with only one solution.
4   Solve quadratic equations with fractions.

Key Terms

Use the vocabulary terms listed below to complete each statement in exercises 1–2.

   quadratic formula   discriminant

1.   The _____ can be used to solve any quadratic equation.

2.   The value of the _____ is used to determine the number of solutions to a quadratic equation.

Guided Examples

Review this example for Objective 1:
1. Write each equation in standard form, if necessary, with 0 on the right side. Then identify the values of a, b, and c. Do not actually solve the equation.

   a. $5x^2 - 5x + 1 = 0$ **b.** $-x^2 + 3x + 1 = 8$

   c. $9x^2 - 16 = 0$

   d. $3(x+5)^2 = 6x - 48$

   a. $5x^2 - 5x + 1 = 0$
   $a = 5$, $b = -5$, $c = 1$

   b. $-x^2 + 3x + 1 = 8$
   Subtract 1 from each side in order to write the equation in standard form.
   $-x^2 + 3x - 7 = 0$
   $a = -1$, $b = 3$, $c = -7$

   c. $9x^2 - 16 = 0$
   The x term is missing, so write the equation as $9x^2 + 0x - 16 = 0$.
   $a = 9$, $b = 0$, $c = -16$.

Now Try:
1. Write each equation in standard form, if necessary, with 0 on the right side. Then identify the values of a, b, and c. Do not actually solve the equation.

   a. $3x^2 - 2x - 4 = 0$

   _____

   b. $5x^2 + 6 = -24x - 9$

   _____

   c. $25y^2 = 49$

   _____

   d. $2(x-2)^2 = 4x - 10$

   _____

d.
$$3(x+5)^2 = 6x - 48$$
$$3\left(x^2 + 10x + 25\right) = 6x - 48 \quad \text{Use FOIL.}$$
$$3x^2 + 30x + 75 = 6x - 48 \quad \begin{array}{l}\text{Distributive}\\ \text{property}\end{array}$$
$$3x^2 - 24x + 123 = 0 \quad \begin{array}{l}\text{Subtract } 6x;\\ \text{add } 48.\end{array}$$
$$a = 3, \, b = -24, \, c = 123$$

Review these examples for Objective 2:	**Now Try:**
2. Use the quadratic formula to solve $5t^2 - 13t + 6 = 0$.	**2.** Use the quadratic formula to solve $6x^2 - 17x + 12 = 0$.

Review these examples for Objective 2:

2. Use the quadratic formula to solve $5t^2 - 13t + 6 = 0$.

$a = 5, \, b = -13, \, c = 6$

$$x = \frac{-b \pm \sqrt{b^2 - 4ac}}{2a}$$

$$x = \frac{-(-13) \pm \sqrt{(-13)^2 - 4(5)(6)}}{2(5)}$$

$$= \frac{13 \pm \sqrt{49}}{10} = \frac{13 \pm 7}{10}$$

$$x = \frac{20}{10} = 2 \text{ or } x = \frac{6}{10} = \frac{3}{5}$$

The solution set is $\left\{\frac{3}{5}, 2\right\}$.

3. Solve $4x(x+1) = 1$.

We must write the equation in standard form.
$$4x(x+1) = 1$$
$$4x^2 + 4x = 1 \quad \text{Distributive property}$$
$$4x^2 + 4x - 1 = 0$$
$$a = 4, \, b = 4, \, c = -1$$

$$x = \frac{-b \pm \sqrt{b^2 - 4ac}}{2a}$$

$$x = \frac{-4 \pm \sqrt{4^2 - 4(4)(-1)}}{2(4)} = \frac{-4 \pm \sqrt{32}}{8}$$

$$x = \frac{-4 \pm 4\sqrt{2}}{8} = \frac{-1 \pm \sqrt{2}}{2}$$

The solution set is $\left\{\frac{-1 \pm \sqrt{2}}{2}\right\}$.

Now Try:

2. Use the quadratic formula to solve $6x^2 - 17x + 12 = 0$.

_____

3. Solve $2x^2 = 2x + 3$.

_____

Review this example for Objective 3:

4. Solve $100p^2 = -60p - 9$.

    First write the equation in standard form.
    $$100p^2 = -60p - 9$$
    $$100p^2 + 60p + 9 = 0$$
    $$a = 100, \, b = 60, \, c = 9$$
    $$x = \frac{-b \pm \sqrt{b^2 - 4ac}}{2a}$$
    $$x = \frac{-60 \pm \sqrt{60^2 - 4(100)(9)}}{2(100)}$$
    $$= \frac{-60 \pm \sqrt{0}}{200} = \frac{-60}{200} = -\frac{3}{10}$$
    The solution set is $\left\{-\frac{3}{10}\right\}$.

Now Try:

4. Solve $25x^2 - 80x + 64 = 0$.

    _____

Review this example for Objective 4:

5. Solve $x^2 + \frac{2}{3}x - \frac{10}{9} = 0$.

    Clear the fraction by multiplying by the LCD, 9.
    $$x^2 + \frac{2}{3}x - \frac{10}{9} = 0$$
    $$9x^2 + 9\left(\frac{2}{3}\right)x - 9\left(\frac{10}{9}\right) = 9(0)$$
    $$9x^2 + 6x - 10 = 0$$
    $$a = 9, \, b = 6, \, c = -10$$
    $$x = \frac{-b \pm \sqrt{b^2 - 4ac}}{2a}$$
    $$x = \frac{-6 \pm \sqrt{6^2 - 4(9)(-10)}}{2(9)}$$
    $$= \frac{-6 \pm \sqrt{396}}{18} = \frac{-6 \pm 6\sqrt{11}}{18}$$
    $$= \frac{-1 \pm \sqrt{11}}{3}$$
    The solution set is $\left\{-\frac{1 \pm \sqrt{11}}{3}\right\}$.

Now Try:

5. Solve $\frac{1}{16}z^2 + \frac{7}{8}z = -\frac{1}{2}$.

    _____

Objective 1 Identify the values of a, b, and c in a quadratic equation.

For extra help, see Example 1 on page 568 of your text, the Section Lecture video for Section 9.3, and Exercise Solution Clip 1.

Write each equation in standard form, if necessary, with 0 on the right side. Then identify the values of a, b, and c. Do not actually solve the equation.

1. $3d^2 = 2d - 4$

1. _____

2. $3p^2 = 12$

2. _____

3. $(z+1)(z+2) = -7$

3. _____

Objective 2 Use the quadratic formula to solve quadratic equations.

For extra help, see Examples 2 and 3 on pages 569–570 of your text, the Section Lecture video for Section 9.3, and Exercise Solution Clips 15 and 17.

Use the quadratic formula to solve each equation.

4. $3x^2 - 7x - 6 = 0$

4. _____

5. $x^2 + 5x - 8 = 0$

5. _____

6. $(2x-5)(2x+7) = 0$

6. _____

7. $5k^2 + 4k + 2 = 0$

7. _____

Objective 3 Solve quadratic equations with only one solution.
For extra help, see Example 4 on page 570 of your text, the Section Lecture video for
Section 9.3, and Exercise Solution Clip 19.

Use the quadratic formula to solve each equation.

8. $16a^2 - 8a + 1 = 0$ 8. _____

9. $16x^2 - 24x = -9$ 9. _____

10. $9r^2 = 6r - 1$ 10. _____

Objective 4 Solve quadratic equations with fractions.
For extra help, see Example 5 on page 571 of your text, the Section Lecture video for
Section 9.3, and Exercise Solution Clip 47.

Use the quadratic formula to solve each equation.

11. $\frac{1}{2}x^2 + 2x - 3 = 0$ 11. _____

12. $\frac{1}{6}y^2 + \frac{1}{2}y = \frac{2}{3}$ 12. _____

13. $\dfrac{1}{4}t^2 - \dfrac{1}{3}t + \dfrac{5}{12} = 0$

13. _____

14. $r^2 + \dfrac{4}{3}r - \dfrac{1}{3} = 0$

14. _____

15. $\dfrac{1}{3}x^2 - \dfrac{1}{2}x = \dfrac{3}{2}$

15. _____

Chapter 9 QUADRATIC EQUATIONS

9.4   Complex Numbers

Learning Objectives
1   Write complex numbers as multiples of i.
2   Add and subtract complex numbers.
3   Multiply complex numbers.
4   Write complex number quotients in standard form.
5   Solve quadratic equations with complex number solutions.

Key Terms

Use the vocabulary terms listed below to complete each statement in exercises 1–6.

complex number   **real part**   **imaginary part**

pure imaginary number   **standard form (of a complex number)**

conjugate (of a complex number)

1.   A complex number $a + bi$ with $a = 0$ and $b \neq 0$ is called a
      _____.

2.   A _____ is a number of the form $a + bi$ where a and b are real
      numbers.

3.   The _____ of $a + bi$ is $a - bi$.

4.   A complex number written in the form $a + bi$ is in _____.

5.   The _____ of $a + bi$ is bi.

6.   The _____ of $a + bi$ is a.

Guided Examples

Review this example for Objective 1:

1.   Write $\sqrt{-24}$ as a multiple of i.

$$\sqrt{-24} = i\sqrt{24} = i\sqrt{4} \cdot \sqrt{6} = 2i\sqrt{6}$$

Now Try:

1.   Write $\sqrt{-72}$ as a multiple of i.

      _____

Review this example for Objective 2:

2. Add or subtract.

   a. $(-2+9i)+(10-3i)$

   b. $(7-9i)-(5-6i)$

   a. $(-2+9i)+(10-3i)$

   $= (-2+10)+(9i-3i)$ Add real parts.
   Add imaginary parts.

   $= 8+6i$

   b. $(7-9i)-(5-6i)$

   $= (7-9i)+(-5+6i)$ Definition of subtraction

   Add real parts.
   Add imaginary parts.

   $= (7-5)+(-9i+6i)$

   $= 2-3i$

Now Try:

2. Add or subtract.

   a. $(1-2i)+(-3+11i)$

   _____

   b. $(-8-i)-(3-5i)$

   _____

Review this example for Objective 3:

3. Find each product.

   a. $6i(8-5i)$

   b. $(3-2i)(2-3i)$

   c. $(1+3i)^2$

   a. $6i(8-5i) = 6i(8)+6i(-5i)$

   Distributive property

   $= 48i-30i^2$

   $= 48i+30 \quad i^2 = -1$

   $= 30+48i \quad$ Standard form

   b. $(3-2i)(2-3i)$

   $= 3(2)+3(-3i)-2i(2)-2i(-3i)$

   Use FOIL.

   $= 6-9i-4i+6i^2$ Multiply.

   $= 6-13i+6(-1)$ Combine terms; $i^2 = -1$

   $= 6-13i-6 \quad$ Multiply.

   $= -13i \qquad$ Add.

   c. $(1+3i)^2 = 1^2+2(1)(3i)+(3i)^2$

   $= 1+6i+9i^2$

   $= 1+6i+9(-1)$

   $= 1+6i-9$

   $= -8+6i$

Now Try:

3. Find each product.

   a. $7i(2-3i)$

   _____

   b. $(-4+6i)(3+2i)$

   _____

   c. $(5-3i)(5+3i)$

   _____

   413

Review this example for Objective 4:

4. Write each quotient in standard form.

   a. $\dfrac{1+i}{2-i}$

   b. $\dfrac{2-i}{4i}$

   a. To divide complex numbers, multiply both the numerator and denominator by the conjugate of the denominator. The conjugate of $2-i$ is $2+i$.

   $\dfrac{1+i}{2-i} = \dfrac{1+i}{2-i} \cdot \dfrac{2+i}{2+i}$

   $= \dfrac{2+i+2i+i^2}{4-i^2}$ Multiply.

   $= \dfrac{2+i+2i+(-1)}{4-(-1)}$ $i^2 = -1$

   $= \dfrac{1+3i}{5}$ Combine like terms.

   b. The conjugate of $4i$ is $-4i$.

   $\dfrac{2-i}{4i} = \dfrac{2-i}{4i} \cdot \dfrac{-4i}{-4i}$

   $= \dfrac{-8i+4i^2}{-16i^2}$ Multiply.

   $= \dfrac{-8i+4(-1)}{-16(-1)}$ $i^2 = -1$

   $= \dfrac{-4-8i}{16}$ Combine like terms.

   $= \dfrac{-4(1+2i)}{16}$ Factor.

   $= -\dfrac{1+2i}{4}$ Divide out the common factor.

Now Try:

4. Write each quotient in standard form.

   a. $\dfrac{2+3i}{2+5i}$

   _____

   b. $\dfrac{3-i}{-2i}$

   _____

Review these examples for Objective 5:

5. Solve $(x-3)^2 = -9$ for complex solutions.

   We will use the square root property.

   $(x-3)^2 = -9$

   $x-3 = \sqrt{-9}$ or $x-3 = -\sqrt{-9}$

   $x-3 = 3i$ $x-3 = -3i$

   $x = 3+3i$ $x = 3-3i$

   The solution set is $\{3 \pm 3i\}$.

Now Try:

5. Solve $(t+4)^2 = -1$ for complex solutions.

   _____

6. Solve $-7x^2 = 5x + 3$ for complex solutions.

Write the equation in standard form as $-7x^2 - 5x - 3 = 0$. Then $a = -7$, $b = -5$, and $c = -3$.

$$x = \frac{-b \pm \sqrt{b^2 - 4ac}}{2a}$$

$$= \frac{-(-5) \pm \sqrt{(-5)^2 - 4(-7)(-3)}}{2(-7)}$$

$$= \frac{5 \pm \sqrt{-59}}{-14} = -\frac{5}{14} \pm \frac{i\sqrt{59}}{14}$$

The solution set is $\left\{-\frac{5}{14} \pm \frac{i\sqrt{59}}{14}\right\}$.

6. Solve $3x = 2x^2 + 5$ for complex solutions.

_____

Objective 1 Write complex numbers as multiples of i.

For extra help, see Example 1 on page 575 of your text, the Section Lecture video for Section 9.4, and Exercise Solution Clip 1.

Write each number as a multiple of i.

1. $\sqrt{-100}$

1. _____

2. $\sqrt{-75}$

2. _____

3. $\sqrt{-128}$

3. _____

Objective 2 Add and subtract complex numbers.

For extra help, see Example 2 on page 576 of your text, the Section Lecture video for Section 9.4, and Exercise Solution Clip 9.

Add or subtract.

4. $(-11 - 5i) + (-7 + 12i)$

4. _____

5. $(7 - 4i) - (-1 + 2i)$

5. _____

6. $(-3+i)-9$

6. _____

Objective 3 Multiply complex numbers.

For extra help, see Example 3 on pages 576–577 of your text, the Section Lecture video for Section 9.4, and Exercise Solution Clip 25.

Find each product.

7. $2i(5-i)$

7. _____

8. $(5-3i)(2+i)$

8. _____

9. $(2+7i)(2-7i)$

9. _____

Objective 4 Divide complex numbers.

For extra help, see Example 4 on page 578 of your text and the Section Lecture video for Section 9.4.

Write each quotient in standard form.

10. $\dfrac{3-4i}{2+3i}$

10. _____

11. $\dfrac{3+5i}{-2i}$

11. _____

12. $\dfrac{i}{4+5i}$

12. _____

Objective 5 Solve quadratic equations with complex number solutions.
For extra help, see Examples 5 and 6 on pages 578–579 of your text, the Section Lecture video for Section 9.4 and Exercise Solution Clip 51.

Solve each quadratic equation for complex solutions. Write the solution in standard form.

13. $(3z - 2)^2 = -72$ 13. _____

14. $(2x - 5)^2 = -8$ 14. _____

15. $k^2 - 2k + 2 = 0$ 15. _____

Chapter 9 QUADRATIC EQUATIONS

9.5   More on Graphing Quadratic Equations; Quadratic Functions

---

Learning Objectives

1   Graph quadratic equations of the form $y = ax^2 + bx + c$ $(a \neq 0)$.
2   Use a graph to determine the number of real solutions of a quadratic equation.

Key Terms

Use the vocabulary terms listed below to complete each statement in exercises 1–4.

   parabola   .   **axis (of symmetry)**   **vertex**   **quadratic function**

1.   A _____ is a graph of a quadratic equation in two variables.

2.   The _____ of a parabola is the lowest point on a parabola that opens up or the highest point on a parabola that opens down.

3.   A function of the form $f(x) = ax^2 = bx + c$ $(a \neq 0)$ is a _____.

4.   The vertical line through the vertex is called the _____.

Guided Examples

Review these examples for Objective 1:

1.   Graph $y = x^2 + 4x - 5$.

Start by finding the x–intercepts. Let $y = 0$ and solve for x.

$$x^2 + 4x - 5 = 0$$
$$(x+5)(x-1) = 0 \quad \text{Factor.}$$
$$x + 5 = 0 \quad \text{or} \quad x - 1 = 0 \quad \text{Zero-factor property}$$
$$x = -5 \qquad\qquad x = 1$$

There are two x-intercepts, $(-5, 0)$ and $(1, 0)$.
Since the x-value of the vertex is halfway between the x-values of the two x-intercepts, it is half their sum.

$$x = \tfrac{1}{2}(-5+1) = \tfrac{1}{2}(-4) = -2$$

Find the corresponding y-value by substituting -2 for x in the original equation: $y = (-2)^2 + 4(-2) - 5 = -9$
The vertex is $(-2, -9)$.

Now Try:

1.   Graph $y = x^2 + x - 2$.

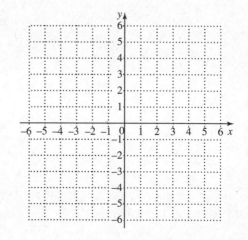

The axis is the line $x = -2$.

To find the y-intercept, substitute $x = 0$ in the original equation:

$$y = (0)^2 + 4(0) - 5 = -5$$

The y-intercept is $(0, -5)$.

Now plot the three intercepts and the vertex. Find additional ordered pairs, if necessary. For example, if $x = -4$, then

$$y = (-4)^2 + 4(-4) - 5 = -5.$$

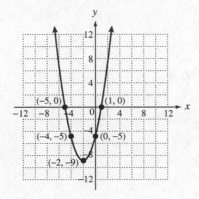

2.   Graph $y = -x^2 + 4x - 1$.

Step 1 Find the vertex. The x-value of the vertex is given by $x = -\dfrac{b}{2a} = -\dfrac{4}{2(-1)} = 2$.

The y-value of the vertex is

$$y = -(2)^2 + 4(2) - 1 = 3.$$

The vertex is $(2, 3)$.

Step 2 Find the y-intercept by letting $x = 0$ in the original equation.

$$y = -(0)^2 + 4(0) - 1 = -1$$

Step 3 Find the x-intercepts by letting $y = 0$ and solving for x, either by factoring or by using the quadratic formula.

$$0 = -x^2 + 4x - 1$$

2.   Graph $y = x^2 - 6x + 11$.

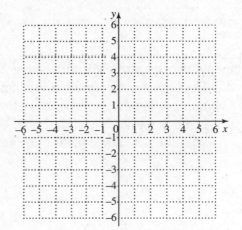

$$x = \frac{-b \pm \sqrt{b^2 - 4ac}}{2a}$$

$$= \frac{-4 \pm \sqrt{4^2 - 4(-1)(-1)}}{2(-1)}$$

$$= \frac{-4 \pm \sqrt{12}}{-2} = \frac{-4 \pm 2\sqrt{3}}{-2}$$

$$= 2 \pm \sqrt{3} \approx 0.3, \ 3.7$$

The x-intercepts are $(0.3, 0)$ and $(3.7, 0)$.
Steps 4 and 5 Additional points are
$(-1, -6)$, $(1, 2)$, $(3, 2)$, $(4, -1)$, and $(5, -6)$.
Plot the intercepts, vertex, and any
additional points. Connect the points with a
smooth curve.

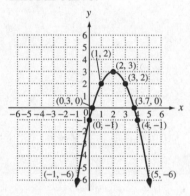

Review this example for Objective 2:

3. Decide from the graph how many real
    number solutions there are of the
    corresponding equation $f(x) = 0$. Give the
    solution set. The equation of the graph is
    $-x^2 + 4x - 3$.

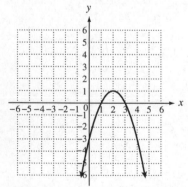

There are two real solutions, $\{1, 3\}$.

Now Try:

3. Decide from the graph how
    many real number solutions
    there are of the corresponding
    equation $f(x) = 0$. Give the
    solution set. The equation of the
    graph is $y = \frac{3}{4}x^2 - 3x + 4$.

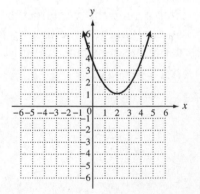

Name: Date:
Instructor: Section:

Objective 1 Graph quadratic equations of the form $y = ax^2 + bx + c$ $(a \neq 0)$.

For extra help, see Examples 1 and 2 on pages 581–583 of your text, the Section Lecture video for Section 9.5 and Exercise Solution Clips 5 and 7.

Give the coordinates of the vertex and sketch the graph of each equation.

1. $y = -(x+1)^2$

1.

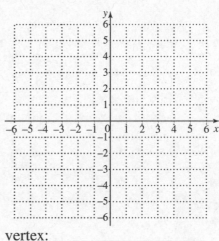

vertex: _____

2. $y = -x^2 + 4x - 1$

2.

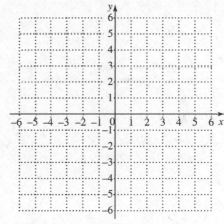

vertex: _____

3. $y = x^2 + 2x - 2$ **3.**

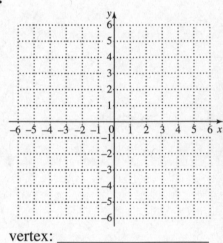

vertex: _____

4. $y = 2x^2 + 4x$ **4.**

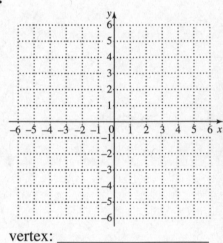

vertex: _____

5. $y = -x^2 - 3x + 1$ **5.**

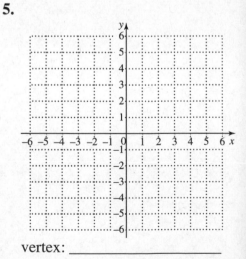

vertex: _____

6. $y = -x^2 + 6x - 9$

6.

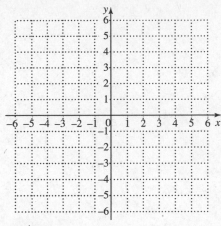

vertex: _____

7. $y = -x^2 + 6x - 13$

7.

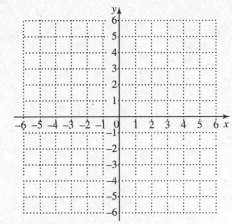

vertex: _____

8. $y = x^2 + 8x + 14$

8.

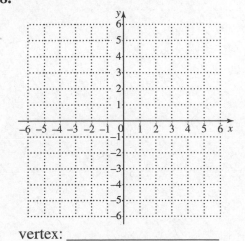

vertex: _____

9. $y = -\frac{1}{2}x^2 + 2x$

9.

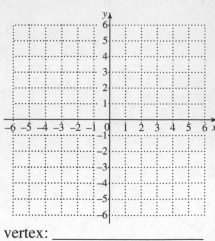

vertex: _____

10. $y = \frac{1}{2}(x+2)^2$

10.

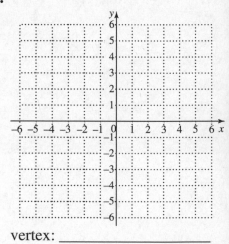

vertex: _____

Name: Date:
Instructor: Section:

Objective 2 Use a graph to determine the number of real solutions of a quadratic equation.

For extra help, see Example 3 on page 584 of your text, the Section Lecture video for Section 9.5 and Exercise Solution Clips 13 and 35.

Decide from each graph how many real solutions f(x) = 0 has. Then give the solution set (of real solutions).

11.

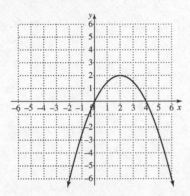

11. _____

12.

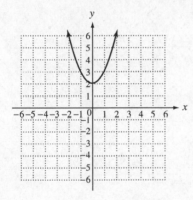

12. _____

13.

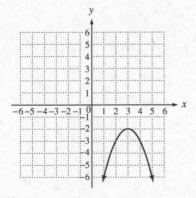

13. _____

14.

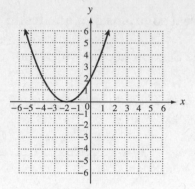

14. _____

15.

15. _____

Chapter 1
The Real Number System

1.1 Fractions
Key Terms

1. proper fraction
2. factor
3. natural numbers
4. least common denominator (LCD)
5. greatest common factor
6. improper fraction
7. whole numbers
8. basic principle of fractions
9. reciprocals 10. sum
11. mixed number
12. numerator 13. composite
14. prime factors
15. difference 16. lowest terms
17. quotient 18. product
19. prime 20. fraction
21. denominator

Now Try

1.a. $5 \cdot 7$ b. $2 \cdot 2 \cdot 2 \cdot 3 \cdot 5$

2.a. $\frac{5}{6}$ b. $\frac{8}{25}$

3.a. $\frac{3}{20}$ b. $\frac{6}{11}$

4.a. $\frac{25}{36}$ b. $\frac{37}{32}$ or $1\frac{5}{32}$

5. $3\frac{1}{4}$ bushels 6. $112,500

Objective 1

1. composite, $2 \cdot 2 \cdot 2 \cdot 2 \cdot 3$

Objective 2

3. $\frac{14}{5}$ or $2\frac{4}{5}$

Objective 3

5. $\frac{95}{4}$ or $23\frac{3}{4}$ 7. $\frac{11}{4}$ or $2\frac{3}{4}$

Objective 4

9. $\frac{1}{6}$

Objective 5

11. 38 face cords 13. $2\frac{5}{8}$ yards

Objective 6

15.a. $156.6 million

b. $2,513,700,000

c. $21,600,000

1.2 Exponents, Order of Operations, and Inequality
Key Terms

1. exponential expression
2. base
3. order of operations
4. exponent 5. inequality
6. grouping symbols

Now Try

1.a. 64 b. $\frac{1024}{243}$ or $4\frac{52}{243}$

2.a. 17 b. 32

c. 49

3.a. 189 b. 3

4.a. false b. true

c. true d. false

5.a. $6 \le 6$

b. $7 \ge \frac{15}{5}$ or $7 \ge 15 \div 5$

6. $5 < 17$

Objective 1

1. 81 3. 0.16

Objective 2

5. 1

Objective 3

7. 68   9. 189

Objective 4

11. True

Objective 5

13. $9 + 13 > 21$

Objective 6

15. $0.922 \geq 0.921$

1.3 Variables, Expressions, and Equations

Key Terms

1. variable   2. equation

3. set   4. solution

5. elements

6. algebraic expression

Now Try

1.a. 18   b. 324

2.a. 16   b. $\frac{8}{16} = \frac{1}{2}$

   c. 129

3.a. $8 + x$, or $x + 8$ b. $\dfrac{8}{x}$

   c. $10x + 21$

4. false   5. $\dfrac{35}{x} = 7; 5$

6. equation

Objective 1

1. 20   3. $\frac{28}{13}$ or $2\frac{2}{13}$

Objective 2

5. $\dfrac{x + 4}{2x}$

Objective 3

7. No   9. Yes

Objective 4

11. $3x = 2 + 2x; 2$

Objective 5

13. expression

15. equation

1.4 Real Numbers and the Number Line

Key Terms

1. coordinate   2. absolute value

3. number line

4. additive inverses

5. set-builder notation

6. real numbers

7. signed numbers

8. integers

9. rational numbers

10. graph

11. irrational number

Now Try

1. 1,465,146

2.a. none   b. $\{-9, 0\}$

   c. $\left\{\sqrt{3}, \pi\right\}$

   d. $\left\{-9, -\frac{4}{3}, 0, 0.\overline{3}, \sqrt{3}, \pi\right\}$

3. false   4. -2.35

5.a. 2   b. -2

   c. 2

6. Philadelphia from 2000–2009

Objective 1

1. rational numbers, real numbers

3.

Objective 2

5. True

Objective 3

7. 15 9. $2\frac{5}{8}$

Objective 4

11. −25

Objective 5

13. Lake 15. 1327 people

1.5 Adding and Subtracting Real Numbers

Key Terms

1. minuend 2. subtrahend

Now Try

1.a. 6 b. −7

2. −23

3. −3 4. −9

5.a. answer is correct

b. answer is correct

6.a. 46 b. −16.1

c. $-\frac{2}{5}$

7.a. −13 b. −21

8. −10 + [20 + (−4)]

9. [10 − (−4)] − 2

10. 1804 meters

11. The CPI increased by 4.2.

Objective 1

1. 15 3. −18

Objective 2

5. $\frac{11}{8} = 1\frac{3}{8}$

Objective 3

7. 22 9. $\frac{47}{8} = 5\frac{7}{8}$

Objective 4

11. −13.8

Objective 5

13. $\left[6+(-2)\right]+8;\,12$

Objective 6

15. 4566 feet

1.6 Multiplying and Dividing Real Numbers

Key Terms

1. dividend 2. divisor

3. multiplicative inverse

Now Try

1.a. −13.12 b. $-\frac{7}{12}$

2. 112

3.a. 6 b. $-\frac{1}{6}$

4.a. 0 b. $\frac{2}{5}$

5.a. 70 b. −15

6. $\frac{25}{9} = 2\frac{7}{9}$

7.a. $-2-(10)(-2)=18$

b. $85 - 0.3[50 - (-10)] = 67$

8. $\dfrac{40(-3)}{5-(-10)} = -8$

9.a. $-2x = 12;\, x = -6$

b. $5 + x = -3;\, x = -8$

Objective 1

1. −28

Objective 2

3. $\frac{4}{5}$

Objective 3

5. $-36, -18, -12, -9, -6, -4, -3, -2,$
   $-1, 1, 2, 3, 4, 6, 9, 12, 18, 36$

Objective 4

7. $-\frac{1}{6}$

Objective 5

9. $\frac{16}{21}$

Objective 6

11. $\frac{3}{4}$

Objective 7

13. $\frac{(-4)(7)}{-3+14}; -\frac{28}{11}$ or $-2\frac{6}{11}$

Objective 8

15. $x + 8 = 6; x = -2$

1.7 Properties of Real Numbers
Key Terms

1. distributive property

2. identity property

3. associative property

4. identity element for multiplication

5. inverse property

6. commutative property

7. identity element for addition

Now Try

1.a. $3 + (-4) = -4 + 3$

b. $-4(p + 9) = (p + 9)(-4)$

2.a. $\left[-4 + (-2)\right] + y = -4 + (-2 + y)$

b. $(2m)(-7) = (2)\left[m(-7)\right]$

3. commutative

4.a. 115   b. -26

5.a. 1   b. 0

6.a. $\frac{3}{7}$ b. $\frac{19}{18} = 1\frac{1}{18}$

7.a. $\frac{3}{2}$ b. 2

8. -12

9.a. $-8x - 36y$ b. $48x - 36y - 44z$

   c. $3(a - b)$

10.a. $6x - 2$ b. $-5x + 3y + z$

Objective 1

1. $ab(2) = 2(ab)$

3. $2 + \left[10 + (-9)\right] = \left[10 + (-9)\right] + 2$

Objective 2

5. $(-r)\left[(-p)(-q)\right] = \left[(-r)(-p)\right](-q)$

Objective 3

7. -7 9. 7

Objective 4

11. $\frac{2}{7} \cdot \frac{7}{2} = 1$; inverse

Objective 5

13. $r(10 - 4); 6r$

15. $-14(x + y)$

1.8 Simplifying Expressions
Key Terms

1. numerical coefficient

2. combining like terms

3. term   4. like terms

5. unlike terms

Now Try

1.a. $45x - 72y$ b. $10x + 3$

2.a. 0   b. $-5a + 25$

c. $-10y^2 + 16y$

3.a. $3y + 2.5$ b. $3x - 4$

4. $(5x - 3) + 4(x + 2); 9x + 5$

Objective 1

1. $6 + 3y$ 3. $-24p - 10$

Objective 2

5. 0.3

Objective 3

7. like   9. like

Objective 4

11. $\frac{3}{10}r - \frac{1}{2}s$

Objective 5

13. $3x - (7x + 2); -4x - 2$

15. $3(-7 + 5x) - 4[2x - (-10)];$
$7x - 61$

Chapter 2
Linear Equations And Inequalities In One Variable

2.1 The Addition Property of Equality
Key Terms

1. linear equation in one variable

2. solution set

3. equivalent equations

4. solution

5. equation

Now Try

1. $\{-3\}$ 2. $\left\{-\frac{21}{2}\right\}$

3. $\{-23\}$ 4. $\left\{\frac{1}{2}\right\}$

5. $\{-10\}$ 6. $\{-2\}$

7. $\{-4\}$

Objective 1

1. Yes   3. No

5. Yes

Objective 2

7. $\left\{\frac{3}{2}\right\}$ or $\left\{1\frac{1}{2}\right\}$ 9. $\{-12.8\}$

Objective 3

11. $\{3\}$ 13. $\{2\}$

15. $\{7.2\}$

2.2 The Multiplication Property of Equality
Key Terms

1. multiplication property of equality

2. coefficient

3. reciprocal

Now Try

1. $\{7\}$ 2. $\left\{-\frac{21}{8}\right\}$

3. $\{-5.4\}$ 4. $\{-42\}$

5. $\{-14\}$ 6. $\{-16\}$

7. $\{8\}$

Objective 1

1. $\{-14\}$ 3. $\left\{-\frac{7}{4}\right\}$

5. $\left\{\frac{7}{9}\right\}$ 7. $\{3.6\}$

Objective 2

9. $\{8\}$ 11. $\{18\}$

13. $\{8\}$ 15. $\{7\}$

2.3 More on Solving Linear Equations

Key Terms

1. conditional equation
2. contradiction   3. empty set
4. identity

Now Try

1. $\{-7\}$ 2. $\{-7\}$
3. $\{5\}$ 4. $\left\{\frac{1}{4}\right\}$
5. $\left\{-\frac{5}{11}\right\}$ 6. $\{4\}$
7. $\{-2\}$ 8. $\{-3\}$
9. {all real numbers}
10. $\varnothing$ 11. $12 - x$

Objective 1

1. $\{-7\}$ 3. $\{-1\}$

Objective 2

5. $\{2\}$ 7. $\{-6\}$

Objective 3

9. $\varnothing$
11. {all real numbers}

Objective 4

13. $36 - m$ 15. $10q$ cents

2.4 An Introduction to Applications of Linear Equations

Key Terms

1. complementary
2. degree
3. supplementary
4. right angle
5. straight angle
6. consecutive integers

Now Try

1. 16   2. 52
3. 160 calories
4. 24 adult tickets, 72 children's tickets
5. Sanchez family: $925, Jones family: $1285, Kwan family $1700
6. 75, 76   7. 13, 15
8. 55°   9. 20°

Objective 2

1. $6(x - 4) = x(-2);\ 3$
3. $-3(x - 4) = -5x + 2;\ -5$

Objective 3

5. 27 inches
7. 16 oz of cranberry juice; 32 oz of orange juice; 128 oz of ginger ale

Objective 4

9. 76, 78   11. 13, 15

Objective 5

13. 49°   15. 59°

2.5 Formulas and Applications from Geometry

Key Terms

1. formula   2. area
3. vertical angles
4. perimeter   5. volume

Now Try

1. 7
2. width: 15 feet; length: 22 feet
3. 91 cm, 192 cm, 277 cm
4. 18 in.   5. 59°, 121°
6. $p = \dfrac{i}{rt}$ 7. $s = \dfrac{P - 2a}{3}$

8.　$b = \dfrac{3\left(V - h^2\right)}{h}$ or $b = \dfrac{3V}{h} - 3h$

9.　$x = \dfrac{Aw}{R} - 1$

Objective 1

1.　28.26　　　3.　12

Objective 2

5.　3052.08 cm^3　7.　4 cm

Objective 3

9.　$\left(3x + 5\right)^\circ = 35^\circ$; $\left(6x - 25\right)^\circ = 35^\circ$

11.　$\left(3x\right)^\circ = 45^\circ$; $\left(9x\right)^\circ = 135^\circ$

Objective 4

13.　$r = \dfrac{S - a}{S}$ or $r = 1 - \dfrac{a}{S}$

15.　$F = \dfrac{9}{5}C + 32$

2.6　Ratios and Proportions

Key Terms

1.　extremes　　　2.　means

3.　ratio

4.　proportion

5.　cross products　6.　terms

Now Try

1.a.　$\dfrac{5}{8}$　　　b.　$\dfrac{69}{10}$

2.　64-oz size; $0.054 per oz.

3.a.　false　　　b.　true

4.　{15}　　　5.　−13

6.　43.2 acres

7.a.　0.32　　　b.　89.3%

8.a.　87.5　　　b.　750

　c.　2%　　　9.　64 questions

Objective 1

1.　$\dfrac{3}{4}$　　　3.　48-oz jar

Objective 2

5.　$\dfrac{10}{3}$　　　7.　$-\dfrac{7}{4}$

Objective 3

9.　15 inches　11.　14 tanks

Objective 4

13.　850 members

15.　discount: $14.25;
sale price: $42.75

2.7　Further Applications of Linear Equations

Key Terms

1.　simple interest

2.　mixture problem

3.　distance problems

4.　denomination

Now Try

1.a.　2.4 L　　　b.　$331.20

2.　$\dfrac{1}{2}$ liter

3.　$4500 at 4%; $9000 at 7%

4.　39 $10 bills; 51 $5 bills

5.　15 hr　　　6.　12 hr

7.　40 mph, 48 mph

Objective 1

1.　22.5 liters

Objective 2

3.　2.67 gallons

5.　7 liters of 60%; 3 liters of 30%

Objective 3

7.　$51,000 at 8%; $17,000 at $9\dfrac{1}{2}$%

Objective 4

9. 185 student tickets

11. 15 nickels; 21 dimes; 30 quarters

Objective 5

13. 1.79 m/sec   15. 6 mi

2.8 Solving Linear Inequalities

Key Terms

1. linear inequality in one variable

2. interval notation

3. three part inequality

4. interval   5. inequality

Now Try

1.a. $(-\infty, -2)$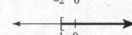

   b. $[-1, \infty)$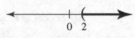

2. $(2, \infty)$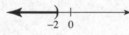

3. $(-\infty, -2)$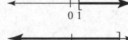

4. $[1, \infty)$

5. $(-\infty, 9]$

6. $315 or more

7. $[-3, 0]$

8. $[-6, 3)$

Objective 1

1. $[3, \infty)$

Objective 2

3. $(-5, \infty)$

Objective 3

5. $(-\infty, -10)$

Objective 4

7. $(4, \infty)$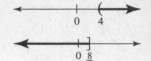

9. $\left(-\infty, \frac{8}{5}\right]$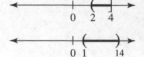

Objective 5

11. $38.84 or less

Objective 6

13. $(2, 4]$

15. $\left(1, \frac{14}{3}\right)$

Chapter 3
Linear Equations And Inequalities In Two Variables; Functions

3.1 Linear Equations in Two Variables; The Rectangular Coordinate System

Key Terms

1. ordered pair   2. quadrant

3. y-axis   4. line graph

5. coordinates   6. x-axis

7. origin   8. scatter diagram

9. rectangular (Cartesian) coordinate system

10. linear equation in two variables

11. plot   12. plane

13. table of values

Now Try

1.a. September doubled July; August doubled October

   b. $40

2.a. no   b. yes

3.a. $\left(-4, \frac{15}{2}\right)$ b. $(2, 0)$

4. $\left(0, \frac{3}{2}\right), (-2, 0), (2, 3), \left(-\frac{14}{3}, -2\right)$

5.

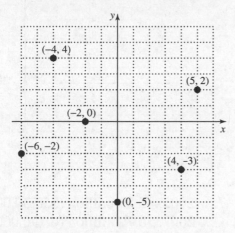

6. 150 miles

Objective 1

1. 1991–1992   3. 1993–1994

Objective 2

5. $\left(0, \frac{1}{3}\right)$

Objective 3

7. No

Objective 4

9.a. $(2, -1)$ b. $(0, -5)$

c. $(4, 3)$ d. $(-1, -7)$

e. $(7, 9)$

11.a. $(-2, -2)$ b. $(-2, 0)$

c. $(-2, 19)$ d. $(-2, 3)$

e. $\left(-2, -\frac{2}{3}\right)$

Objective 5

13.

x	0	$\frac{4}{3}$	4
y	2	0	−4

Objective 6

15.

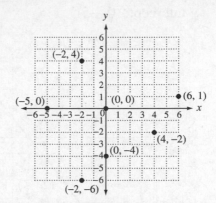

3.2 Graphing Linear Equations in Two Variables

Key Terms

1. graph   2. x-intercept

3. y-intercept

Now Try

1.

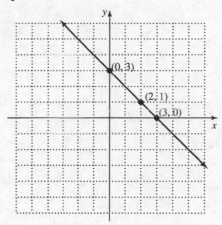

2.

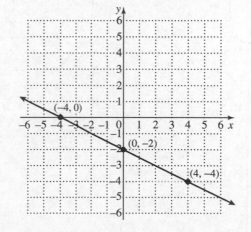

3.   *x*-intercept: (4, 0)
     y-intercept: (0, −5)

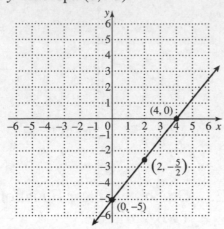

4.

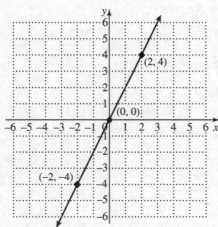

5.

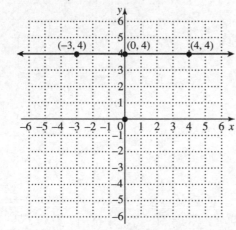

6.

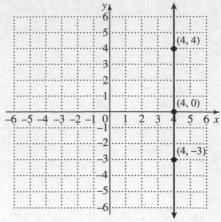

7.   2005: $5.53 million ($5,530,000)
     2007: $6.79 million (6,790,000);
     2008: $10.57 million ($10,570,000)

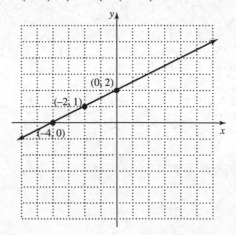

Objective 1

1. $(0, 2)$, $(−4, 0)$, $(−2, 1)$

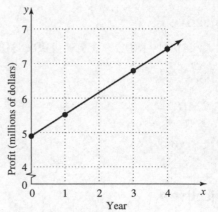

3. $(0,2)$, $(3,0)$, $(-3,4)$

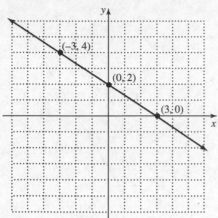

Objective 2

5. x-intercept: $\left(\frac{9}{4},0\right)$

    y-intercept: $(0,3)$

Objective 3

7.

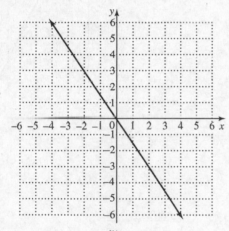

9.

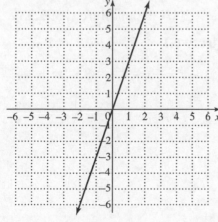

Objective 4

11.

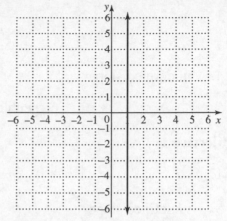

Objective 5

13. $(2000, 2435)$, $(2001, 2350)$,
    $(2002, 2265)$, $(2003, 2180)$,
    $(2004, 2095)$, $(2005, 2010)$

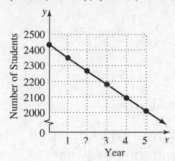

15. $(2003, 325)$, $(2004, 367)$,
    $(2005, 409)$, $(2006, 451)$

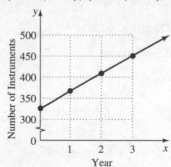

3.3   The Slope of a Line
Key Terms

1. slope   2.   run

3. subscript notation

4. parallel   5.   rise

6. perpendicular

437

Now Try

1. $\frac{5}{8}$ 2. $-\frac{3}{5}$

3. 0

4. undefined slope

5. $\frac{4}{3}$ 6. perpendicular

Objective 1

1. –2   3. $\frac{1}{5}$

5. undefined slope

Objective 2

7. $\frac{3}{4}$ 9. 0

Objective 3

11. $4; -\frac{1}{4}$; perpendicular

13. $-2; -\frac{1}{4}$; neither

15. 0; undefined slope; perpendicular

3.4   Writing and Graphing Equations of Lines

Key Terms

1. slope-intercept form

2. point-slope form

3. standard form

Now Try

1.a. slope: $\frac{2}{3}$; y-intercept: $\frac{5}{3}$

   b. slope: $-\frac{1}{4}$; y-intercept: $-\frac{5}{8}$

2. $y = \frac{3}{2}x - \frac{2}{3}$

3.

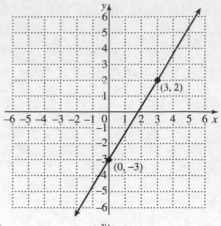

4.

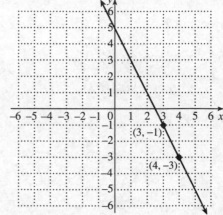

5. $y = \frac{3}{4}x - 4$

6. $y = \frac{3}{2}x - 11$

7. $y = -\frac{10}{3}x + \frac{4}{3}; 10x + 3y = 4$

8. $y = \frac{208}{3}x + \frac{532}{3}$

   2007 expenditure: $593.33

Objective 1

1. $y = \frac{2}{3}x - 4$ 3. $y = -4$

Objective 2

5.

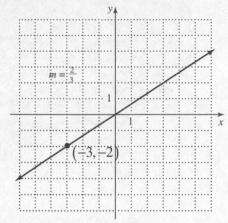

Objective 3

7. $y = \frac{1}{3}x + \frac{7}{3}$ 9. $y = -\frac{2}{3}x - 3$

Objective 4

11.a. $y = -11x + 29$

b. $11x + y = 29$

Objective 5

13. $m = 19;\ y = 19x + 82$

15. $m = \frac{58}{3};\ y = \frac{58}{3}x + \frac{253}{3}$

3.5 Graphing Linear Inequalities in Two Variables

Key Terms

1. test point   2. boundary line

3. linear inequality in two variables

Now Try

1.

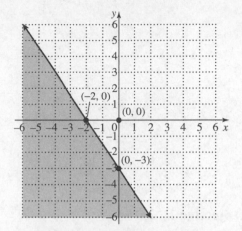

2.

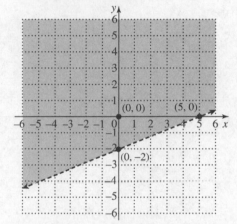

3.

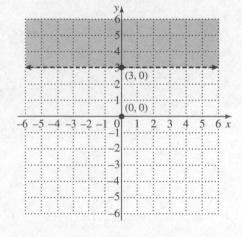

4.

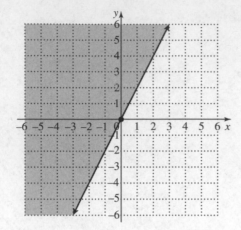

5.

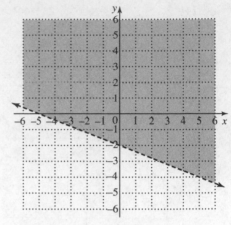

Objective 1

1.

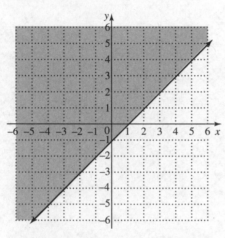

7.

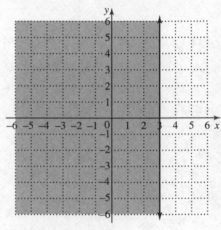

3.

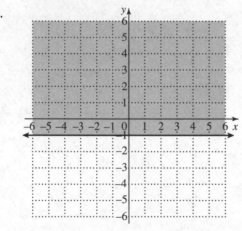

9.

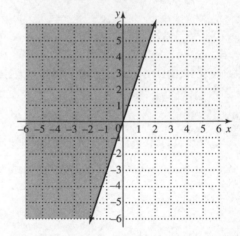

Objective 2

11.

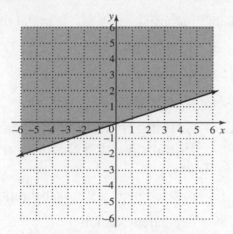

13.

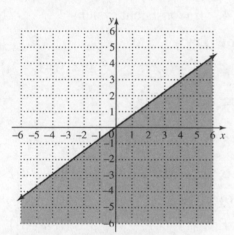

15.

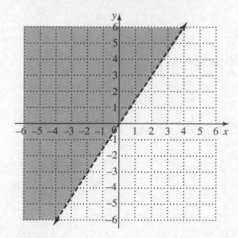

3.6   Introduction to Functions
Key Terms

1. function   2.   relation

3. function notation

4. range   5.   components

6. domain

Now Try

1. domain:$\{-2, 1, 3, 7\}$; range: $\{4\}$

2.a. not a function b.   function

3.a. function   b.   not a function

4. domain: $(-\infty, \infty)$; range: $[4, \infty)$

5. 18

6.a. $\{(2001, 210\}, (2002, 294),$
   $(2003, 316), (2004, 378),$
   $(2005, 455), (2006, 524), (2007,$
   $608)\}$

   b. domain: $\{2001, 2002, 2003, 2004,$
   $2005, 2006, 2007\}$
   range: $\{210, 294, 316, 378, 455,$
   $524, 608\}$

   c. $455   d.   2004

Objective 1

1. domain: $\{1,2,3,4\}$
   range: $\{3,4,7,9\}$

Objective 2

3. function   5.   not a function

Objective 3

7. function

Objective 4

9. domain: {all real numbers}
   range: {all real numbers}

Objective 5

11.a. 10   b.   −11

Objective 6

13. domain:
   $\{1089, 2581, 2991, 3223, 3665\}$
   range: $\{3.0, 9.9, 16.2, 16.7, 41.1\}$

15. 3665 theaters

Chapter 4
Systems of Linear Equations and Inequalities

4.1 Solving Systems of Linear Equations by Graphing

Key Terms

1. set-builder notation
2. solution set of a system
3. dependent equations
4. inconsistent system
5. solution of a system
6. independent equations
7. consistent system
8. system of linear equations (linear system)

Now Try

1.a. yes   b. no
2. $\{(-1, -2)\}$
3.a. $\varnothing$
  b. $\{(x, y) \mid x + 2y = 4\}$
4.a. one line; infinite number of solutions
  b. parallel lines; no solution
  c. neither parallel nor the same; one solution

Objective 1

1. Yes   3. No

Objective 2

5. $\{(-3, 4)\}$ 7. $\{(-2, -4)\}$

Objective 3

9. dependent; $\{(x, y) \mid 2x + 3y = 0\}$
11. inconsistent; $\varnothing$

Objective 4

13.a. inconsistent   b. parallel lines
  c. no solution
15.a. dependent   b. one line
  c. infinite number of solutions

4.2 Solving Systems of Linear Equations by Substitution

Key Terms

1. ordered pair   2. substitution
3. dependent system
4. inconsistent system

Now Try

1. $\{(-1, 3)\}$ 2. $\{(-1, -7)\}$
3. $\left\{\left(3, -\frac{2}{3}\right)\right\}$ 4. $\varnothing$
5. $\{(x, y) \mid 2x - y = 4\}$
6. $\{(-12, 0)\}$ 7. $\{(7, -2)\}$

Objective 1

1. $\{(1, 6)\}$ 3. $\{(1, -5)\}$
5. $\left\{\left(\frac{1}{2}, -\frac{1}{2}\right)\right\}$

Objective 2

7. $\{(x, y) \mid 36x + 20y = 12\}$
9. $\{(x, y) \mid 3x - 4y = 8\}$

Objective 3

11. $\{(6, 2)\}$ 13. $\{(-16, 0)\}$
15. $\{(x, y) \mid 3x + 4y = 5\}$

4.3 Solving Systems of Linear Equations by Elimination

Key Terms

1. opposites
2. elimination method
3. true

Now Try

1. $\{(2, -4)\}$ 2. $\{(2, -4)\}$
3. $\{(3, -2)\}$
4. $\left(-\dfrac{13}{8}, \dfrac{11}{8}\right)$

5.a. $\varnothing$

  b. $\{(x, y) \mid 5x + 2y = 3\}$

Objective 1

1. $(5, 0)$ 3. $\left(\dfrac{1}{2}, -\dfrac{1}{6}\right)$

Objective 2

5. $(2, -2)$ 7. $(0, -4)$

Objective 3

9. $\left(\dfrac{51}{16}, -\dfrac{69}{32}\right)$ 11. $\left(-\dfrac{21}{29}, -\dfrac{38}{29}\right)$

Objective 4

13. $\{(x, y) \mid 2x - 4y = 1\}$
15. $\{(x, y) \mid 2x - 3y = 1\}$

4.4   Applications of Linear Systems

Key Terms

1. assign variables
2. check   3. $d = rt$

Now Try

1. Carla's class, 28 students; Linda's class, 21 students
2. 396 children's tickets; 327 adult tickets
3. 28 liters of 75% solution; 42 liters of 55% solution
4. Rick, 48 mph; Hilary, 78 mph

Objective 1

1. 8 and 12
3. 32 cm, 32 cm, 52 cm

Objective 2

5. 6 fives; 5 twenties
7. 30 fives; 60 tens

Objective 3

9. 20 pounds at $1.60 per pound
   10 pounds at $2.50 per pound
11. 5 liters of 20% solution
    10 liters of 5% solution

Objective 4

13. 8 mph; 12 mph
15. Plane A, 400 mph
    Plane B, 360 mph

4.5   Solving Systems of Linear Inequalities

Key Terms

1. solution set of a system of linear inequalities
2. test point   3.   boundary line
4. system of linear inequalities

Now Try

1.

2.

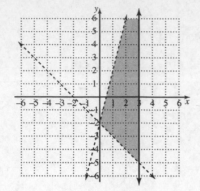

Objective 1

1.

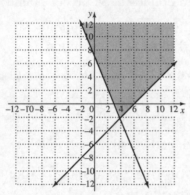

3.

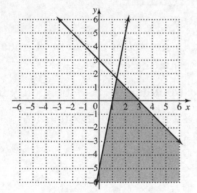

5.

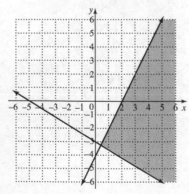

7.

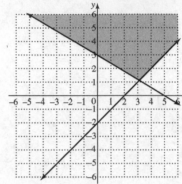

9.

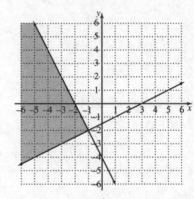

11.

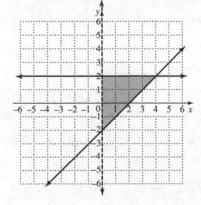

13.

15.

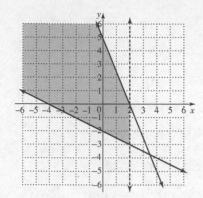

Chapter 5
Exponents and Polynomials

5.1 The Product Rule and Power Rules for Exponents

Key Terms

1. exponent
2. exponential expression
3. power   4. base

Now Try

1. $\left(-\dfrac{2}{5}\right)^4 = \dfrac{16}{625}$

2. -36; base 6; exponent, 2

3.a. $4^7 = 16,384$ b. $(-3)^5 = -243$

c. m^{10}

d. Product rule does not apply; 3087

e. a^{10}

f. Product rule does not apply; -100

g. $60n^9$

4.a. $(-2)^{35}$ b. b^{10}

5.a. $8a^{12}b^3$ b. $-27w^9z^{21}$

6.a. $\dfrac{x^8}{y^8}$ b. $\dfrac{256}{81}$

7.a. $\dfrac{32}{3}$ b. $\dfrac{3^4 b^8}{11^4}$

c. $32a^{13}b^{26}c^5$

8. $28q^{11}$

Objective 1

1. $(-2y)^5$

Objective 2

3. 4^{11}

Objective 3

5. 3^{12}

Objective 4

7. $5a^3b^9$

Objective 5

9. $-\dfrac{8x^3}{125}$

Objective 6

11. $5^{11}x^{18}y^{37}$ 13. $\dfrac{k^7 m^{28} p^{14}}{3^7 n^{28}}$

Objective 7

15. $16\pi a^{10}$

5.2 Integer Exponents and the Quotient Rule

Key Terms

1. negative exponent
2. zero exponent
3. quotient rule for exponents

Now Try

1.a. 1   b. -1
   c. 1   d. 2

2.a. $\dfrac{1}{216}$ b. 625

c. $-\dfrac{27}{125}$ d. $\dfrac{17}{64}$

e. $\dfrac{1}{x^{10}}, \; x \neq 0$

3.a. $\dfrac{25}{256}$ b. $\dfrac{3yz^3}{4a^5}$

c. $\dfrac{243x^{15}}{32y^{10}}$

4.a. $\dfrac{1}{12}$ b. $(3+a)^2$

c. $\dfrac{yz^2}{15^6 x^2}$ d. $\dfrac{y^3}{25x^{12}z^2}$

5.a. 7^6 b. $\dfrac{1}{(3+a)^{10}}$

c. $\dfrac{216x^{18}z^3}{125}$ d. $\dfrac{256x^{10}}{25y^4 z}$

Objective 1

1. -1 3. 0

Objective 2

5. $\dfrac{1}{20}$ 7. $\dfrac{2y^7}{3x^4}$

Objective 3

9. $a^6 b^6$ 11. $8^5 b^4 c^7$

Objective 4

13. $\dfrac{3^2}{2^4 y^2}$ 15. $\dfrac{1}{k^4 t^{20}}$

5.3 An Applications of Exponents: Scientific Notation

Key Terms

1. scientific notation

2. right 3. left

Now Try

1.a. 2.3651×10^{10} b. -4.7×10^{-4}

c. 5.03×10^{-5}

2.a. $72,000,000$ b. 0.04007

c. -0.000045

3.a. $2.53 \times 10^2 = 253$

b. $5.0 \times 10^4 = 50,000$

4. $5.56 \times 10^{12} \text{ kg/km}^3$

Objective 1

1. 4.579×10^6 3. 2.46×10^{-1}

5. 4.26×10^{-3}

Objective 2

7. $-2,450,000$ 9. 0.4752

Objective 3

11. $210,000,000$

13. 0.0313

15. 9.5×10^{20} meters

5.4 Adding and Subtracting Polynomials; Graphing Simple Polynomials

Key Terms

1. monomial

2. degree of a term

3. degree of a polynomial

4. numerical coefficient

5. parabola 6. binomial

7. term 8. axis

9. trinomial

10. descending powers

11. like terms 12. vertex

13. polynomial

14. line of symmetry

Now Try

1. −4, 10, −1

2. $12c^2 - 6c$

3. $3m^5 + 3m^4 - 5m^2$
   degree: 5; trinomial

4. 16

5.a. $5m^3 - 2m^2 - 4m + 4$

  b. $4x^2 + 6x - 18$

6. $3x^3 + 7x^2 - 12x + 2$

7. $4m^2 - 2$

8. $-4m^2n - 4mn - 6m - 2n$

9.

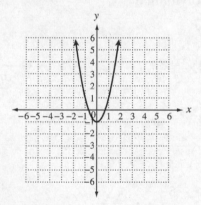

Objective 1

1. 4 terms; 9, 3, −4, 2

Objective 2

3. $-3z^3$ 5. $-\frac{1}{4}r^3$

Objective 3

7. $n^8 - n^2$; degree 8; binomial

Objective 4

9.a. 410   b. 55

Objective 5

11. $2z^5 - z^4 - 2z^3 + 7z^2 - 8z + 7$

13. $7p^2 + 10p - 12$

Objective 6

15.

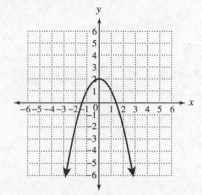

5.5   Multiplying Polynomials

Key Terms

1. FOIL method   2. inner product

3. outer product

Now Try

1. $-35z^6 + 28z^5 - 14z^3$

2. $x^3 - 2x^2 - 12x + 27$

3. $6x^4 + 19x^3 - 9x^2 - 2x$

4. $2x^2 - 15x + 27$

5.a. $-6m^2 - mn + 12n^2$

  b. $8a^4 - 10a^3 + 3a^2$

Objective 1

1. $-4x^6 - 12x^5 - 6x^4$

3. $-6y^5 - 9y^4 + 12y^3 - 33y^2$

5. $-8r^6 + 12r^5 - 8r^4$

Objective 2

7. $2r^3 + 3r^2 - 4r + 15$

9. $6y^4 - 5y^3 - 8y^2 + 7y - 6$

Objective 3

11. $20a^2 + 11ab - 3b^2$

13. $8a^4 + 10a^2 + 3$

15. $x^2 - \frac{5}{3}x + \frac{4}{9}$

5.6 Special Products
Key Terms

1. product of the sum and difference of two terms

2. square of a binomial

Now Try

1. $z^2 + 12z + 36$

2.a. $a^2 + 4ab + 4b^2$

b. $4m^2 - 12mp + 9p^2$

c. $9x^2 + 4xy + \frac{4}{9}y^2$

d. $4x^4 + 20x^3 + 25x^2$

3.a. $49 - x^2$

b. $25m^2 - 16$

c. $9m^2 - 49n^2$

d. $36x^2 - \frac{4}{9}y^2$

e. $64x^4 - 9x^2$

4.a. $343 - 147x + 21x^2 - x^3$

b. $125m^3 - 300m^2n + 240mn^2 - 64n^3$

c. $81a^4 - 216a^3 + 216a^2 - 96a + 16$

Objective 1

1. $4m^2 + 20m + 25$

3. $4p^2 + 12pq + 9q^2$

5. $9a^2 + 3ab + \frac{1}{4}b^2$

Objective 2

7. $16p^2 - 49q^2$

9. $49m^2 - \frac{9}{16}$

Objective 3

11. $8x^3 - 36x^2 + 54x - 27$

13. $-64k^4 - 96k^3 - 48k^2 - 8k$

15. $256s^4 + 768s^3t + 864s^2t^2$
$$+ 432st^3 + 81t^4$$

5.7 Dividing Polynomials
Key Terms

1. dividing a polynomial by a monomial

2. long division

3. dividing a polynomial by a polynomial

Now Try

1. $8y^3 - 3y$ 2. $7b^2 - 3 - \dfrac{1}{9b^2}$

3. $4w^3 - 2w + \dfrac{3}{w}$

4. $-y^2 + 6x^2y - 6x^3$

5. $2r - 5$

6. $3m - 2 + \dfrac{8}{3m - 4}$

7. $3x^3 - 3x^2 - 3x + 6$

8. $3x^2 - 6x + 2 + \dfrac{13x - 7}{2x^2 + 3}$

9. $4x^2 + \dfrac{2}{3}x - \dfrac{4}{3} - \dfrac{5}{6x + 3}$

Objective 1

1. $4m^2 - 3m + 2$

3. $1 - \dfrac{1}{m^2}$

5. $\dfrac{1}{2}m + \dfrac{3}{2} - \dfrac{6}{m}$

7. $-2 + \dfrac{2}{y} - \dfrac{10}{y^2}$

Objective 2

9. $a - 7 + \dfrac{37}{2a+3}$

11. $z^2 - 5z + 9 + \dfrac{-25}{2z+3}$

13. $3x - 11 + \dfrac{40x - 80}{x^2 - 5}$

15. $2x^4 + x^3 - 3x^2 + 2x + 1 + \dfrac{2}{3x+2}$

Chapter 6
Factoring and Applications

6.1 Greatest Common Factors; Factoring by Grouping

Key Terms

1. factored form

2. greatest common factor (GCF)

3. factor

4. common factor

Now Try

1.a. 12   b. 14

2.a. $8y^2$ b. $15ab^2$

3.a. $3y(-5y + 6)$ or $-3y(5y - 6)$

b. $12a\left(8a^2 - 4a + 5\right)$

c. $x^4 z^2\left(1 + x^2\right)$

d. $7x^2 y\left(2xy + 1 - 3x^3 y^2\right)$

4.a. $(r - 4s)\left(x^2 + z^2\right)$

b. $(3x + 1)(b + 4)$

5.a. $(3c - d)(4b - 5x)$

b. $(3v - 8)\left(2v^2 + 7\right)$

c. $(x - 6)(4y - 1)$

d. $(7xy + 4)\left(4x^2 y - 3\right)$

6.a. $(6u + 5v)(6u + 5)$

b. $(3z + c)(3m - 4n)$

Objective 1

1. 3   3. $2ab^3$

5. $15a^2 y$

Objective 2

7. no common factor except 1

9. $-(x + y)$

Objective 3

11. $(1 - x)(1 - y)$

13. $\left(x^3 - 3\right)\left(1 + y^2\right)$

15. $(7w + 3x)(2w - 5x)$

6.2 Factoring Trinomials

Key Terms

1. prime polynomial

2. FOIL

3. common factor

Now Try

1. $(x + 2)(x + 12)$

2. $(x - 3)(x - 12)$

3. $(x + 7)(x - 4)$

4. $(t + 4)(t - 5)$

5. prime

6. $(p-3m)(p-4m)$

7. $2t(s-10)(s+2)$

Objective 1

1. $(z+3)(z+2)$

3. prime

5. $(k+5)(k+4)$

7. $(a-7)(a+3)$

9. $(a-8b)(a+2b)$

Objective 2

11. $2m(m-2)(m+1)$

13. $3x(y-6)(y-2)$

15. $2x(x-2y)(x-5y)$

6.3 More on Factoring Trinomials
Key Terms

1. inner product   2. FOIL

3. outer product

Now Try

1.a. $(4b+3)(2b+3)$

b. $(3m+4)(m-3)$

c. $(3n+4p)(2n+p)$

2. $4(x+5)(4x-5)$

3. $(3y+1)(y+4)$

4. $(x-3)(5x-3)$

5. $(3z-4)(z+2)$

6. $(10z-3y)(4z+3y)$

7. $-3a^2b(2a-1)(5a+2)$

Objective 1

1. $(2x+5)(3x+2)$

3. $(3y-1)(8y-3)$

5. $s(2-t)(1+3t)$

Objective 2

7. $(3s+1)(7s+2)$

9. $(5q-4)(3q+1)$

11. $(3p-5)(2p+3)$

13. $(6x+y)(x-y)$

15. $y^3z^2(y-3z)(2y+z)$

6.4 Special Factoring Techniques
Key Terms

1. difference of squares

2. sum of cubes

3. perfect square trinomial

4. difference of cubes

Now Try

1.a. $(y+4)(y-4)$

b. cannot be factored

2.a. $(5a+6)(5a-6)$

b. $(13m+15p)(13m-15p)$

3.a. $9x^2(3x+10)(3x-10)$

b. $(m^2+25)(m+5)(m-5)$

c. $(10x^2y-9z)(10x^2y+9z)$

4. $(x+2)^2$

5.a. $(x-4)^2$ b. $(4x-5)^2$

c. cannot be factored

d. $3x^2(3x-1)^2$

6.a. $(x-7)(x^2+7x+49)$

b. $(3x-2)(9x^2+6x+4)$

c. $-3(3y-5)(9y^2+15y+25)$

d. $2x(5x-4y)(25x^2+20xy+16y^2)$

7.a. $(3x+5)(9x^2-15x+25)$

b. $8(x+2b)(x^2-2bx+4b^2)$

c. $(uv^2+6)(u^2v^4-6uv^2+36)$

Objective 1

1. $(3x+1)(3x-1)$

3. $(12x+5y)(12x-5y)$

Objective 2

5. $(y+11)^2$ 7. $(4t+7)^2$

Objective 3

9. $(4a-7b)(16a^2+28ab+49b^2)$

11. $(2a-5b)(4a^2+10ab+25b^2)$

Objective 4

13. $8(a+2b)(a^2-2ab+4b^2)$

15. $3(4x^2+7)(16x^4-28x^2+49)$

6.5 Solving Quadratic Equations by Factoring

Key Terms

1. standard form

2. quadratic equation

Now Try

1.a. $\left\{\frac{2}{5},-\frac{1}{3}\right\}$ b. $\{0,-5\}$

2.a. $\{-3,-6\}$ b. $\{-5,4\}$

3. $\left\{\frac{1}{2},-\frac{3}{4}\right\}$

4.a. $\left\{\frac{7}{3},-\frac{7}{3}\right\}$ b. $\{0,-3\}$

c. $\left\{-\frac{2}{3},3\right\}$

5. $\left\{\frac{5}{6}\right\}$

6.a. $\left\{0,\frac{9}{7},-\frac{9}{7}\right\}$ b. $\{-5,-4,-2\}$

7. $\{-5,3\}$

Objective 1

1. $\{0,2\}$ 3. $\left\{-\frac{5}{2},-\frac{2}{3}\right\}$

5. $\left\{-\frac{5}{4},\frac{1}{3}\right\}$ 7. $\left\{-\frac{3}{2},-\frac{1}{3}\right\}$

9. $\left\{-\frac{5}{4},\frac{5}{4}\right\}$

Objective 2

11. $\left\{-\frac{3}{2},0,\frac{3}{2}\right\}$ 13. $\left\{-\frac{3}{2},\frac{3}{2},2\right\}$

15. $\left\{-5,\frac{3}{2},7\right\}$

6.6 Applications of Quadratic Equations

Key Terms

1. legs   2. hypotenuse

Now Try

1. base: 12 cm; height: 7 cm

2. 11, 13, 15

3. The ladder is 16 ft up the side of the building.

4. $1\frac{3}{8}$ sec or $1\frac{1}{2}$ sec

5. 5.6%

Objective 1

1. width: 8; length: 24

3. width: 7 in.; length: 21 in.

Objective 2

5. −4 and −3 or 3 and 4

7. 6, 8

Objective 3

9. 60 mi   11. 20 mi

Objective 4

13. a. $\frac{3}{2}$ or $1\frac{1}{2}$ sec; b. 1 sec or 2 sec

15. 236 ft

Chapter 7
Rational Expressions and Applications

7.1 The Fundamental Property of Rational Expressions

Key Terms

1. rational expression

2. fundamental property of rational expressions

3. lowest terms

Now Try

1.a. 5   b. $-\frac{11}{2}$

2.a. 0

b. never undefined

c. −2, 2

3. $\frac{3y^2}{7}$

4.a. $\frac{1}{2}$ b. $\frac{3y+2}{2y+1}$

5. −1

6.a. −x − 5

b. already in lowest terms

7. $\dfrac{-(2x+6)}{x-5}$; $\dfrac{-2x-6}{x-5}$

$\dfrac{2x+6}{-(x-5)}$; $\dfrac{2x+6}{-x+5}$

Objective 1

1.a. $\frac{64}{7}$ b. undefined

3.a. $-\frac{3}{10}$ b. $\frac{11}{10}$

Objective 2

5. 0, 4   7. −5, 5

Objective 3

9. $2r-s$ 11. $\dfrac{x+5}{x+2}$

Objective 4

13. $\dfrac{-(2x-3)}{x+2}$, $\dfrac{-2x+3}{x+2}$,

$\dfrac{2x-3}{-(x+2)}$, $\dfrac{2x-3}{-x-2}$

15. $\dfrac{-(2x-1)}{3x+5}$, $\dfrac{-2x+1}{3x+5}$,

$\dfrac{2x-1}{-(3x+5)}$, $\dfrac{2x-1}{-3x-5}$

7.2 Multiplying and Dividing Rational Expressions

Key Terms

1. rational function

2. rational expression

Now Try

1. $\dfrac{10m^3n}{3}$ 2. $\dfrac{8y^2(y+4)}{27}$

3. $\dfrac{x+4}{2(x-4)}$ 4. $\dfrac{15y^3}{2}$

5. $\dfrac{14w}{z^4}$ 6. $\dfrac{2(m-5)}{m+3}$

7. $-\dfrac{6(a+1)}{a}$

Objective 1

1. $-\dfrac{3}{2}$ 3. $-\dfrac{x-2}{4(x+1)}$

5. $\dfrac{a-1}{a+1}$ 7. $\dfrac{3-x}{2x+3}$

Objective 2

9. $\dfrac{4a+3}{a-4}$ 11. $\dfrac{z-8}{2z-3}$

13. $\dfrac{-3(6k-1)}{3k-1}$ 15. $\dfrac{2k-3}{k-1}$

7.3   Least Common Denominators
Key Terms

1. equivalent expressions
2. least common denominator

Now Try

1.a. 216   b. 96y

2. $24x^3y^4$

3.a. $15r^3(r-5)$

b. $(p-4)(p-3)(p+3)$

c. $-(a-b)(a+b)$

4.a. $\dfrac{15}{18}$ b. $\dfrac{16r^2}{28r}$

5.a. $\dfrac{4(11a+1)}{8a-24}$ b. $\dfrac{4p^2}{14p^3-70p^2}$

Objective 1

1. 420

3. $-42(t-4)$ or $-42(4-t)$

5. $z^2(z-2)(z+4)^2$

7. $m^2(m-2)(m+7)$

Objective 2

9. $-3y(y+3)$ or $-3y^2-9y$

11. $9(y+2)$ or $9y+18$

13. $3(k+7)$ or $3k+21$

15. $2p^2(p+9)$ or $2p^3+18p^2$

7.4   Adding and Subtracting Rational Expressions
Key Terms

1. different denominators
2. parentheses
3. same denominator

Now Try

1.a. $\dfrac{11}{3w^2}$ b. $\dfrac{1}{b-2}$

2.a. $\dfrac{41}{60}$ b. $\dfrac{46}{45z}$

3. $\dfrac{2m+1}{(m-3)(m+3)}$

4. $\dfrac{3y^2+10y-12}{(y+4)(y+2)(y-2)}$

5. $\dfrac{7r-3s}{s-4r}$ or $\dfrac{3s-7r}{4r-s}$

6. 5   7. $\dfrac{30}{s(s-6)}$

8. $\dfrac{x+24}{x-8}$ or $\dfrac{-24-x}{8-x}$

9. $\dfrac{2z^2+5z+4}{(z+2)(z+3)(z+4)}$

Objective 1

1. $\dfrac{2}{m-5}$ 3. $\dfrac{1}{2y+1}$

Objective 2

5. $\dfrac{6h-4}{(1-h)(1+h)}$

7. $\dfrac{16s^2+19s-1}{(3s-2)(s-4)(2s+3)}$

9. $\dfrac{3x^2-18x+7}{(2x+1)(2x-1)(x+2)}$

Objective 3

11. $\dfrac{16}{3(x+4)}$

13. $\dfrac{q-22}{(2q+1)(q+2)(q-2)}$

15. $\dfrac{11x^2-x-11}{(2x-1)(x+3)(3x+2)}$

7.5 Complex Fractions
Key Terms

1. complex fraction

2. LCD

Now Try

1.a. $\dfrac{1}{3}$ b. $\dfrac{x(9x+1)}{2x+9}$

2. $\dfrac{7m^2}{2n^3}$ 3. $\dfrac{2(2a+3)}{3a+5}$

4.a. $\dfrac{23}{6}$ b. $\dfrac{24}{y}$

5. $\dfrac{8x-18}{30x^2+3x}$ or $\dfrac{2(4x-9)}{3x(10x+1)}$

6.a. $\dfrac{3(3t+2)}{2t-5}$ b. $\dfrac{4(x-4)}{3}$

Objective 1

1. $\dfrac{2}{7}$ 3. $\dfrac{2y-5}{3y-8}$

5. $\dfrac{24}{w-3}$ 7. $\dfrac{5(3a+4)}{2a+5}$

Objective 2

9. $\dfrac{x\left(2x-y^2\right)}{x^2+y^2}$ 11. $\dfrac{(x-2)^2}{x(x+2)}$

13. $\dfrac{2(1-4h)}{h(1+4h)}$ 15. $\dfrac{4m-3}{2(3-2m)}$

7.6 Solving Equations with Rational Expressions
Key Terms

1. extraneous solution

2. proposed solution

Now Try

1.a. equation; –2 b. expression; $\dfrac{3x}{10}$

2. $\{-14\}$ 3. $\varnothing$

4. $\{6\}$ 5. -7

6. $\left\{-\dfrac{3}{2}, \dfrac{3}{5}\right\}$ 7. $\{-24, 1\}$

8.a. $G = \dfrac{d^2 F}{m_1 m_2}$

b. $R_r = \dfrac{R_1(R_2 - A)}{A}$ or

$\quad R_r = \dfrac{R_1 R_2}{A} - R_1$

9. $R = \dfrac{R_1 R_2}{R_1 + R_2}$

Objective 1

1. equation; 4

Objective 2

3. All real numbers

5. $\{4, 6\}$ 7. $\{4\}$

9. $\varnothing$ 11. $\{4\}$

Objective 3

13. $T_1 = \dfrac{V_1 P_1 T_2}{V_2 P_2}$

15. $b_2 = \dfrac{2A}{h} - b_1$ or $b_2 = \dfrac{2A - b_1 h}{h}$

7.7   Applications of Rational Expressions

Key Terms

1. rate of work   2. check

3. read   4. smaller

Now Try

1. 2   2. 2 mph

3. $\dfrac{10}{7}$ or $1\dfrac{3}{7}$ hr

Objective 1

1. $\dfrac{3}{4}$ or $\dfrac{1}{3}$ 3. $\dfrac{2}{3}$ or $\dfrac{3}{2}$

5. $-\dfrac{2}{3}$ or 1

Objective 2

7. 24 mph   9. 8 mph

Objective 3

11. $\dfrac{24}{11}$ or $2\dfrac{2}{11}$ hr

13. $\dfrac{12}{11}$ or $1\dfrac{1}{11}$ hr

15. 2 hr

7.8   Variation

Key Terms

1. constant of variation

2. direct variation

3. inverse variation

Now Try

1. 28   2. 153.86 sq cm

3. $\dfrac{20}{3}$

4. 90 revolutions per minute

Objective 1

1. 50   3. 25 lbm

5. 379.94 sq cm

7. 24,000 watts

Objective 2

9. $\dfrac{1}{3}$ 11. 200 cu ft

13. 15 foot-candles

15. 213.3 ohms

Chapter 8
Roots and Radicals

8.1   Evaluating Roots

Key Terms

1. fourth root   2. radical

3. principal root

4. radicand   5. perfect square

6. negative square root

7. index (order)   8. cube root

9. perfect fourth power

10. square root   11. perfect cube

12. radical expression

Now Try

1. ± 20

2.a. 22   b. -18

   c. $-\dfrac{17}{100}$

3.a. 484   b. 324

   c. $3x^2 + 4$

4.a. irrational   b. rational

   c. not a real number

5.a. 5.916   b. -11.747

6.a. 61   b. $\sqrt{288} \approx 16.971$

7. 48 ft   8. 12

9.a. 6   b. 4

   c. 4

Objective 1

1. $\pm \dfrac{7}{12}$ 3. $3x^2 + 2x$

Objective 2

5. not a real number

Objective 3

7. -162.351

Objective 4

9. 24 ft

Objective 5

11. $\sqrt{13}$

Objective 6

13. -4 15. 2

8.2 Multiplying, Dividing, and Simplifying Radicals

Key Terms

1. quotient rule for radicals

2. simplified form

3. product rule for radicals

Now Try

1.a. $\sqrt{14}$ b. 13

   c. $\sqrt{21x}$

2.a. $2\sqrt{21}$ b. $9\sqrt{2}$

   c. cannot be simplified

3.a. $4\sqrt{3}$ b. $60\sqrt{2}$

4.a. $\dfrac{6}{7}$ b. 9

   c. $\dfrac{5\sqrt{5}}{8}$

5. $\dfrac{\sqrt{2}}{3}$ 6. $\dfrac{2}{9}$

7.a. $10y^6$ b. $4x^3\sqrt{3x}$

   c. $\dfrac{\sqrt{10}}{7x^4}$

8.a. $3\sqrt[3]{6}$ b. $5\sqrt[4]{3}$

   c. $\dfrac{8}{3}$

9.a. x^6 b. $2n^2\sqrt[3]{4n}$

   c. $\dfrac{y\sqrt[3]{y}}{5}$

Objective 1

1. $\sqrt{35ab}$

Objective 2

3. $5\sqrt{10}$ 5. $3\sqrt{10}$

7. $12\sqrt{3}$

Objective 3

9. 30

11. $\dfrac{\sqrt{5}}{2}$

Objective 4

13. $\dfrac{\sqrt{6}}{x}$

Objective 5

15. $\dfrac{4}{3}$

8.3 Adding and Subtracting Radicals

Key Terms

1. like radicals   2. unlike radicals

Now Try

1.a. $8\sqrt{2}$ b. $-4\sqrt{13}$

   c. cannot be added

2.a. $7\sqrt{2}$ b. $-10\sqrt{3}$

   c. $40\sqrt{2}+9\sqrt{3}$ d. 0

3.a. $10\sqrt{2}$ b. $4x\sqrt{5}-15x\sqrt{3}$

   c. $39w\sqrt{6}$ d. $13r\sqrt[3]{r^2}$

Objective 1

1. $2\sqrt{2}$

3. cannot be simplified further

5. $-5\sqrt{5}$

Objective 2

7. $24\sqrt{2}$ 9. $-2\sqrt{3}$

Objective 3

11. $4\sqrt{35}$

13. $6\sqrt{3y}-5\sqrt{6y}$

15. $-4\sqrt{3y}$

8.4 Rationalizing the Denominator

Key Terms

1. rationalizing the denominator

2. simplified form

Now Try

1.a. $\dfrac{2\sqrt{15}}{3}$ b. $\dfrac{3\sqrt{7}}{7}$

2. $\dfrac{4\sqrt{6}}{9}$ 3. $\dfrac{\sqrt{5}}{2}$

4.a. $\dfrac{2\sqrt{3xz}}{z}$ b. $\dfrac{3\sqrt{10qt}}{5t}$

5.a. $\dfrac{\sqrt[3]{18}}{2}$ b. $\dfrac{\sqrt[3]{5}}{5}$

   c. $\dfrac{a\sqrt[3]{147ab}}{7b}$

Objective 1

1. $\sqrt{5}$ 3. $\dfrac{-2\sqrt{3}}{3}$

5. $\dfrac{\sqrt{6}}{6}$

Objective 2

7. $\dfrac{3\sqrt{7}}{35}$ 9. $\dfrac{ab\sqrt{30b}}{6}$

Objective 3

11. $\dfrac{\sqrt[3]{50}}{5}$ 13. $\dfrac{3\sqrt[3]{7}}{7}$

15. $\dfrac{\sqrt[3]{5x^2y^2}}{5y}$

8.5 More Simplifying and Operations with Radicals

Key Terms

1. FOIL   2. conjugates

3. special products

4. rationalize the denominator

Now Try

1.a. $2\sqrt{15} + 4\sqrt{35}$ b. $-55 - 25\sqrt{6}$

c. $\sqrt{15} + \sqrt{10} - 2\sqrt{6} - 4$

2.a. $28 + 10\sqrt{3}$ b. $9 - 12\sqrt{x} + 4x$

c. -38

3.a. $-4\sqrt{3} + 8$ b. $5 + 2\sqrt{6}$

c. $\dfrac{3\sqrt{x} + 9}{x - 9}$ 4. $\dfrac{6 + 3\sqrt{6}}{4}$

Objective 1

1. $2 - \sqrt{10}$

3. $4\sqrt{10} - 4\sqrt{35} + \sqrt{6} - \sqrt{21}$

5. $97 - 56\sqrt{3}$

Objective 2

7. $\dfrac{4\sqrt{5} + 5}{5}$

9. $\sqrt{6} + 2 + \sqrt{10} + \sqrt{15}$

Objective 3

11. $\dfrac{\sqrt{7} - \sqrt{2}}{3}$ 13. $3 + \sqrt{2}$

15. $27\sqrt{3} + 5$

8.6 Solving Equations with Radicals

Key Terms

1. squaring property

2. radical equation

3. extraneous solution

Now Try

1. $\{10\}$ 2. $\dfrac{2}{5}$

3. $\varnothing$ 4. $\varnothing$

5. $\{5\}$ 6. $\{4, 9\}$

7. $\{4\}$

8.a. $\{1\}$ b. $\{1, 4\}$

Objective 1

1. $\{81\}$ 3. $\left\{\dfrac{1}{3}\right\}$

Objective 2

5. $\varnothing$ 7. $\{-3\}$

Objective 3

9. $\{0\}$ 11. $\{-6, 3\}$

Objective 4

13. $\{-1\}$ 15. $\left\{\dfrac{1}{3}, -\dfrac{3}{2}\right\}$

8.7 Using Rational Numbers as Exponents

Key Terms

1. $a^{m/n}$ 2. $a^{-m/n}$

3. rational exponent

4. $a^{1/n}$

Now Try

1.a. 25 b. 5

c. 3

2.a. 9 b. 8

c. 216

3.a. $\dfrac{1}{49}$ b. $-\dfrac{1}{625}$

4.a. 125 b. $\dfrac{1}{5^{1/7}}$

c. $-\dfrac{8}{125}$ d. 6

5.a. $\dfrac{x^{4/5}}{y^2}$ b. $\dfrac{1}{x^9}$

c. $\dfrac{y}{x^{11/3}}$

6.a 10

b. $\sqrt[4]{x}$ or $x^{1/4}$

Objective 1

1. 13

3. 8

Objective 2

5. 25

7. $\dfrac{1}{100}$

Objective 3

9. $6^{7/5}$

11. $r^{1/3}s^{1/3}$

13. a

Objective 4

15. $\sqrt[4]{a}$

Chapter 9
Quadratic Equations

9.1 Solving Quadratic Equations by the Square Root Property

Key Terms

1. quadratic equation

2. square root property

Now Try

1.a $\{-9, 7\}$

b. $\{-9, 9\}$

2.a. $\{\pm 15\}$

b. $\left\{\pm\sqrt{47}\right\}$

c. $\varnothing$

d. $\left\{\pm\dfrac{\sqrt{3}}{2}\right\}$

3. $\left\{-\dfrac{3}{2}, \dfrac{5}{2}\right\}$

4. $\left\{\dfrac{-5\pm 4\sqrt{2}}{2}\right\}$

5. $\varnothing$

6. 69.28 ft/sec

Objective 1

1. $\{7, 9\}$

3. $\left\{-\dfrac{7}{3}, 4\right\}$

Objective 2

5. $\left\{\pm 2\sqrt{6}\right\}$

7. $\left\{\pm\dfrac{3\sqrt{10}}{17}\right\}$

Objective 3

9. $\left\{-3\pm 3\sqrt{2}\right\}$

11. No real number solutions

Objective 4

13. 7.07 feet

15. 5.86 in.

9.2 Solving Quadratic Equations by Completing the Square

Key Terms

1. perfect square trinomial

2. completing the square

3. square root property

Now Try

1.a. 64, $(x+8)^2$

b. 144, $(x-12)^2$

2. $\left\{-4\pm 2\sqrt{3}\right\}$

3. $\left\{3\pm 2\sqrt{2}\right\}$

4. $\left\{-\dfrac{4}{3}, \dfrac{2}{3}\right\}$

5. $\left\{-\dfrac{9}{4}, \dfrac{23}{4}\right\}$

6. $\varnothing$

7. $\left\{\dfrac{-1\pm 3\sqrt{5}}{2}\right\}$

8. 1.2 sec

Objective 1

1. $\{-9, 7\}$

3. $\left\{\dfrac{-5\pm 3\sqrt{5}}{2}\right\}$

Objective 2

5. $\left\{\dfrac{5}{2}, 4\right\}$

7. $\left\{-\dfrac{3}{2}, \dfrac{5}{3}\right\}$

Objective 3

9. $\left\{\dfrac{3\pm\sqrt{69}}{6}\right\}$

11. $\varnothing$

Objective 4

13. 1 sec and 3 sec; both are reasonable

15. 2 units or 8 units

9.3 Solving Quadratic Equations by the Quadratic Formula

Key Terms

1. quadratic formula

2. discriminant

Now Try

1.a. $a = 3, b = -2, c = -4$

b. $a = 5, b = 24, c = 15$

c. $a = 25, b = 0, c = -49$

d. $a = 2, b = -12, c = 18$

2. $\left\{\frac{4}{3}, \frac{3}{2}\right\}$ 3. $\left\{\frac{1 \pm \sqrt{7}}{2}\right\}$

4. $\left\{\frac{8}{5}\right\}$ 5. $\left\{-7 \pm \sqrt{41}\right\}$

Objective 1

1. $a = 3, b = -2, c = 4$

3. $a = 1, b = 3, c = 9$

Objective 2

5. $\left\{-\frac{5 \pm \sqrt{57}}{2}\right\}$ 7. $\varnothing$

Objective 3

9. $\left\{\frac{3}{4}\right\}$

Objective 4

11. $\left\{-2 - \sqrt{10}, \, -2 + \sqrt{10}\right\}$

13. $\varnothing$ 15. $\left\{-\frac{3}{2}, 3\right\}$

9.4 Complex Numbers

Key Terms

1. pure imaginary number

2. complex number

3. conjugate

4. standard form

5. imaginary part

6. real part

Now Try

1. $6i\sqrt{2}$

2.a. $-2 + 9i$ b. $-11 + 4i$

3.a. $21 + 14i$ b. $-24 + 10i$

c. 34

4.a. $\frac{19}{29} - \frac{4}{29}i$ b. $\frac{1}{2} + \frac{3}{2}i$

5. $\left\{-4 \pm i\right\}$ 6. $\left\{\frac{3}{4} \pm \frac{\sqrt{31}}{4}i\right\}$

Objective 1

1. $10i$ 3. $8i\sqrt{2}$

Objective 2

5. $8 - 6i$

Objective 3

7. $2 + 10i$ 9. 53

Objective 4

11. $-\frac{5}{2} + \frac{3}{2}i$

Objective 5

13. $\left\{\frac{2}{3} \pm 2i\sqrt{2}\right\}$ 15. $\left\{1 \pm i\right\}$

9.5 More on Graphing Quadratic Equations; Quadratic Functions

Key Terms

1. parabola   2. vertex

3. quadratic function

4. axis of symmetry

Now Try

1.

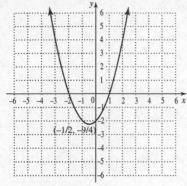

Vertex: $\left(-\frac{1}{2}, -\frac{9}{4}\right)$

x-intercepts: $(-2, 0)$, $(1, 0)$
y-intercept: $(0, -2)$
additional points: $(-1, -2)$, $(2, 4)$,
$(-3, 4)$

2.

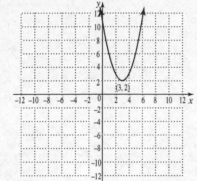

Vertex: $(3, 2)$; y-intercept: $(0, 11)$
additional points: $(1, 6)$, $(2, 3)$,
$(4, 3)$, $(5, 6)$

3.   no real solutions; $\varnothing$

Objective 1

1.

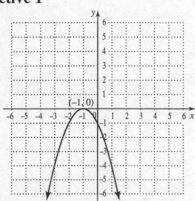

Vertex: $(-1, 0)$

3.

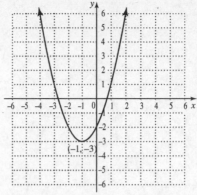

Vertex: $(-1, -3)$

5.

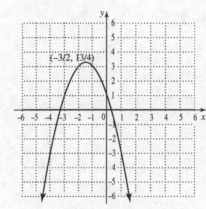

Vertex: $\left(-\frac{3}{2}, \frac{13}{4}\right)$

7.

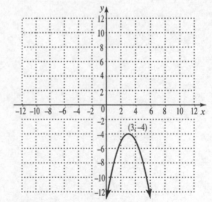

Vertex: $(3, -4)$

9.

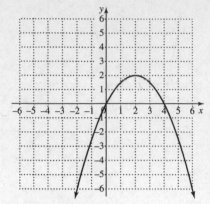

Vertex: (2, 2)

Objective 2

11. two solutions; {0, 4}

13. no real solutions; ∅

15. two solutions; {−2, 3}